D0846120

LONDON ESSAYS IN GEOGRAPHY

LONDON
ESSAYS IN GEOGRAPHY

RODWELL JONES MEMORIAL VOLUME

EDITED BY

LAURENCE DUDLEY STAMP

AND

SIDNEY WILLIAM WOOLDRIDGE

Published for
THE LONDON SCHOOL OF ECONOMICS
AND POLITICAL SCIENCE
(*University of London*)

Essay Index Reprint Series

BOOKS FOR LIBRARIES PRESS
FREEPORT, NEW YORK

First published 1951 for the
London School of Economics and Political Science
by Longmans, Green and Co., Ltd.

Reprinted 1969 by arrangement with
London School of Economics and Political Science

STANDARD BOOK NUMBER:
8369-1050-8

LIBRARY OF CONGRESS CATALOG CARD NUMBER:
76-80399

PRINTED IN THE UNITED STATES OF AMERICA

This Volume of Essays
is published
in honour of the memory of
LLEWELLYN RODWELL JONES
1881–1947
by a group of his
former students and colleagues

CONTENTS

Contents

INTRODUCTION

THE JOINT SCHOOL OF GEOGRAPHY AT KING'S COLLEGE AND THE LONDON SCHOOL OF ECONOMICS

The complex collegiate structure of the University of London has resulted in the establishment of several schools of geography in the colleges. At first sight this might appear to involve unnecessary duplication, but it has positive advantages in limiting the size of classes, strengthening the bond between teacher and student and encouraging collegiate specialization. Nevertheless, the University disposes of so great a teaching strength in many subjects, that there are equally manifest advantages in collaboration and inter-collegiate work and many such arrangements have naturally been made. In one case at least two colleges have formed a Joint School and it was in this School that the late Professor Rodwell Jones did his major work.

To trace the origin of the Joint School it is necessary to look at its antecedents. It was in 1895 that a small group headed by Sidney Webb (afterwards Lord Passfield) determined to found a college for the study of economics, politics and the social sciences. They recognized the importance of including the geographical as well as the historical, economic and political approaches to the study of social problems, with the result that geography was included in the curriculum of the London School of Economics from its foundation and H. J. Mackinder—a young man, still in his thirties, although already Principal of University College, Reading—was invited to organize the teaching of the subject for the new school. Mackinder became Reader in Geography in 1900, later becoming Professor and, for the years 1903–08, Director of the School. Hilda Rodwell Jones (afterwards Mrs. Ormsby) was appointed to a Lectureship in Geography in 1912 and the Department not only provided courses for students for the degree of B.Sc.(Econ.), in which geography is compulsory in the First Year and optional as a Special Subject, but also for entrants to the Academic Diploma in Geography, then of full Honours Degree standard.

In the meantime the University had drawn up the syllabus for an Honours Degree in Geography in the Faculty of Arts (shortly afterwards also in the Faculty of Science). The first Honours examination was held in 1921 and entrants began their training in October, 1918.

It was in the latter month that King's College registered the first student for the new degree, though she was compelled to do First Year work at the London School of Economics and Final work (except subsidiary geology) at University College, where L. W. Lyde held a College Professorship of Geography.

To meet the needs of the post-war influx of students in 1919 Major Llewellyn Rodwell Jones, M.C., after five years of active service, was appointed as Lecturer in Geography at the School of Economics, joining there his sister, some years his senior. After some years as a schoolmaster he had lectured in railway geography at Leeds University in 1913–14. It was in October, 1919, also, that Lieutenant L. D. Stamp, demobilized from the Royal Engineers, was appointed by Dr. W. T. Gordon as Demonstrator in Geology at King's College—in the Department from which he had graduated two years before. It was Dr. Gordon, with a student registered in geography already working in his Department, who saw the need for King's to provide for the new degrees.

The degrees in Arts and Science called for a wider basis of instruction, particularly in physical, mathematical and historical geography, as well as in biogeography, than was provided at L.S.E. It would no doubt have been possible for L.S.E. to have increased its staff to cover these aspects, leaving its near neighbour King's College to take its own decision, whether or not to provide teaching for the new degree. King's College necessarily taught many of the subjects which furnish the geographer's background, especially geology and history, but it would have needed to make new provision for the teaching of what has been termed " the core and culmination of the subject "—regional geography and its closely associated economic aspects. In place of such an arrangement the heads of the two colleges, Sir William (now Lord) Beveridge and Dr. (now Sir) Ernest Barker, and a group of their colleagues devised a much better one. Thus the Joint School came into being in 1922, leaving Mackinder and his staff to continue the teaching of the regional and economic aspects of the subject, while the Departments of Geology, History, Botany, Zoology and Civil Engineering made their appropriate contributions at King's College. Professor W. T. Gordon (Geology) and Professor A. P. Newton (Imperial History) were the chief architects, with Dr. Barker, of the King's College side of the structure, whilst Rodwell Jones with his chief and Sir William Beveridge shared in the early discussions on behalf of L.S.E.

The proposed arrangements were rendered possible by the fact that the two colleges are nearer neighbours than any others in London. They are separated, it is true, by a historic and traffic-ridden thoroughfare, the hasty crossing of which has endangered the lives of geography staff and students all too often. But it proved reasonably practicable to attend lectures in both, and the small measure of inconvenience was amply off-set by the obvious academic advantages of a large specialist staff. None the less, such an arrangement has its dangers ; geography by its nature is prone to centrifugal tendencies in aim and method and these at first seemed likely to be intensified by lack of co-ordination between the collegiate contributions. A further unbalancing factor was the possession by L.S.E. of definite premises for its department, while King's College had none. The geology department at King's soon became the natural student headquarters of the geography students, for Professor Gordon was the administrative head of department on this side and made generous provision for his new charges. Nevertheless, they had no place they could really call their own, and though many of them were registered as King's College students they tended to be little more than birds of passage, visiting the college for lectures, but living and working at L.S.E. Such difficulties increased during the first three years, with greatly increased student numbers. In 1925 Mackinder resigned his academic post under the increasing pressure of his parliamentary duties. He had done much in the founding of the School which benefited beyond measure by his world-wide prestige as a geographer. He was succeeded by his chief assistant, Dr. Rodwell Jones, and the latter's first task was the easing of the early stresses in the new structure and its consolidation and development. Among his greatest contributions to his subject and to his University were the patience and wisdom which he brought to these problems. His staff at first included his sister, Mrs. Hilda Ormsby, one of Mackinder's students, Mr. L. G. Robinson (in Historical Geography), Miss Winefride Hunt and Mr. D. H. Montgomerie. Professor Sargent, of the Department of Commerce, who had shared in the teaching of geography to economics students, continued as a friend and active collaborator of the department. Geography remained an important subject in the Faculty of Economics and this added appreciably to the weight and complexity of Rodwell Jones's charge. He was joined in 1926, a year after his own appointment, by Dr. Dudley Stamp, who had left King's College in 1921 for geological exploration in Burma and returned from the Chair of Geology and Geography at Rangoon to the newly created Cassel Readership in Economic Geography. In the following years Mr. W. G. East came to carry on the work in Historical Geography when Mr. Robinson took up his studies in Diplomatic History and Mr. S. H. Beaver took the place of Miss Hunt.

At King's College the greater part of the increasing work fell upon Professor Gordon's department. His staff was increased in 1928 by the appointment of Mr. D. L. Linton, jointly in Geology and Geography, while Dr. S. W. Wooldridge (who had succeeded Dr. Stamp as Demonstrator in 1921) became increasingly employed in teaching the latter subject. In 1930, with the opening of the new East Wing, provision was made for a geographical laboratory at King's College and the staff was strengthened by the appointment of Dr. H. J. Wood, a graduate in the Faculty of Economics at L.S.E.

The years 1925–30 thus marked the essential period of building and consolidation in the history of the Joint School and during this period Rodwell Jones was indefatigable in guiding its work and maintaining its intellectual standards. A retiring and modest man, he was always considerate and approachable as a chief. Both colleagues and students came quickly to look upon him with the deepest affection and respect and to depend upon his unobtrusive leadership and high integrity of mind and character. Those of us who came to know him well found him a strong and distinctive, but not overbearing, personality, numbering among his most lovable qualities a sense of dry humour and a most kindly and generous nature. His younger colleagues of former days will not readily forget his manner of organizing the annual field-class. He had, in high degree, that prized attribute of successful military commanders, the power to direct, and to delegate without interference. At a preliminary meeting the scheme was sketched in outline and "jobs" allotted. Then, often at his own expense, he would take some of the staff over the ground and comment upon their proposals for carrying out the work. He took his share in this, but left to them most of the credit. It was he who introduced that invaluable practice, never now to be omitted from a King's-L.S.E. field-class, of sending groups of students out alone on one of the last days of the class to make their own observations and report them in the evening. When he took the Chair at the evening meeting the Staff listened, sometimes with alarm and despondency, but often with appreciation, to the "reports" of the parties, excellent evidence of both what they had taught and what they had not succeeded in teaching.

The most active years of teaching and administration saw also Rodwell Jones's chief geographical writing, his text-books on North England (1921) and North America (with P. W. Bryan in 1925), his papers on Canada and Africa, based upon work in the field, and what is perhaps his major work, *The Geography of London River* (1931). He was not a prolific writer, but all his work bore the stamp of an acute and sensitive mind imbued with the highest standards of scholarship and aware that a University is a society dedicated above all else to the search for knowledge. He gave much thought to the content and teaching of economic geography and particularly to the principles which may serve to synthesize and illuminate its great bulk of descriptive fact. The revised regulations for the geography degree in arts and science, finally formulated in 1947, owe much to the fruits of his experience. Very prominent, however, in any view of his work must be his major contribution to the fashioning of the Joint School. With the retirement of Mrs. Ormsby in 1939, he and Professor Gordon alone remained of the original Joint Staff, though the rest of the staff, Dr. Stamp, Mr. East, Mr. Beaver, Dr. Wooldridge and Dr. Wood, had all worked with them for ten years or more. The War brought evacuation and separation, for King's College with the majority of the Arts and Science students in geography repaired to Bristol, while L.S.E. found refuge at Cambridge. If anything was needed to demonstrate to us how much we depended upon one another the separation did so ; it was almost complete through the bleak and ominous winter of 1939–40 and it was very cheering to the spirit when we were able temporarily to rejoin forces for the field-class at Bradford-on-Avon in the March of the latter year. In the following session began one of the most generous and self-sacrificing of Professor Rodwell Jones's services. He periodically went to Bristol to strengthen the slender teaching resources of King's College and to lecture daily for a week or more on those aspects of the subject in which he was an acknowledged master. This he continued to do when King's College returned to London in 1943, with L.S.E. still at Cambridge, and thus the spirit of joint work was maintained. At Cambridge he held together the work of the Economics students in geography, together with that of the Arts students registered at L.S.E. Most of his staff were largely or wholly engrossed in war-time service, but were able to give some help and Dr. Ormsby returned for a time to lecture. But he carried much of the burden alone, assisted latterly by Dr. Church, a graduate of the School and a member of the present staff. The personal debt of the war-time geography students in economics was strongly and sincerely expressed at the presentation they made to him on his retirement in 1945. Characteristically it was his desire to hand over to a younger man before he need have done by University regulations —he was then only sixty-four. He was to enjoy all too short a respite of leisure, for he contracted an obstinate form of jaundice and died after a short illness in August, 1947.

We cannot here trace in detail all that Rodwell Jones did for the Joint School. Its best expression was the steadily increasing coherence of the whole. The chief fault of the earlier days was the clean-cut separation between the teaching of specialist or ancillary aspects at King's and the central and more distinctively geographical aspects at L.S.E. Such an arrangement must always embody the risk of a false and misleading antithesis. It was with the active encouragement and full concurrence of both Professor Rodwell Jones and Professor Gordon that this cleavage was broken down and King's College teachers began to participate in the regional work. A minor advance, but an important one, was registered when, avoiding any possible shadowy administrative barricades, the teacher and not the class made the perilous passage of the Strand and Aldwych, using the available rooms freely and inter-changeably as time-table convenience might direct. But all such were mere "structural" adaptations. The greater problem and the greater triumph, spiritual and intellectual, was to make a team of senior and junior members—a society—of staff and students working together cordially and effectively with clearly organized aims and methods.

To the writers of this introduction, the present conjoint heads of the department, together with their colleagues, falls the task of continuing the work established in earlier years. Of the former staff, Mr. Beaver and Dr. Wood are still serving. No fewer than eight former members of staff or old students occupy senior University posts at home or abroad. They, and the other contributors to this volume and the large number of graduates whose names are here listed, will join with us in treasuring the memory of a loved and respected friend and teacher.

L. D. S.
S. W. W.

January, 1948.

I

APPLIED GEOGRAPHY[1]

By L. DUDLEY STAMP, C.B.E., B.A., D.Sc.

It is fortunately quite unnecessary in this institution to apologize for the word " applied " which appears in the title of my address this afternoon. In other circumstances it might have been necessary to combat the idea that, where a distinction has been made between the " pure " and " applied " branches of a subject, there is, by inference, something impure in the application of knowledge otherwise pure. What has, indeed, become almost a tradition dies hard—that the word " applied " denotes something inferior and belonging to the sphere of lesser beings than those who remain unsullied by the consequences of the application of their higher thoughts. In the period between the wars more than one President of the British Association found it necessary to deal with the impact of science on society, to stress the unhappy result of technological progress outpacing social adjustments. There are, now, perhaps, few scientists who would subscribe wholeheartedly to the older view that the scientist has no responsibility for the consequences of his findings, yet that was a view very widely held less than a generation ago.

I am not concerned this afternoon with the oft-essayed and, to me, somewhat fruitless task of attempting to define geography. For my particular purpose I am not concerned as to whether it should be regarded as one of the liberal arts, as one of the natural sciences or as one of the social sciences, though I appreciate and applaud the established decision of this great University that it can properly be considered as all three. In the University of London it is possible to graduate with Honours in Geography in the Faculties of Arts, Science and Economics. Thus the subject occupies the remarkable position of a Cinderella in three homes at once. Whilst this may serve to stress its many unappreciated merits, there is no need to pursue the analogy further when the qualities and characters of so many sisters are involved.

Since it is my purpose to demonstrate the application of geographical principles and methods to the solution of the problems of town and country planning, it is perhaps not inappropriate to recall the almost parallel careers of geography and town planning. We are apt, very

[1] Inaugural Address delivered at the London School of Economics, May 9, 1946, under the chairmanship of Lord Reith of Stonehaven.

wrongly it is true, to consider both as modern developments. In a British Act of Parliament town planning is, I believe, first mentioned in the Housing Act of 1909, with a very restricted application to the proper layout of land already partly developed or in train of being developed. With the growth of democracy and the remarkable post-1914–18 increase in the numbers of owner-occupiers of both urban and rural land, it became self-evident that entirely unplanned and unregulated growth of building was leading to some very unsatisfactory results—as undesirable æsthetically as they were economically. There followed the further attention to town planning in the Town Planning Act of 1925 and then the Town and Country Planning Act of 1932. The preparation of schemes under that Act was vigorously pursued until the outbreak of war, but it was an experiment which had already proved that series of local plans, however good, do not necessarily add up into one good national plan. It is because we are still groping towards a satisfactory marriage of an overriding scheme of co-ordinated national development with the maintenance of local initiative that town and country planning is in vital need of any help which can be rendered by a detached investigation of its problems.

To me it is a very special pleasure to be honoured by the presence of Lord Reith in the Chair today. When, early in 1941, the Office of Works became the Ministry of Works and Buildings, with Lord Reith as Minister, it was one of his first acts to set up a Consultative Panel. The Panel only met as a body on one occasion—because Lord Reith explained that he hoped it would work as a series of committees, each in an advisory capacity, for he was clearly aware of the many and new problems to be faced. Two of the members of the Panel were my colleague, Professor E. G. R. Taylor, at that time Professor of Geography at Birkbeck College, and myself, and we were at once requested to join a Research Committee. It was thus Lord Reith who first afforded the opportunity to geographers to try out their methods on a national scale. In due course the Planning Division of the Ministry of Health was transferred, and for a brief time Lambeth Bridge House housed the Ministry of Works and Planning. I am not sure what work was carried out by other committees of the Panel, but the Research Committee, ably abetted by Mr. H. G. Vincent, C.B. (then Principal Assistant Secretary), carried on under Lord Reith's successor in office, Lord Portal, and was reconstituted as the Research Maps Committee when the separate Ministry of Town and Country Planning was set up. The Research Maps Office (under Dr. Willatts, formerly of this School), had already been established and in due course the headquarters Research staff and the Regional Research staffs followed ; parallel arrangements were made in Scotland and, to give help and advice on the agricultural and rural aspects of land use and planning, the Central Planning Branch of the Ministry of Agriculture, with a regional organization, was set up. All this work owes its initial inspiration to Lord Reith.

Because these developments have all taken place since 1941 ; because town and country planning legislation is relatively new ; and because the Ministry was only set up in 1943, the idea is current that land planning itself is something new. Nothing could be further from the truth, especially in this country. Nearly two thousand years ago the Romans practised national planning not only by laying out that magnificent network of main roads which is still the essential framework of our road system, but also by deliberately siting their main settlements. So accurately did they judge the influence of site factors—they must have had a very effective New Towns Committee—that only two or three of their selections have failed to develop into great modern towns and cities. Silchester and Uriconium are the exceptions which prove the rule exemplified by Newcastle, Carlisle, Lancaster, Manchester, York, Lincoln, Chester, Bath, Exeter, Winchester, Chichester, Rochester, Dover and a dozen others. Incidentally this was an age when the science of geography occupied the attention of the greatest brains of the time—from Herodotus, Plato and Aristotle to Strabo and Ptolemy.

If the Romans practised national land planning in a country of which they knew but little, it was the Anglo-Saxons, the Jutes, Danes and Norwegians who showed by their settlements how strong were the geographical factors governing the behaviour of those who would settle on and live by the land. The siting of villages reflects essential accessibility combined with availability of a reliable water supply : the spacing of villages reflects the quality of land—the poorer the land the more needed to support a community and so the wider the spacing —and the layout of the manor and village reflected local differences of soil and land fertility. It is a bold generalization, but in the majority of cases one would not be far wrong if one said that the common arable fields can be taken to indicate the best soils of the neighbourhood.

The Anglo-Saxon and Celtic countrymen often turned their backs on the urban framework of the Romans : it was the Norman overlords who looked again to that old framework. In the middle ages Britain just grew on the twin foundations already laid—deliberate planning was almost forgotten. In a way there was comprehensive planning in that a large proportion of the land was in the hands of a relatively small number of great landowners. It was they who, especially in the latter part of the eighteenth century, developed the conscious art of landscape architecture in the layout of their parks, so that, with enclosure, the face of Britain began to take on its present-day pattern. In towns, unity of ownership made it possible for the great architects to be great town planners too, backed as they were by landowners of vision. Unity of ownership unfortunately also produced the industrial slums of Victorian England. The rise of democracy has coincided with an ever-increasing number of owners of land. Those owners have yet to learn to plan and act in unison. The Universities, notably of London and Liverpool, afforded scope to such pioneers in the field as Adshead, Reilly and Abercrombie : the link was with architecture though it would be wrong to suggest that the practical experiments of Ebenezer Howard or the penetrating analyses of Patrick Geddes and Frederic Le Play had no influence. The Town Planning Institute was founded in 1914.

Meanwhile the geographers and cosmographers of the ancient world and their successors in Tudor and Stuart times gave place in the nineteenth century, on the one hand to explorer-empire builders, and on the other to uninspired schoolmasters whose dreary recitation of oddly selected facts matched the dreariness of the slum homes of their pupils. The foundations of a new subject which should seek cause and effect in the world pattern were laid laboriously in the latter part of the nineteenth and early twentieth century by such men and women of vision as Reclus, Ellen Churchill Semple, Herbertson and Vidal de la Blache. We honour especially in this connexion our own distinguished ex-Director and my predecessor as Head of the Geography Department here, Sir Halford Mackinder. The consequences of our geographical ignorance were thrown into prominence by the war of 1914–18 and the subsequent peace conferences. The next ten years saw the establishment of honours schools of geography in all the chief British Universities. With an earlier training in geology and engineering, I graduated in geography in the same year as my wife, in 1921, in the year in which honours examinations in the subject were first held in this University.

Surely now the time is ripe to examine the contribution which geographers are making, or can make, to the solution of practical problems.

The Geographical Approach

The geographical approach may be best understood by reference to a specific example. Perhaps I may be forgiven for choosing as my example the work of the Land Utilisation Survey of Great Britain.

This Survey was established in 1930 with the help of this School and was made possible by the forbearance of my most tolerant chief, Professor Rodwell Jones ; it owes its completion largely to the untiring efforts of successive generations of graduates.

All physical planning and development must of necessity start from the present position. Whether eventually the decision be to eliminate and recreate or to restore and develop, a fundamental need must always be to know and to understand what already exists. The first step must therefore be a survey. It must be emphasized that a scientific survey seeks to discover and record what exists : the character and emphasis of the survey may vary according to the use which is to be made of the work but it is utterly wrong to confuse a scientific survey with a *policy* which may afterwards be developed in consequence of its findings. I desire to emphasize this point since, not infrequently, the urge to investigate is interpreted either as a determination to maintain the *status quo* or just as frequently as an indication of the desire to change what exists. It is true that we tend to study pathological cases in an attempt to find out what is wrong more often than we study healthy cases to find out how those happier conditions have developed.

The primary aim of the Land Utilisation Survey was to record the then existing use of every acre of England, Wales and Scotland. The use of volunteers, often young and little trained, necessitated a simple scheme—an eightfold classification of land into arable, permanent grass, heathland with moorland and rough pasture, forest and woodland, orchards, nurseries, houses with gardens (roughly less than 12 to the acre), and land agriculturally unproductive. These facts were recorded uniformly on the 22,000 separate sheets of the 6-inch Ordnance map required to cover the country. Various safeguards to ensure accuracy were used but the fact that each field or piece of land on the margin of two sheets was recorded independently by two surveyors affords a statistically exacting check. Between 10 and 20 per cent of the sheets were covered by fully trained observers, for the most part graduates, and much greater detail was recorded—crops on cultivated land, sometimes the character and condition of pasture, the type, character and arboreal composition of woodland, the character of each tract of land recorded as agriculturally unproductive, and so on. This afforded a large random sample within the general framework. Time was an important factor : changes in land use, though less rapid than commonly imagined, were taking place. The bulk of England was covered in 1931, the bulk of Wales and Scotland in 1932 and the whole was absolutely finished before the oatbreak of war.

The editing and checking of the field sheets, the reduction of the work to the 1-inch scale and the publication of the 1-inch maps—now complete for the whole of England and Wales and for the more populous parts of Scotland—proved a colossal and expensive task and this is not the place to record either the difficulties or how they were surmounted.

The Survey affords a snapshot picture of the face of Britain in the nineteen-thirties. Nine-tenths of Britain is rural, for towns cover less than a tenth of the surface ; in other words nine-tenths of the people occupy in their homes and workplaces less than a tenth of the surface. The emphasis thus appears to be on rural Britain. But the survey is an example of a purely objective survey : the emphasis was on accurate observation and the recording of facts, not on any use which might be made of the work.

The use of contrasted colours for the different types of land-use suggests at once certain broad and striking differences between one part of the country and another. In detail there is almost everywhere an extraordinary complexity of pattern.

The next stage is, accordingly, to attempt the interpretation of both regional differences and local patterns. The natural approach for the geographer is to seek correlations by comparing the land-use maps with those showing such features as relief of the land, solid and surface geology, soils, rainfall and so on. A tremendous amount can be revealed by so doing, but is this the correct procedure ? Is not the land-use pattern a kaleidoscopic one and is it not possible that the snapshot has caught the victim at an unusual moment and one that is far from typical ? One is led to enquire into land use changes so that the snapshot though it does not become part of a continuous cinematograph screen picture at least becomes one of a series of " stills." The history of land-use changes was attempted in two ways simultaneously : the use of the annual official statistics collected since 1866 which gave the general trend of changes in each county, and the reconstruction of land-use maps for sample areas (usually parishes) at certain intervals of time. Tragically the records of land-use changes of the 1917–18 agricultural drive have usually been destroyed and there is little data of the requisite accuracy and detail till one gets back to the 'seventies. Fortunately, just as the nineteen-thirties marked a certain *nadir* in British agriculture (with the lowest acreage under the plough recorded for at least a hundred years), so the eighteen-seventies marked a so-called golden age with ploughland at a maximum. For much of the country the Ordnance Surveyors preparing the detailed 25-inch and 6-inch surveys recorded in manuscript in their survey book, in a special column, the use of each parcel of land. The practice was discontinued

" for reasons of economy " but enough exists to afford some interesting samples. Still more valuable are the manuscript maps (of which three copies should exist—in the parish, in the diocesan centre and at London headquarters) and schedules prepared about 1845 by the Tithe Redemption Commission. The use of every parcel of titheable land was recorded and one can get samples from most parts of the country. Prior to that records are scanty. Moorland and woodland are faithfully recorded on the first edition 1-inch Ordnance Survey maps (1801 onwards) and some fine survey maps exist of certain counties for the latter half of the eighteenth century. Here and there are estate and manorial maps of earlier date and one gets at least a sketch of the conditions of the times by mapping the Domesday data.

Are there any general deductions which can be drawn from this historical review? The answer is " Yes "; and they are ones which caused and may still cause considerable surprise. The chief is the remarkable *stability* of land use in many areas. One can understand a certain stability of forest and woodland areas since a tree of lumber timber size takes 60 to 120 years to mature. The requirements of charcoal for smelting and timber for ship and house building had left Britain almost as deforested two centuries ago as it is today. Very remarkable is the stability of the great moorland and heathland areas. But more surprising is the stability of many of the important tracts of arable land. Under the plough in the difficult days of the nineteen-thirties, they were so used in 1870 and 1845 and in the preceding centuries. Elsewhere great land reclamation schemes, as the draining of the Fens or of the Lancashire mosses, tell their own story in major changes. But in other parts, especially where the land-use pattern of 1931–3 is an intimate and complex one, there seem to have been constant changes. In other words the changes from year to year recorded in the official statistics have definitely taken place mainly in certain areas which can be defined and mapped.

Our problem becomes one of explaining not only the present pattern but the evolution of that pattern.

So, gradually, we can disentangle the influence of the various factors and we may group these as geographical and non-geographical.

Two of the characteristics of the geographical factors are their permanence, and, contrary to general belief, their *increasing* importance. One of the commonest fallacies of the present day is the belief that the progress of science and technology has emancipated man from the influence of his environment. It is true that advances in transport and communications have largely overcome disadvantages of position and distance but it is this very fact which has emphasized the overriding importance of other geographical factors. Our ancestors built their villages and towns in an evenly-spaced pattern over the countryside: only where geographical factors proved favourable has there been great expansion and development of the early settlements. London has developed as a great port because of the dredgability of Thames mud; Bristol has given place to Avonmouth because the Avon gorge interposes a barrier where the river is neither dredgable nor capable of being widened. Tyneside has developed its great shipbuilding industry for the same reason: the industry on the Wear has not extended above Sunderland because of the river's rocky bed. The growth of towns in the Industrial Revolution on the coalfields—economy in transport costs—is paralleled in later days by Corby and Scunthorpe on the low-grade iron-ore fields. In simple language it costs money, much money, both in capital expenditure and in current expenditure, to overcome physical difficulties. In a world of fierce competition it is increasingly important to seek sites where the physical factors are as favourable as possible. Why is it, broadly speaking, uneconomic to cultivate wheat over a large part of the British Isles? Our ancestors if they wanted wheaten bread were forced to grow the wheat: today we know that our climate is less favourable through its excessive moisture and unreliable sunshine than the climate of many of the "newer" lands, whilst the form of our terrain is less suited for large-scale cultivation. Facile generalizations are rarely wholly true, but for almost any commercial crop it is possible to lay down extreme or geographical limits of cultivation—the limits being, broadly, inadequate moisture on the one hand, excessive moisture on the other, with north and south limits of inadequate or excessive total heat—within which much narrower limits of economic cultivation can from time to time be laid down. For wheat the greater part of Britain lies within the geographical limits but only a much smaller part within the economic. Nature complicates the issue by rewarding the adventurous—Nature loves a gambler—for near the limits of cultivation one season will see a complete crop failure, another a crop of exceptional quality and quantity. Why else are English apples of unbeatable flavour: why do we rely on Scottish seed oats and seed potatoes?

The geographical factors are thus (*a*) the elevation of the lands and the form of the ground; (*b*) the structure of the earth's crust and especially the disposition of minerals of economic importance; (*c*) climate and weather; and (*d*) the soil, to which we may add the obvious factor of space relationship—location and site.

By and large we see the land pattern of Britain reflecting these factors. Land over a thousand feet—lower as we go northwards—

lacks both settlements and cultivation and is occupied by moorland; country of rugged or varied relief and steep slopes presents permanent difficulties to settlement and development. We see the familiar concentration of population on the coalfields and note the problems created by the exhaustion of the raw material which was their *raison d'être*; we note in detail the localizing influence of iron-ore fields, salt-fields, limestone for cement and clay for bricks and recall the once overriding influence on the economic development of this country exercised by the ores of gold, tin, lead and copper. We are brought right to the present day by Wentworth—gardens or coal? Though we know well enough the connexion between moorlands and excessive moisture; and the suitability of the wetter west for grass rather than cereal crops and of the drier east for arable farming, we are only just beginning to appreciate the more subtle influences of climate. When we have really understood the lessons waiting to be learnt from the detailed studies of local climate, especially of the behaviour of cold air in spring, we shall perhaps change our wildly fluctuating annual apple crop into a steady production and realize that three-quarters of our orchards are wrongly sited. What of the incidence of fog and flying: what also of the effect of snow cover on road haulage over the Scottish Southern Uplands or the Pennines? Soil we know to be the joint product of its raw material, its site, climate and cultivation. We can boast soils as fertile as any in the whole world, as well as others so intractable as to be useless. The former are, to the nation, as precious as they are limited—we can upgrade our poorer soils but only by increasing greatly the costs of production of crops grown thereon.

We have in Britain a land of infinite variety in which the differences are very marked even in short distances. In essaying a classification of land, one embarks on a thorny and difficult path. After some three years of investigation a classification of land into ten types was evolved, the types being numbered 1 to 10 inclusive, and this is the scheme shown on the Land Classification Map, now published by the Ministry of Town and Country Planning, used also by the Ministry of Agriculture, and agreed with the Soil Survey. It is very difficult to ensure recognition that the ten types are *types* and that it is not a case of No. 1 the best and No. 10 the poorest—this raises the question of best for what? Though categories 1 to 4 are the good agricultural lands, 5 and 6 the medium and 7 to 10 the poor, the categories 1 and 3 may each be described as "best"—No. 1 by virtue of its character for arable farming and intensive cultivation, No. 3 by virtue of its water conditions for grass. Emphasis on type avoids the vexed question of measuring fertility (actual or potential) or productivity (actual or potential). The ingenious "ranking coefficient" evolved by Mr. M. G. Kendall is based on the yield of ten common crops: it ignores grass and so when applied to pre-war conditions in English counties, showed Leicestershire as low down the scale as Cumberland. I disagree with certain of my scientific colleagues that one cannot use a classification of land unless one *starts* by indicating for what purpose the classification is required. It is true that this scheme provides the first provisional answer to the questions, " What is good agricultural land and where is it found "—which Sir Montague Barlow and his Royal Commission found unanswered in 1940—but it does not presuppose any particular use being made of the facts.

This classification of land throws light on the historical development of land-use. We can establish the general principle that there has been relative stability of land-use on the poorest lands—the " wastes " of older writers and the heaths and commons of today, as well as the mountain moorlands—and also on the best lands. Where site and soil are alike good, the farmer has been able to stand up to the whole range of fluctuation of agricultural prices through the years. The maximum of change has taken place and is still taking place on land of intermediate quality. More than a century ago Cobbett recognized land submarginal in times of agricultural depression which it " paid to plough in dear-corn times." So we have the now familiar feature of the moorland margin retracting—moving uphill—when prices are good; moving downhill when prices fall. During the recent war the plough-up campaign affected but little the land use pattern of the arable eastern counties, whereas it completely changed the face of the former grassy shires in the Midlands and west.

Thus the purely economic factors exercise their influence within general limits set by geographical factors. It is true that the primary geographical factors are reflected in secondary factors—for example, space relationships may be echoed in transport costs. But there seems to me great danger in reliance on an analysis of secondary results—as in a theory of location of industry based on transport costs.

Whilst the broad pattern of land-use is clearly in response to the geographical factors, even a casual study reveals anomalies. A map of farm boundaries (such as now exists for official use as a result of the Farm Survey of 1940) reveals quite surprising differences between farming practice and consequent land-use even on neighbouring farms on the same type of land. Without entering into the technical question of what constitutes, for any given type of farming, an economic holding, we may note that it is even more important for the well-balanced mixed farm to include different *types* of land than for the holding to be a simple continuous area. The lopping off of one or two fields may completely destroy a whole farm as an economic unit. From district to district there are different farming traditions—Stapledon and Davies's Grassland Survey of 1940 is to a considerable extent one of grassland management—and particularly striking is the contrast in land-use pattern between Scotland and England in the nineteen-thirties due to the reliance on a six-course ley system in Scotland and on the management of permanent grass in many parts of England. Some anomalies in the land-use pattern are resolved by an ownership map. The amenities were just as safe in some hands as they were liable to destruction in others. The great mediæval landowners, incidentally, were far too wise to waste good land and for their parks they nearly always selected the poorest land in the area. Their successors in land planning, the local authorities, too often choose and despoil the best—to the nation's disadvantage.

It will be noticed that the non-geographical factors which have helped to determine the present use of land are largely non-permanent in the sense that they are alterable by human action. This is true of ownership, rentals, price levels, farming fashions, road and rail access, adequacy or otherwise of such capital equipment as buildings, provision or absence of electricity, water supply and adequate drainage. The changing influence of these factors too often blinds one to the permanence of the less obtrusive physical factors.

Survey and Planning

So far the geographer has, following accepted methods, carried out a survey and evaluated the operative causes. Treating his investigations historically, he will have established trends. In so far as those trends are due to geographical factors they will continue unless positive action is taken to counteract them.

He is in the position of a scientist who has investigated his facts and established his scientific laws. Should he leave their application to practical affairs to others or should he himself take a hand? Should the geographer play a part in town and country planning when there are professional planners? It is not without relevance to recall that membership of the Town Planning Institute is gained by a stiff professional examination but exemption from part is granted *only* to those who have previously qualified as (*a*) architects, (*b*) surveyors or (*c*) engineers. An honour which I value very highly is my election to Honorary Membership of the Institute and I hope that it may portend the acceptance of other basic qualifications than the ones acceptable to date. As Lord Justice Scott, whom I am very pleased to see here today, said recently " Town Planning is the art of which geography is the science." In land planning we are dealing with something which is *alive*—whether it be the people for whom we plan or their ever-changing environment—and natural laws and sciences must play their part. The whole question of type and size of houses required must depend both on the changing age-composition of the population, to quote but one example—in this case involving the demographer. It is, in fact, essential to regard every town plan not as something fixed and definite but flexible and changing with need or evolving to meet changing conditions almost as a living organism.

I am not attempting to decide whether the use of land should be determined in the future, as it has been in the past, by a continuation of *laissez faire* or whether we should develop a policy of planned use of land. The Government of the day has decided that issue in favour of planning. But we must steer a course which will avoid the danger of departmental planning on the one hand and regional or local planning unrelated to national needs on the other. We are in imminent danger of losing sight of two truths : (*a*) that physical planning is in essence the determination of the right use of land—every acre of it—in the national interest, and (*b*) that land is the one ultimate physical asset of the nation, practically fixed in its strictly limited area (about 1¼ acres per head of population in Britain) and consequently very precious.

What principles are we adopting, or should we adopt, in planning the nation's estate ?

One principle is surely that of the optimal and multiple use of land. Optimal use—in the national and not in sectional or local interests—implies a careful balancing of the claims of competing uses and users. Because so much of the use of land in this old-settled country has been determined by centuries, indeed millenia, of trial and error, we are in an entirely different position from, say, the United States or Russia, and the *onus of proof* should be on the change of user, not on the sitting tenant. Having determined optimal use, we should seek to encourage, not to forbid, other uses. Some waterside meadows near a town may have their optimal use as a " lung " or open space : they can well be kept in good condition by using them as grazing for dairy cattle. Why we do not lay out our cemeteries as parks after the charming manner of Forest Hills in California I can never understand. Public utility authorities—recently given still more powers—are amongst the worst offenders in unnecessary restriction on use of land. A tract of poor hill land may well serve as grazing for sheep or be afforested, as the gathering ground for an urban water supply, and at the same time an area for fresh air and exercise for human beings. Water supply companies, to save themselves the trouble of employing an analytical chemist, have only to wave a banner labelled " pure water at any cost " to " get away " with outrageous restrictions on the use of vast areas of land. Londoners bathe in and row on as well as drink the Thames and are healthy and happy.

If we insist on the principle of multiple use it is quite clear that no part of our land surface should be lying idle. There is, in fact, far too much derelict land lying completely idle and too often an eyesore into the bargain. When attention was called by the Ministry of Agriculture to the possibility of using such land instead of taking virgin agricultural land, the Ministry of Town and Country Planning instructed my colleague, Mr. Beaver, to carry out his survey of derelict land in the Black Country—a survey which has shown the possibilities of creating in the old industrial tract one of the finest and most beautiful and efficient housing-industrial regions in Britain.

The second principle is that we should use our land so as to satisfy as many of the rightful needs of our people and as adequately as possible. What, in fact, are those needs?

We may name five as fundamental—work, homes, food, recreation and mobility. The key to the whole is work, and I would put the location of industry into the forefront. The substitution of electric power for coal has tended to overemphasize a certain increased mobility of industry and has suggested an emancipation from geographical factors of location which is not borne out by the facts. The extractive industries are fixed (though where the raw material is in sufficient abundance, as in limestone for cement, there can be choice of location) and so are the heavy industries based on them. The need for a location by tidewater fixes many industries dependent on imported, bulky and heavy raw materials or on exports of bulky and semi-manufactured products. A locational factor often overlooked is water, of which some modern industries require fantastic quantities, whilst the effluent they return may be contaminated. It is an almost incredible fact that no comprehensive survey of our water resources nor of drainage has ever been made. Then whole groups of industries are tied because of " linkage," and so are servicing industries, such as gas supply or repair works. The number, even of light industries, really to be described as mobile or " footloose " is far more limited than is commonly realized. All industrial plant suffers obsolescence and on *a priori* grounds there is a national advantage in the re-use of industrial sites in areas where housing and a full range of services already exist. Unfortunately there is here a conflict between national and sectional interests : the clearing for *re*development of an old site involves capital expenditure which can be avoided by the use of a virgin site—especially in a new area where land is cheap. Later comes unnecessary expenditure of public money in the provision of housing and services. The danger is recognized by the present Board of Trade policy in regard to the Development Areas—the direction of industry to or encouragement of industry in the older industrial regions. Unfortunately the boundaries of the Development Areas (following administrative divisions) are drawn so widely that new industry may well settle in the Development Areas but on virgin sites, so that existing towns or their people may suffer rather than benefit. The policy of the Ministry of Agriculture in seeking to avoid unnecessary use of rural land thus powerfully reinforces the intention of the Board of Trade. During the war many new factories have been sited in remote areas for strategic reasons : because public money has thus been spent in war needs there is real danger of totally false economy in seeking to perpetuate factories whose location is wholly unsuitable in peace time.

A careful study of industrial location was made by the Scott Committee. We showed in our majority report, signed by eleven of the twelve members, that the factors were against the siting of new factories in the open country or villages. According to the industry the large or small town—we accepted in advance the findings of Lord Reith's committee that some new towns were needed—was the normal location.

The second primary need of mankind is a *home*, so that adequate housing, at a lower density than was acceptable to our Victorian ancestors, has a prior claim on land. We are short of three or four million houses, so they are needed for the moment everywhere, but otherwise their siting must be governed by industrial location and development. In the siting of housing, however, modern tendencies are most regrettable. We seem to have lost all sense of vision and imagination. I do not exempt either the Ministry of Planning or the Ministry of Health ; I blame especially the local authorities. A " good housing site " has become almost synonymous with flat land where the new creations cannot be otherwise than deadly dull. I doubt if there is a local authority functioning today which would allow that the sites of Durham, or Lincoln, or Bath or Edinburgh were even " possible." The suburban sprawl, so far from being tidied up, is in great danger of extension. Incidentally, the demands of industry, housing and airfield construction for level land of the same type places in jeopardy some of the finest agricultural land in this country, often specially suited for intensive market gardening. Again, in attempting to save for the nation's food-needs this type of land, agriculture is forwarding the interests of good planning by suggesting the building of homes on agriculturally poorer but æsthetically much more attractive land. Some of the poorest agricultural land, such as that on which Bournemouth is built, or which covers much of Surrey, makes the most attractive gardens where children can play even shortly after rain without wallowing in mud and catching cold.

The third great need to be satisfied from our land is *food.* We can boast some of the finest soils in the world (as in Fenland) as well as some of the poorest. For certain purposes (such as the growth of grass for all-the-year-round feeding of cattle) our climate approaches the ideal. We have the natural advantages for a farming economy capable of competing successfully with any country in the world. The war has to a large extent put the industry, which with its million workers and dependants is our largest employer of labour, on its feet again. Two essentials in the future success have already been granted in advance—fixation of prices and guaranteed markets. There is no doubt that a balanced mixed farming based on a four- or six-course or slightly longer rotation is the one fitted to the natural conditions of a large part of Britain. There is a universal trend towards this system in the newer countries which after robbing their land by monocultural methods must now seek to maintain soil and soil fertility. In normal times production of grain, especially wheat, is localized in the drier east of Britain and elsewhere is incidental rather than basic to an economy directed to milk and meat, with development of intensive market gardens, fruit farming and poultry farming locally. Though the organization of agriculture is a huge subject, a national nutritional policy with a balanced diet within reach of all is a starting-point. Right allocation of land with security of tenure to ensure this is the second step. In the third place we cannot afford *idle* land, we can use even the poorest by integration with the needs of other large requirements of land, such as forestry, sport and the fighting services. The recent truce between agriculture and forestry is an important step in the right direction. Incidentally, very large areas of our land, at present almost idle, should bear a crop of useful, beautiful timber trees.

A fourth basic need of mankind is *recreation* for mind and body. Parks, open spaces and playing-fields in town planning schemes are still too often regarded by local authorities as luxuries : the recent announcement of financial support for national parks, coastal preservation and footpaths by the Chancellor of the Exchequer is a welcome sign that the central Government is taking the right view. Fortunately the most suitable land for such purposes has a relatively low value in agricultural production and most is unsuitable for industry. As holidays with pay enable more and more of our people to enjoy a break away from home we must face the dangers of the spoliation of the countryside by increasing numbers of shacks and bungalows. To counteract this we should certainly encourage and not hinder the provision of holiday camps.

There is little use providing land for the four preceding purposes unless there is the essential mobility for men and materials. If our railway system is adequate, our patched-up Roman-Mediæval road system certainly is not. We need to canalize fast traffic on a small number of high-speed motorways, completely new, and to provide a few scenic routes. We can then avoid constant expensive bits of unsatisfactory road widening and so effect efficiency and economy, as well as saving intact the beauties of our countryside. Where is our boldness and imagination? Where our business sense when we attempt to measure future needs by taking a census of traffic on existing inadequate roads?

If we accept these as the five basic requirements of land, both local and national planning become the allocation of every acre of the land surface to its right use or uses in the national interest.

The Sieve Method

That which has become generally known as the "sieve method" affords an interesting example of a purely geographical technique. It is purely geographical in that the "sieves" or "filters" which are used are maps (the ordinary method is the use of an opaque black ink on transparent paper). The advantages and dangers of the method may be illustrated by an application to one or two current problems.

Let us first take a national problem—the search for a site for a new town of, say, 30,000 to 50,000 people. Whether or not a new town is needed is a matter of *policy* outside the geographer's sphere: for the sake of illustration I am presuming that the need is regarded as proved. If one looks at a map the first reaction is one of the many apparently wide-open spaces which offer themselves. How many sites are in fact suitable will depend on the factors which are regarded as essential —in other words on the nature of the sieves employed. We may sieve out all land over 1,000 feet elevation; we may sieve out land where the relief is so rugged or with so many and varied changes of slope that development would be difficult. These are unalterable factors. We may sieve out areas where such minerals as coal and iron ore are known to occur and are still to be worked. If it is policy so to do we may sieve out areas of the higher qualities of agricultural land. If we regard access by mainline railway essential and that an extension of the main rail network is unlikely, we may sieve out areas not already served by rail. This introduces us to the *alterable* factors and to a certain danger of treating geographical and non-geographical factors as of equal weight. A manufacturer seeking to establish a new factory might, in contrast to the requirements for a new town, require a site within two miles of an existing town of, say, at least 20,000 inhabitants —in this case the " sieve " to be applied would be quite different.

The result of applying successive sieves is often to leave curious and unexpected areas which can then be examined positively and a final choice made. Professor Sargent Florence, following his work on towns with ill-balanced industry, has combined statistical analysis with the sieve method in planning for the location of industry. Incidentally the statistical and geographical analyses of the same data often yield particularly valuable results, the geographical stressing spatial relationships lost in the statistical.

Conclusion

It has been my aim in this discourse to show that a real contribution can be made to the solution of the problems of physical planning by the use of geographical methods. I should regret if I had left the impression that the geographer could provide the whole answer. Such a gigantic task as town and country planning, within the framework of a national policy and consistent with world economy, can only be attacked by those methods by which we came successfully through the war—combined operations. It is essentially the work for a team which will provide the requisite geographical, statistical and economic analysis of the facts for the planner, the architect and the builder to translate into practical shape. Along these lines a good start has been made in the Ministry of Town and Country Planning. Much of the work needed is spade work—research work in the true sense of the term. Is this work appropriate to a Government Department or to a University ? If it is carried out in a Government Department there is right of access to all essential material and, with the sometimes much-delayed sanction of the Treasury, funds, personnel and equipment can be provided. But much research is of necessity unfruitful : one comes to dread the completely unanswerable question : " What will the proposed research show ? " Promising lines of investigation are stultified because they transgress inter-departmental boundaries. Whilst work carried out within a Government department is more likely to be directly considered in framing policy, there is also danger that half-completed results may become absorbed into the administrative machine and be given official backing before they have been fully tested and tried. The ban on publication and public discussion limits exchange of ideas largely to official circles.

During the war years I have much enjoyed my work as Chief Adviser on Land Utilization to the Ministry of Agriculture and other advisory work. I have seen the development of ideas, their permeation to all parts of the country and their valuable results in practice. I have come to realize the many practical difficulties which face the adminis-tator in the application of even simple principles and how his work is often hindered by lack of basic data. But I have learnt to value more than ever before my academic freedom and believe that basic research is best undertaken in the freer atmosphere of a University.

I say unhesitatingly that there is an immense field of work awaiting investigation : with increased resources of money, equipment and of personnel the Universities have a great task before them in the research field. We need especially two things : a constant influx of first-class men and women to relieve the present desperate shortage ; we need to maintain the closest liaison with local and central government administration so that, on the one hand we may be permitted to use the full range of information available, and on the other hand we may prepare our results in a form in which they can be directly used. There remain barriers. One is an opposition to research and survey actuated sometimes by the fear of what may be discovered and a consequent demand for action. Often this opposition comes from areas about which we know least—such as some of the overseas parts of the British Empire. Another is a petty jealousy which prohibits combined operations—we find it hard to believe that the range of studies covered by a colleague can be as important as our own. Unfortunately neither Government Departments nor University Departments are blameless on this score.

We need combined operations in the Social Sciences and integration with the natural sciences. Is it too much to hope that this institution will lead the way both within its own walls and in its relations with sister institutions ?

II

THE ROLE AND RELATIONS OF GEOMORPHOLOGY[1]

By S. W. WOOLDRIDGE, D.Sc.

In choosing a subject for this lecture I have been somewhat embarrassed by the fact that, not many years since, I delivered another such lecture at a sister college in our federal University. On that occasion it seemed appropriate to deal with certain aspects of the philosophy of geography in the large—a subject rich in differences of opinion and, therefore, of perennial interest to geographers. In the years between, I might perchance have changed my mind on this subject ; there is at least ample warrant in example for such fluidity in opinion. Alternatively, since no one has formally attacked the propositions which I then presented, I might, in a truly Greek spirit, seize this occasion to answer my own pronouncement, pointing out its evident heresies and shortcomings. But though the latter are not in doubt, I have not in fact changed my mind and conclude that I cannot profitably return to the topic. I prefer rather to pass from the whole to the part, from geography as a compendious discipline to geomorphology, that special division of the field which has engaged a great part of my own interest.

There is, indeed, another good reason for my choice of subject. This session marks the departmental separation at King's College of geography from geology, with which it had been grouped for a quarter of a century. As a member of the staff of the old double department, I inevitably inherit its traditions. I am not ashamed of my geological parentage, nor in any way inclined to sever the intellectual liaison between the two earth sciences. That liaison is most clearly seen in the field of geomorphology.

Geomorphology has, naturally indeed, been regarded as a borderline study falling between the fields of geology and geography. Its role and relations, however, have been the subject of some doubt and dispute. Now that geography has attained full recognition in the Universities beside geology, well established much earlier, it is desirable that the real nature and contribution of geomorphology should be more closely examined.

It has, of course, never been doubted that the roots of geomorphology lie and must always lie in geology, but the story of its emergence as a

[1] Inaugural Lecture delivered at King's College, March 19, 1948.

co-ordinated subject is worthy of a much closer historical scrutiny than it has yet received. The great nineteenth-century age of geology has now receded sufficiently in perspective to make possible adequate historical study of the growth of ideas. Britain played a cardinal role in the fashioning of these ideas and the views of British geologists are thus highly significant. In any case, a geomorphologist may be forgiven for believing that, in the sociology of knowledge as in the study of land-forms, the past is the key to the present.

We may start our enquiry by noting that Humboldt and Ritter, the founders of modern geography, though familiar with the geology of their days, were necessarily inheritors of the traditions of Werner and the Freiburg School. This afforded no evolutionary viewpoint in respect of the face of the earth ; its attitude was static and " structural." It was Hutton and Playfair who sowed the seed of geomorphology, but though Playfair's *Illustrations of the Huttonian Theory* was published five years before Humboldt and Ritter first met as young men in Berlin, there is little evidence that they ever learned the geographic force of the Huttonian doctrine. Their geology was, indeed, not only pre-Huttonian, but pre-Lyellian, though this is less significant than it might, at first sight, appear. There has been too great a tendency to regard Lyell as the literal disciple of Hutton. So far as concerns geomorphology their points of view differed radically. The ideas of the " normal cycle of erosion " are implicit in the pages of Hutton and Playfair, but Lyell long retained something of the " structuralist " point of view in regard to valleys. As late as 1838 we find him treating the Wealden valleys as essentially fissures, slightly modified by sub-aerial action, and he was certainly no whole-hearted convert to " erosionist " views during those critical years in which geology began to take formal philosophical shape. The reality of the antithesis between the viewpoints of Lyell and Hutton is shown by the publication, as late as 1857, of George Greenwood's *Rain and Rivers*, with the significant sub-title, *Hutton and Playfair against Lyell and all comers.* It is difficult to realize today that the efficacy of sub-aerial denudation, and particularly of river erosion, remained in dispute in Britain until the years 1860–70. These saw the publication of the work of Jukes on the rivers of Southern Ireland,[2] and Whitaker's paper on cliffs and escarpments.[3] The latter, even at so late a date, contained a doctrine too radical for the Geological Society and was published in the *Geological Magazine.* Topley's great memoir on the Weald came early in the following decade and succeeding years witnessed the labours of Dutton, Powell and Gilbert in

[2] *Quart. Journ. Geol. Soc.*, Vol. XVIII, 1862, p. 378.
[3] *Geol. Mag.*, Vol. IV, 1867, p. 474.

America, which are all agreed in regarding as the mainspring of later thought in the field of land-sculpture.

It was well-nigh quarter of a century before these views became fully current in Britain. The well-known papers by W. M. Davis on "The Development of certain English Rivers"[4] and "The Drainage of Cuestas"[5] marked the appearance of the new mode of thought in British journals. Nor were British geologists wholly unreceptive of the methods and their results. Marr's *Scientific Study of Scenery* appeared in 1900, and Archibald Geikie's great text-book included a final twenty-five pages on physiographical geology which, though it is little more than an appendix to a major work, gives adequate references to the whole field as then known.

It was at this stage that the first signs appeared of a desire to relegate geomorphology to geography, or at least to treat it as a borderline field. Marr's preface refers to geomorphology as "a subject which has sprung from the union of geology and geography," and Geikie's opening paragraph on "Physiographical Geology" concludes with the words: "The rocks and their contents form one subject of study, the history of their present scenery, another." Equally significant is James Geikie's well-known work on "Earth Sculpture" (1898). This gives adequate attention to denudation in the large, but retains a certain emphasis derived from older "structuralist" views. It gives a condensed and somewhat misleading account of river development but, in the preface, it is made clear that "geographical evolution," a subject "studied with assiduity by Professor W. M. Davis and others in North America," does not come within the compass of the book. We may note also that H. B. Woodward's *History of Geology* (1911), which necessarily reflects the point of view of British nineteenth-century writers, includes only three index references to the word "denudation" and refers to "the fascinating subject of earth sculpture" only in its concluding remarks and then only to note that "the subject is by no means a wholly geographical one," since fragments of ancient land-scapes are occasionally revealed "fossil," as beneath the Torridon Sandstone or the Trias at Charnwood Forest.

In all this and much more that might be quoted there is an implied distinction between the study of land-forms and geology proper. It is evident that some British geologists, at least, thought of land-form study as "geography." Nor were they alone in this view, for Mackinder has admitted that in his earlier days he was attracted by Geikie's antithesis and was inclined to regard the laying down of the

[4] *Geog. Journ.*, 1895, p. 128.
[5] *Proc. Geol. Assoc.*, Vol. XV, 1901, p. 75.

rocks as geology and their shaping into a surface as geography. Yet we note that this is not what Geikie said ; his grip of his subject was far too sound to permit any such exclusion of land-sculpture from geology ; it was physiographical geology. Nor need we suppose that all other geologists wished for such exclusion. There was, indeed, no systematic or organized subject of geography to which they could relegate this topic ; it was an intra- not an inter-departmental partition which they were seeking to erect. No useful purpose whatever is served by criticising their attitude on grounds of logic ; it is better to enquire why and how the situation arose. Our conclusion must be that it arose from the necessary character of geology itself combined with an historical accident.

Geology in Britain perfectly exemplifies the principle expressed by Mill when he pointed out that the definition of a science "like the wall of a city . . . has usually been erected not to be a receptacle for such edifices which might afterwards spring up, but to circumscribe an aggregate already in existence."[6] The study of land-sculpture, beyond a very broad and elementary stage, was doubtless excluded from formal geological instruction in earlier years because it was a disputable topic. This and the late closure of the controversy constituted the historical accident. If Hutton had written less obscurely and Playfair's *Illustrations* been more widely known, events might have taken a different course. By the time that geomorphology attained to vigorous growth, geology had fashioned a full curriculum of formal training. Archibald Geikie's text-book, representing the Edinburgh courses, and the regimen laid down by J. W. Judd at South Kensington, which became the basis of the first London degree in geology, sufficiently indicate both the character and the problems of such formal training. If stratigraphical or historical geology be the culmination of the subject it must necessarily rest upon the prior disciplines concerning the materials and processes of earth-history. Petrology carries its necessary prelude of mineralogy and crystallography. Palæontology must be, if necessary, an *ad hoc* study, with or without some measure of biological background. Physical geology is the forerunner of stratigraphy ; it can be, and perhaps must be, in large part a study of rock-making. The traditional geological curriculum necessarily ranges these separate and special studies in series rather than in parallel ; there is a long preparatory apprenticeship to be served before the fullness of geology is attained and this is to say nothing of the ancillary scientific disciplines. To some it has, no doubt, always seemed, with H. G. Wells, that geology is, for these reasons, "an ill-assembled

[6] *Unsettled Questions of Political Economy*, p. 120.

subject."[7] But it is difficult to see what alternative method could have been followed by its early teachers. What here concerns us most is, that in this scheme, geomorphology must figure either at an early or a late stage. In an early stage it forms an admirable introductory vehicle in training, and this is where it has generally been placed in courses. As a field for research it comes after stratigraphical geology in logical order, since it is answerable for the last brief chapters of the geological record. From this point of view Geikie's treatment is wholly sound; physiographical geology completes and complements the stratigraphical story. Yet it can be omitted from the story without loss to its epic sweep; it is little more than an appendix to the whole.

There are several aspects of this fact at which we may look more carefully and which go far to explain and even, in part, to justify the attitude of some geologists to geomorphology. Land-forms, as such, necessarily play a relatively small part in geological history as a whole. The present marks a mere stage in the geologist's series of superimposed pictures. Though similar diversified landscapes have existed before, of them "nothing stands," their hills have indeed been shadows which have gone beyond recall, save where chance fragments have been buried and exhumed. There is nothing that the geologist can really compare with the present physical landscapes of the earth and a close scrutiny of them throws comparatively little direct light on general geological history. In the narrower sense, at least, geology is chiefly an affair of sea-bottoms not of land surfaces, even when the full tale of the great continental formations is told. The substance of geological science is locked up in the cumulative pile of sedimentation and the records it embodies, including those of earth-movement. If the mind of man had been brought to bear at a later stage of the geological cycle when the continents were far advanced towards base-levelling, the progress of geology might well have been hindered by paucity of natural exposures, but it would still have been possible from the evidence of outcrops and borings, supplemented perhaps by geophysical methods, to discern the structure of the crust and reconstruct its history. But for geomorphology there would have been little scope, save for the fact that the nature and reality of peneplains might then have been all too plainly evident! Such considerations as these must obviously and rightly affect the amount of attention which the geologist gives to land-sculpture as such, within his crowded programme of study. They undoubtedly led to some over-emphasis of deposition at the expense of denudation in the older geology. Hutton was clear that the two are related like the halves of an hour-glass, but not all the nine-

[7] *Experiment in Autobiography*, Vol. I, p. 226.

teenth-century geologists had views equally sound. The point is an elementary one, yet there is no doubt that the teaching of stratigraphical geology was for long unbalanced by an imperfect view of the source of sediments. The tendency is expressed to the point of cariacature by a quite ludicrous diagram which Professor Hogben has allowed to appear in his *Science for the Citizen.* It is entitled " Denudation and Deposition " and shows first a group of mountains rising above a coastal plain, next the submergence of the coastal plain and the unconformable accumulation of a marine cover upon it, and finally its re-emergence. Throughout the whole of the period of this considerable episode of deposition the mountains of the background stand immobile and unaffected in a single feature of their outline by denudation ! It is needless to say that no geologist worthy of the name could perpetrate such an error as this, yet it does represent, in some sense, the curious lack of balance between knowledge of the two co-ordinate phases of the geological cycle. Deposition is self-registering, denudation, so it has been thought, is not.

The last point brings us at once to another question of importance responsible for a certain amount of misunderstanding between American geomorphologists and European geologists. By the end of the nineteenth century the stratigraphy of Western Europe, and Britain in particular, was known in far greater detail than that of North America. The physical history or " palæogeography " of the British Isles had been written by Jukes Browne with a completeness not then attainable for any other region in the world. To form an acceptable sequel to such a history it was necessary that British geomorphology should be "chronological." The system of geomorphology propounded by W. M. Davis *was* in a sense chronological or sequential, but it could not be grafted on to a stratigraphical system, nor was it designed for such a purpose. The cycle of erosion was of indeterminate length and dealt in stages, not ages as Davis always insisted. It enabled one to sketch in broad terms the physical history of regions of which the stratigraphical history was unknown or imperfectly known and hence it was and is a powerful tool in reconnaissance. But Britain and much of Western Europe did not need reconnaissance. It is interesting to note some of the first applications of Davis's work by British geologists. Cowper Reed, in the Sedgewick Prize Essay for 1900, dealt with the rivers of Eastern Yorkshire, dividing their history into six cycles. It was impossible then and is still impossible to relate this cyclical chronology to the stages of Tertiary stratigraphical history. On the other hand A. E. Salter [8] saw in Davis's theories a means by which he sought

[8] *Proc. Geol. Assoc.*, Vol. XIX, 1905, p. 1.

to clarify the tangled history of the superficial deposits of the London Basin. He classified the various deposits in " drift series " arranged in descending order, each marking a stage in the history of a consequent stream entering the Basin by one of its marginal gaps. This classification cut right across the stratigraphical grouping, which treated the deposits as marking various stages of the Pliocene or Pleistocene periods. Davis himself put forward his " two-cycle theory " of the evolution of the Southern British drainage.[9] He ignored the evidence of the marine Lenham Beds (L. Pliocene), finding evidences of two successive cycles of sub-aerial denudation. Subsequent work has shown that for a great part of the area this theory is completely in error. The Lenham Beds cannot be ignored, and from our knowledge of their former extent we can now perceive that the drainage of a great part of the area is superimposed from a sheet of Pliocene deposits and is in its first cycle. But the basic evidence for this conclusion is stratigraphical. In brief, one may say that British stratigraphy was too well known to render it possible or useful to think and speak in terms of cycles. The geologist tended to feel that, where he had any evidence at all, it was both more tangible and more precise than that offered by morphology alone. From another standpoint it could, of course, be said that the stratigraphy was not well-enough known to afford any chance of correlation with morphological cycles. What is wanted as the complement of the stratigraphical story is the denudation chronology of Tertiary times. The Tertiary deposits are limited in area, small in thickness and concentrated almost entirely in the South-East Province. Over much of the British Isles even the Chalk is missing as a datum formation. In the west and north of the country stratigraphical perspective is thus inevitably lost. The summit-plane of Wales has been variously identified as an exhumed sub-Triassic or sub-Cretaceous surface and as a Tertiary peneplain, and there is no direct geological evidence, in the narrower sense, to decide between the alternatives. The demonstration, in later years, throughout Britain of a large number of planation surfaces over a wide range of height, the recurrent suspicion that they are marine, rather than sub-aerial, and the suggestion of a eustatic interpretation of the evidence, bids fair to make it even more difficult to think in terms of " cycles " in the orthodox sense.

The foregoing argument is intended to indicate some of the reasons why geologists tend to ignore geomorphology, especially in Britain. It is not implied that the reasons are entirely sound or that no counter case could be put. It may be conceded, however, that the geologist,

[9] Davis, op. cit., 1895.

confronted with the need of specialist training in many different fields, can only be expected to accord place to geomorphology in so far as it pays dividends in illuminating earth-structure and earth-history. Regarded as a special branch of latter-day growth, it is only one of several; geophysics and micro-palæontology suggest themselves as modern developments affording powerful tools in elucidating earth-history, quite apart from their immediate economic applications. It is easy to say that the whole range of possible methods of investigation should be applied to the problems of geology, but foolish to ignore that they are of very different powers and values. Geomorphology is a part only of physical geology and a smaller part still of the whole science. Nor can we expect it to be cultivated solely at the behoof and for the service of geographers.

Our brief review of the relations of geology and geomorphology illustrates a principle of great importance in what I have referred to as the "sociology of knowledge." A subject is best understood as a whole, not by defining its logical boundaries, but by considering the methods or tools by which its aim is pursued. It is in the light of this principle that we shall find it profitable to examine the relations of geomorphology and geography.

Historically, these relations are essentially simpler than those we have traced in the case of geology. Geomorphology emerged as something like a coherent subject towards the end of the nineteenth century at the very time that academic geography was in process of being re-born. In terms of Mill's analogy, geomorphology stood to geology in the relation of "an edifice which had afterwards sprung up." To the new geography it was one of a number of "aggregates already in existence," which, it seemed, might conveniently and profitably be circumscribed by the geographical wall. Considerable efforts were made thus to include it by Peschel, Richthofen and Penck in Germany and by Davis and Tarr in America. No similar situation can be said to have arisen in Britain; geography here was initially too weak to circumscribe anything and in any case its British founders had other ideas. One has only to compare three volumes in the Oxford Regions of the World Series, Mackinder's *Britain and the British Seas* (1906), Partsch's *Central Europe* (1903) and Russell's *North America* (1904) to see the varying emphasis on geomorphology fully illustrated. Mackinder had, no doubt, the most compact and manageable region to survey. His book deals adequately with the geomorphology; indeed it is not too much to say that it embodies original contributions in the field, notably the chapter on the rivers of Britain. Yet the British study accords to geomorphology a relatively subordinate role. It was perhaps his experience in seeking unavailingly to teach his " regional method " to other contributors to his series which led Mackinder, many years later, to aver that geographers " in escaping from servitude had robbed the Egyptians, the geologists, and had been cursed for the possession of ill-gotten goods by a generation spent in the wilderness."[10] Yet by the time that it was uttered the reproof had lost much of its force. Geographers were looking increasingly askance at the attractions of geomorphology and were seeking, by ignoring it, to throw it back, as unwanted and irrelevant lumber, into the geologist's back garden.

I shall not attempt to trace fully here the rambling and indecisive controversy on the place of geomorphology in geography. The clearest recent statement is that of Ogilvie in his Herbertson Memorial lecture.[11] He affirms the need of close study by geographers of land-forms and declares in favour of " explanatory " rather than empirical description. He notes that geologists in Britain present, as a rule, only vague and elementary statements on the origin of relief and in effect abandon the last chapter of earth-history. He feels it will be necessary for some time yet for geographers to be trained in morphological methods, so that they can make good, for their own purposes, the omissions of geologists, but he urges them to remember that in so doing they will tend to write evolutionary studies, essentially chapters in earth-history, which, in his view, " ought to come from the pen of the geologist."

There is much in Ogilvie's treatment which commands assent from all geographers of reasonable mind, but a number of criticisms may fairly be made. The " genetic " difficulty is, indeed, an old one. An extreme interpretation of the nature of geography would fix its whole attention on being rather than becoming, but this represents an ideal impossible of attainment, and of curiously little intellectual attraction when attained. If to describe land-forms is geography and to discuss their origin is geology, then a similar limitation must apply throughout the field of systematic geography. Hartshorne enforces his recent discussion of " The Nature of Geography "[12] by a diagram in which the " geographical plane " intersects the plane of the systematic sciences, each of which has its geographical counterpart—meteorology—climatology, botany—plant geography, economics—economic geography and so on. Yet in each case the " genetic problem " exists. Plant communities are subject to " succession " and human societies to development. Climates are the result of genesis

[10] Presidential Address to Section E of the British Association, London, 1931, p. 4.
[11] *Geography*, Vol. XXIII, 1938, p. 1.
[12] *Ann. Assoc. Amer. Geog.*, Vol. XXIX, 1939, p. 323.

not less than land-forms. One notes that Hartshorne omits geomorphology from the geographical plane altogether, substituting the simple words "land-forms." This no doubt expresses his conviction that geomorphology is really geology, and had, therefore, better not be mentioned. Reflection assures us that this is an impossible and ultimately a disastrous position to take up. To pursue it logically would be to concede that as climatology, economic geography and the rest become genetically coherent they too will pass out of the field of geography and that it will be impious to mention them! The truth of the matter surely is that it is quite impossible for geography to achieve an *instantaneous* cross-section of the phenomena of the systematic sciences. All the phenomena are subject to "becoming" and their inter-relation cannot be perceived if their evolutionary tendencies are ignored. The element of truth in the implied objection is that if the geographer moves too far in the dimension which lies, as it were, at right angles to his own, parts of his field will disappear over the curve of the mental horizon and his synthesis will become unbalanced. But this is only to say that he must limit himself to consideration of proximate, not ultimate, genesis. Whether as a whole, or in any of its branches, geography must remain, to this extent, a two-dimensional not a one-dimensional subject.

Geomorphology is the historical geography of the physical landscape. The only alternative to the genetic approach is to accept land-forms as "given," which would be in effect to see them with dull and unimaginative eyes and in all probability to misread them and misconceive their relations. Davis tersely and finally vindicated the genetic approach when he wrote: "To look upon a landscape . . . without any recognition of the labour expended in producing it or of the extraordinary adjustments of streams to structures and of waste to weather, is like visiting Rome in the ignorant belief that the Romans of today had no ancestors." It is the idea of development or evolution in the sense here implied by Davis which manifestly constitutes the major philosophical contribution of geomorphology to the bare bones of physical geography. This is undeniable—yet it has been denied, at least by implication, by certain geographers. It is with some feeling of surprise that I find myself standing here, in the College where Lyell first professed his subject, attempting to re-affirm a principle so self-evident. For geography to reject the intellectual concept of developing land-forms would be stupidly retrograde, the self-imposition of limitations inevitable in the days of Humboldt and Ritter—but quite unnecessary in our own. Geography has, indeed, shown from time to time a tragic propensity to simplify its domestic regimen by "throwing the baby away with the bath-water." It has shown this not in regard to geomorphology alone ; some of its reformers propose thus to serve a veritable platoon of babies with their appropriate *quanta* of bath-water. Given their way they would zealously proceed to empty geography of most of its content, value and significance. I can only conclude that, in the language of modern psychology, they have become possessed, in respect of their subject, by a psychotic death-urge ; indeed, their persecuting zeal is such that, if uncurbed, its life would very rapidly become extinct.

The more immediately practical relevance of geomorphology to the geographer is seen in two distinct, but not unrelated, contexts. In the first place, though processes of landscape evolution are slow in comparison with those of human development, the physical evolution of what are essentially human regions is still proceeding. The past becomes the key to the present in conceiving of these regions as wholes. The great Aquitaine Lowland lying between the Pyrenees and the Massif Central of France, was once a gulf of the late Tertiary sea, almost literally an extension of the present Bay of Biscay ; the gulf was gradually filled up with rock-waste from the surrounding uplands. This process is still actually in progress by virtue of the heavy river deposition of the Garonne and its tributaries ; it affects the life of the present inhabitants of the region, leaving its stamp upon land-use, settlement and economy in general. The physical history in this and many other cases is more than mere background to the human development ; it overlaps with and is continued in it. While geography as practised by geographers retains its root meaning of a " description of the Earth " it can no more be blind to physical than to human history.

But again, when we come to subdivide areas in detail, to examine the actual differentiation of the Earth's surface, the first of our considerations is generally its morphological diversity. Physical " regions " may be distinguished on many different bases, climate, structure, lithology, soil or hydrology. Where a cover of natural or semi-natural vegetation exists this is, no doubt, the best index to the natural variation of the surface. Where a true cultural landscape has replaced the primæval scene this method fails us ; it fails us, indeed, in varying degree throughout the greater part of the world. Morphology is, then, not only the factor which, taken alone, most readily integrates the others, but the morphology, even the micro-morphology, generally shows very close correlation with the warp and woof of the cultural landscape. The student is too apt to assume that when we distinguish, say, the Cotswold terrain, we are dealing essentially with a physical region. It is certainly this, but it is evidently more than this. The detailed texture of the human landscape is a faithful reflex of the form of the surface. So is it likewise, in the familiar and classic French "*pays*." Again, we may traverse a Yorkshire dale and find all the road and rail crossings on morainic bridges, the settlements on moraines or late glacial deltas, and the intervening lake flats forming natural farm units. The Chalk cuestas of the South Country offer their recurrent pattern of scarp-foot and dip-foot "sites" using the term "site" as the pedologists use it, but extending it in the wider geographical context. The dip-slopes vary in form, soil cover and utilization according as they have been exhumed from beneath an Eocene or a Pliocene cover, or have been exposed since Mid-Tertiary times. It is in the light of such examples as these that the geographer studies geomorphology, not because, or not only because, it is a register of Earth-history but because it is a guide to land quality in the widest sense. Every physiographic fact is, indeed, the starting-point of two trains of inference, one leading back into the geological past and one into the geographical present. No doubt it is true that a geographer's choice of tools and the very philosophy of his subject are conditioned in some measure by his environment; certainly no British geographer can readily deny the relevance of geomorphology. But there are in fact very few areas of the Earth's surface in which the form of the land is not the best and most immediate indicator not only of its recent history but of its structure, variations of soil and vegetation cover, and the mode of circulation of water above or within it. When human occupancy and economy are also found to show significant relationship with the facts of form surely the geographer may conclude that morphological analysis is fundamental to his final synthesis and value it accordingly.

Again there may come the query—why cannot we accept the facts of form without studying their antecedents? To this question we have already given, in the words of Davis, the general philosophical reply. But it prompts other reflections. I have heard it innocently and plaintively suggested that if only geologists would really bend their minds to the description in detail of the geomorphology of the world, regional geography would be much easier to study. It is not long since Mrs. Barbara Wooton [13] made effective play, in another connexion, with the nursery poet's lines:

> If all the world were apple pie,
> And all the sea were ink,
> And all the trees were bread and cheese,
> What should we do for drink?

The format of the argument fits our case. If all geologists were to

[13] *Lament for Economics.*

turn into geomorphologists, all botanists into ecologists, if all historians were knowledgeable cartographers and all economists were preoccupied with the geographical implications of applied economics, geographers might well be in search of an occupation ; they could adopt a pose of delightful ease, but one that would soon cease to retain any dignity or credit whatsoever. But none of these admirable labour-saving devices is at all likely to avail the geographer. He cannot sit back and receive his data like some pre-digested intellectual breakfast food. He must needs go out and fetch it himself and adapt it to his purposes. In order to do so he must at least be trained in the necessary ancillary disciplines. It is at this point that so much confusion seems to be engendered. However regrettable it may seem to some when the research geographer is lured into evolutionary studies, one cannot artificially break the chain of relationships in training and return a pedantic methodological negative if the student asks " why ? " Indeed, my humanist colleagues among the geographers never dream of such a negative within the social field. They blandly trespass at will, as fancy directs, and I for one will not say them nay if they will be logical enough and tolerant enough to extend a comparable freedom to my co-workers and myself.

The argument, as I have sought to present it, may be summed up by saying that as a contribution to earth-history the evidence of land-forms is vastly more incomplete and fragmentary than that of stratified deposits, but as constituents of landscapes, land-forms manifestly take a primary place. There are innumerable areas in which the record of land-sculpture can add little to geological history, but which for the geographer are important regions or parts of regions. In describing the land-forms of such an area he will necessarily indicate how they have developed and are developing and in so doing he need not be deterred by the pedantic objection that he has for the nonce become a geologist. If in the same area he traces changes recently occurring, and perhaps still continuing, in the distribution of cultivation or settlement, no one in his senses will accuse him of becoming a historian !

It cannot surely be an unreasonable or improper hope that geologists may not turn their backs on geomorphology as too " geographical " nor geographers shun it as " geological," but that both may cultivate the field for their proper purposes. If in so doing they on occasion help one another this will be a matter neither for surprise nor regret. Sovereign subjects are perhaps as outmoded in the world of today as sovereign states and above the jealousies and disputation of the separate sections of the learned world stands the truth that knowledge is one and indivisible.

III

THE DEVELOPMENT OF THE NORTHAMPTONSHIRE IRON INDUSTRY, 1851–1930

BY S. H. BEAVER, M.A., F.G.S.

Introduction

The iron-ore field in the Northampton Sands formation of the East Midlands, and the iron and steel industry based upon it, are important and well-known features of the economic geography and cultural landscape of that part of England. The ironstone output reached an all-time record in 1942 with 10½ million tons, representing just over a half of the entire British output ; this ore would have yielded about 3⅓ million tons of pig-iron, about 43 per cent of the national total. The steel output from the Basic Bessemer works of Messrs. Stewarts and Lloyds at Corby was about ½ million tons in the same year.[1]

The effect on the landscape of this vast ironstone output, almost all of which is obtained by opencast methods, has of course been very striking, and has caused much public concern, which resulted in two Government enquiries being held into the methods of restoring damaged areas and preventing the spread of devastation.[2], [3]

The present-day geography of the ironstone industry is a response to several factors : (i) the physical circumstances, i.e. the geology of the ore-bed and of the rocks which overlie it ; (ii) the "technical" environment, i.e. the nature of the machinery which can at present be employed for excavating overburden and ore ; (iii) the location of the blast-furnace plants which need the ore, and of the railway lines which lead thereto ; (iv) the physical and chemical nature of the ore, which control the use which is made of it, e.g. in relatively small blast-furnaces for making foundry iron, or in large blast-furnaces for producing steel-making iron, which is converted either by the basic open-hearth process (at various localities outside the ore-field) or by the basic Bessemer process (at Corby, Northants, and at Ebbw Vale,

[1] British Iron and Steel Federation, "Statistics of the Iron and Steel Industries of the U.K. for the years 1939–44," London, Sept., 1945.

[2] "Report of the Committee on the Reconstruction of Land Affected by Iron Ore Working," Min. of Health, H.M.S.O., 1939.

[3] "Report on the Restoration Problem in the Iron and Steel Industry in the Midlands : Summary of Findings and Recommendations," H.M.S.O., 1947, Cmd. 6906.

S. Wales) into steel. These factors are by no means permanent in themselves or fixed in their relations to one another, and the economic geography of the industry is thus not static, but constantly changing.

A brief sketch of the industry based on the Northampton ironstone has already been published.[4] It is the purpose of the present essay to trace in some detail the development of quarrying and smelting in the different parts of the ore-field, as a study of the relationship of a great human activity to its geographical environment.

Early Records

The records of early ironworking within the Northampton ironstone field in Northamptonshire, Rutland and South Lincolnshire are scanty. Iron weapons and slag heaps have been found in a settlement of Celtic age at Hunsbury Hill,[5] about a mile south-west of Northampton, and numerous records are available of slag dumps in association with Roman pottery.[6] These occurrences presumably indicate smelting activities. At Colsterworth, in South Lincolnshire, an actual Roman furnace has been discovered, datable to between A.D. 75 and 150.[7] In the Domesday survey, ironworks—" ferraria "—are recorded as having existed in the time of Edward the Confessor at Gretton and Corby (though it is by no means certain that they were smelting furnaces ; they may have been forges) ; and in the twelfth century the royal forest of Rockingham contained a number of furnaces.[8] From this time until 1850, however, the existence of the iron ores seems to have been forgotten, and John Morton, writing on Northamptonshire in 1712,[9] and marvelling at the extent of the former smelting industry, states definitely that there is no ore in the county.

1851–63. Re-Discovery of Ore and Early Development of Quarrying and Smelting

The re-discovery of the ore took place in 1851.[10], [11] The ironstone was quickly found to underlie a large part of the county, and early in 1852 the first consignments were forwarded to Staffordshire. The Staffordshire smelters, however, did not take kindly to the new ore, which was different from anything they had previously employed,

[4] S. H. Beaver. " The Iron Industry of Northamptonshire, Rutland and South Lincolnshire," *Geography*, 1933, 102–17.

[5] Victoria County History, Northamptonshire, Vol. I, 1902, p. 152.

[6] Ibid., p. 206.

[7] C. E. N. Bromehead, *Proc. Geol. Assoc.*, LVIII, 1947, p. 362.

[8] Bridges, *History of Northamptonshire*, 1791, Vol. II, p. 309.

[9] J. Morton, *Natural History of Northamptonshire*, 1712, p. 549.

[10] S. H. Blackwell, article in *Lectures on Results of the Great Exhibition*, 1852.

[11] J. Percy, *Metallurgy*, Vol. II, 1864, p. 225.

and not until the end of the decade did it begin to move in any quantity into the Black Country.

Although Northamptonshire has no coal, the leanness of the ore and the necessity to use far more ore than fuel to produce a ton of pig-iron, meant that local smelting was a fairly obvious corollary of the working of ironstone. The credit for the establishment of the industry in the county is due to Mr. William Butlin. After experimenting in 1852, he erected a small cold-blast furnace at Wellingborough in the summer of 1853, and by perseverance and careful management he solved most of the problems attaching to the smelting of the remarkable new ironstone. His works occupied a site alongside the Midland Railway. A second furnace plant made its appearance in 1855 at Heyford, not far from Weedon on the main line of the London and North Western Railway. Meanwhile, quarries in the ironstone were springing up all along the outcrop of the bed where it overlooked or lay alongside an existing line of railway—e.g. Blisworth, Gayton, Duston, Wellingborough, Desborough and Stamford (Fig. 5). The Blisworth quarries also had the advantage of access to the Grand Junction Canal, by which and the Coventry Canal, ore could be sent to the Midlands.

By about 1857, then, the industry was really on its feet and both ore-extraction and smelting were proceeding apace. But expansion was not rapid. The difficulties of the smelters took some time to overcome, and it was not until 1866 that a third furnace plant made its appearance ; whilst the prejudice which existed in South Staffordshire and elsewhere against the Northamptonshire ore was not easily eradicated.

The output of both ore and pig-iron (Figs. 1 and 4) remained steady for some years, though at a level almost insignificant when compared with the rest of Britain. The ore production was only about 2 per cent of the total (Fig. 3), whilst the pig-iron output was a mere 10,000 tons per annum out of a total of almost 4 million. It was a period of experiment, and of the creation of a favourable public opinion towards the ores and the iron. There was no doubt that the Northampton Ironstone could be used in the local furnaces to produce good pig-iron for most purposes, and the three Northamptonshire furnaces were certainly efficient, with a fuel consumption of under 30 cwt. per ton of pig-iron,[12] a figure which remained virtually unaltered for over sixty years. Moreover, the proximity of the works to the quarries, and the excellent rail facilities available for the import of relatively cheap fuel (even though it came from Durham), rendered the price

[12] *Mining Journal*, July 16 ; Nov. 26, 1859.

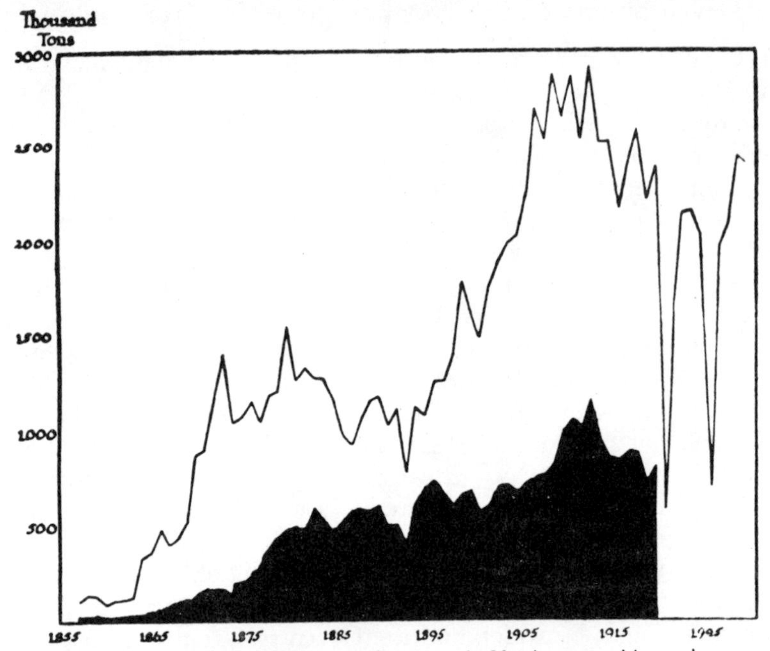

Fig. 1.—Graph showing production of iron ore in Northamptonshire and amount sent out of the county (in black).

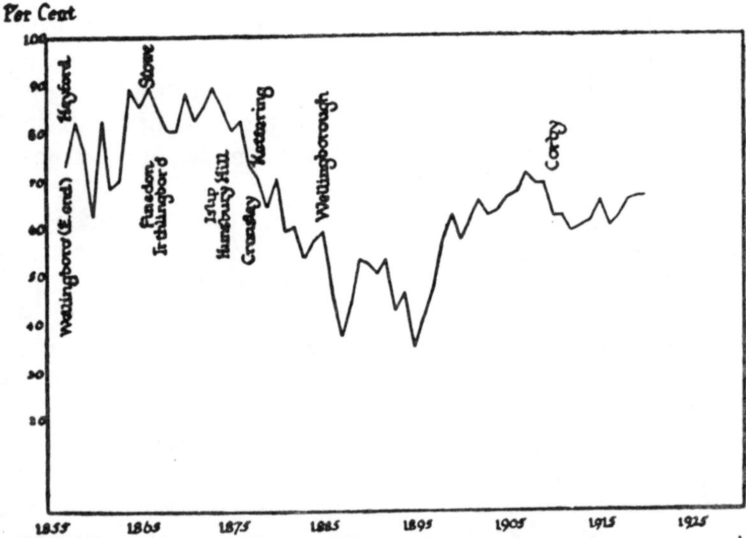

Fig. 2.—Graph showing percentage of Northamptonshire ore sent away, and dates of opening of smelting plants.

Note : There are two forces at work in controlling the proportions of the ore output used in and outside the county. First, the growth of blast furnaces tends to reduce the proportion exported, and second, the growth of the ore output tends to raise that proportion. Although this is true in a general way, Fig. 2 should be carefully compared with the graph showing actual production (Fig. 1), since years of low output tend to coincide with years of small export (e.g. 1887, 1892), the reasons being that the blast furnaces continue to require approximately the same amount, and in consequence the surplus left over for export is considerably reduced. This was especially the case before 1880, in which period there were few or no outside ironworks owning quarries in Northamptonshire. Subsequently, especially after 1900, the matter becomes rather more complicated, because these " foreign " blast furnaces also needed a constant supply of ore. No data are available of the amount of ore ent out of the county after 1920.

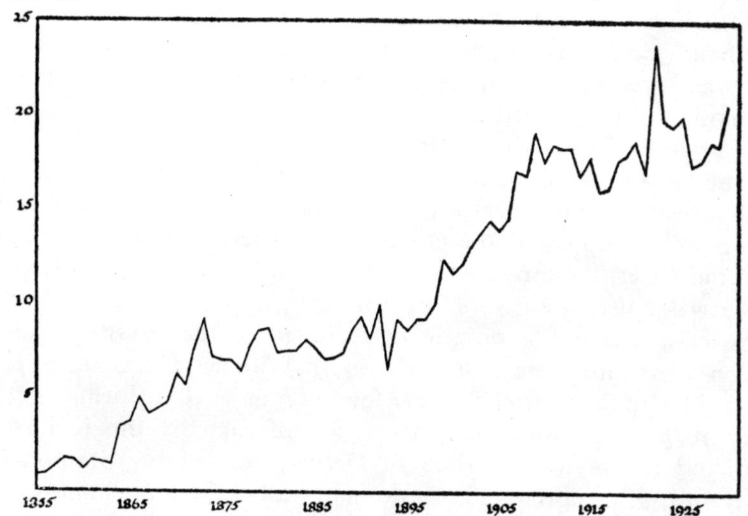

Fig. 3.—Graph showing percentage of British output of iron ore obtained from Northamptonshire.

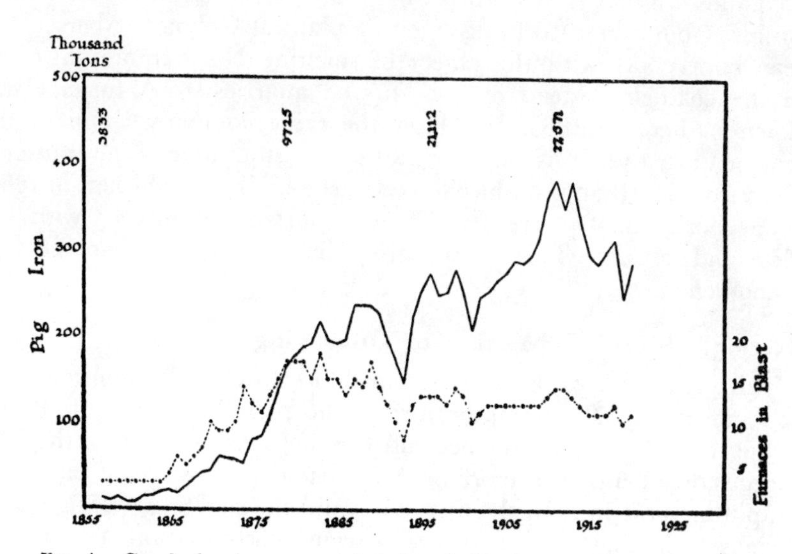

Fig. 4.—Graph showing output of pig-iron in Northamptonshire and number of furnaces in blast.

Figures on graph are average output per furnace per annum in selected years. County output figures for pig-iron are not available after 1920.

of pig-iron very much lower than in Staffordshire (45*s.* as against 64*s.*), where the increasing cost of working the thin Coal Measures clayband ores was beginning to be felt. Most of the pig-iron apparently went into Staffordshire, where for a long time it was surreptitiously employed by puddlers in the face of a marked prejudice against its use for wrought iron.[13]

The localities which received Northamptonshire ore were far more widespread. Much certainly went into South Staffordshire, but the same prejudice existed there against the ore as against the pig-iron, and moreover, the mixture of the clayband ores and the Northampton stone was not easy to work. Until 1860, in fact, the South Staffordshire area was not a prominent recipient of the ironstone, most of which went into Derbyshire and South Wales. The detailed figures given in Hunt's *Mineral Statistics* for 1857 show that during that year over 40,000 tons were sent, about a half each by the L.N.W. and Midland railways, to stations in Derbyshire, notably Stanton Gate, Clay Cross, Eckington and Chesterfield—all of them the home of works which still use large quantities of the Northampton ironstone. South Wales began to receive a considerable amount of Northamptonshire ore owing to the increased cost of working the almost exhausted local claybands. What is more surprising, however, is to find a Northamptonshire ironmaster acquiring smelting plant at Golynos (Abersychan, near Pontypool) with the object of smelting Northamptonshire ore on the coalfield instead of close to the quarries.[14] A furnace was blown in here in 1863, and later the same company acquired the nearby Varteg plant as well.[15] Still more astonishing is the transport of ore from Northamptonshire to Newcastle-on-Tyne, whither, in 1860, some consignments were sent for the purpose of mixing with the Cleveland ores.[16] This must surely have been only an isolated phenomenon.

1864–73. Rapid Expansion of Quarrying

This period marks the first real stride forward in the ironstone and iron industry of Northamptonshire. The prejudices in Staffordshire having been largely overcome, and the methods of smelting the lean siliceous ores being now more or less perfected, the way was paved for a rapid increase in both the quarrying and the smelting, which is very obvious from Figs. 1, 3 and 4. It seems fairly certain, too, that a series of articles on the Northamptonshire iron industry which appeared

[13] *Mining Journal*, June 18, 1859.
[14] *Mining and Smelting Magazine*, 1863, p. 303.
[15] Ibid., 1864, p. 232.
[16] *Mining Journal*, July 21 and 28, 1860.

in the *Mining Journal* between 1868 and 1871 had much to do with this development, by bringing to the notice of ironmasters all over the country the nature and extent of the Inferior Oolite ironstone and the quality of the furnace products, and thus giving the county a wide advertisement.[17] Moreover, the market for ore and for pig-iron in the Midlands was expanding rapidly, with the general increase in the demand for iron of all kinds. Thus we find that between 1863 and 1873 the annual output of ore increased from just over 100,000 tons to nearly 1½ million tons, the latter figure representing about 9 per cent of the total British production ; whilst the pig-iron output rose from about 15,000 tons to 60,000 tons a year (Figs. 1, 3 and 4).

The all-round expansion of trade which characterized the early 'seventies, culminating in the boom of 1873, was especially evident in the iron industry, and it gave a remarkable stimulus to the quarrying of iron ore in Northamptonshire. Not only did the output of the existing quarries increase materially, but the period 1871–3 witnessed the opening of at least seventeen new workings. This expansion of quarrying is very evident from Figs. 5 and 6, which show the location and output of workings in 1866 and 1873. The control exercised by the three principal lines of railway (L.N.W. main line, Midland main line, and Northampton and Peterborough line) is still very clear ; numerous new quarries were opened in the Finedon and Desborough districts, and on both sides of the Nene valley between Wellingborough and Thrapston, whilst the construction of the Northampton and Market Harborough line in 1859 had permitted the commencement of quarrying at Brixworth, and the completion of the Kettering and Huntingdon branch in 1866 opened up the Cranford-Islip area. This increase in the output of ore was not balanced by quite so extensive a development of blast furnaces, and in consequence we still find the greater proportion of the ore being sent to other iron-smelting districts. Close on 90 per cent of the ore raised went out of the county via the Midland and L.N.W. railways and the Grand Junction Canal. A good deal was sent into Staffordshire, but it seems that the Staffordshire ironmasters preferred importing the pig-iron rather than the ore. Derbyshire and South Wales are constantly mentioned in the contemporary press as the destination of Northamptonshire stone, whilst South Yorkshire began to receive large quantities. In 1868 the single firm of Butlin, at Wellingborough, was sending as much as 2,500 tons of ore per week into Derbyshire and Yorkshire.[18] An important feature of this trade

[17] *Mining Journal*, Dec. 21, 1867 ; Jan. 11, Mar. 14, Sept. 19, Oct. 10, 1868 ; June 5, 1869 ; May 28, Dec. 31, 1870 ; Mar. 11, Sept. 30, 1871, etc.
[18] *Ibid*. Jan. 11, 1868.

was the fact that the one thing which Northamptonshire lacked, i.e. coal, could be brought as a return freight from the coalfields to which the ore was sent.[19] Another noteworthy feature was the regional differentiation in the ore export, the Blisworth and Duston area, served by the L.N.W. railway, sending mainly to Staffs. and South Wales,

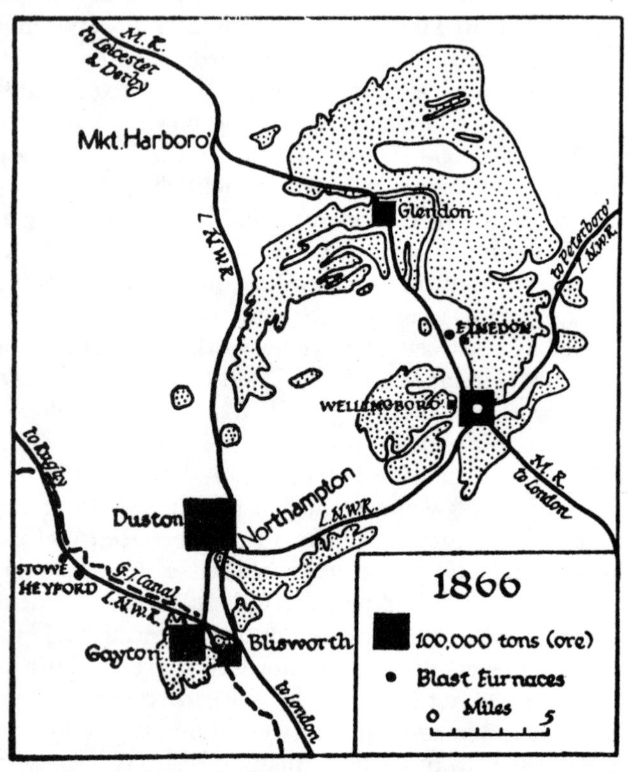

FIG. 5.—Quarries and blast furnaces in relation to railways and canals, 1866.

Note on Orefield (stippled) *:* The outline of the orefield given on this and subsequent maps is substantially that shown in the 1920 Mineral Resources volume (Vol. XII) of the Geological Survey. The results of the re-survey conducted during the early 1940's have, at the time of writing, not been made public except on a very small scale—the 1 : 625,000 " Coal and Iron " map published by the Ordnance Survey ; and it seemed appropriate therefore, since the present essay refers to the pre-1930 period, to retain the ore field as it was known at that time. The major alterations (additions) are in the area between central Northamptonshire and south Lincolnshire, where the re-survey has revealed considerable and hitherto unsuspected ore reserves.

whilst the Ise valley region, served by the Midland railway, sent almost exclusively to Derbyshire and Yorkshire.[20]

There was naturally, in the development of blast furnace plants, a considerable lag behind the rapid expansion of the ore-raising industry, since smelting works require a large capital expenditure and need a year or two for their construction ; so that whereas the period 1869–73

[19] *Mining Journal,* Dec. 21, 1867. [20] Ibid., Sept. 19, 1868.

marked the rise of the quarries, it was in the succeeding decade that most of the blast furnaces commenced operations. There were, however, three plants set up in the earlier period, which may be briefly commented upon (Figs. 5 and 6). In 1866 the Glendon Iron Company

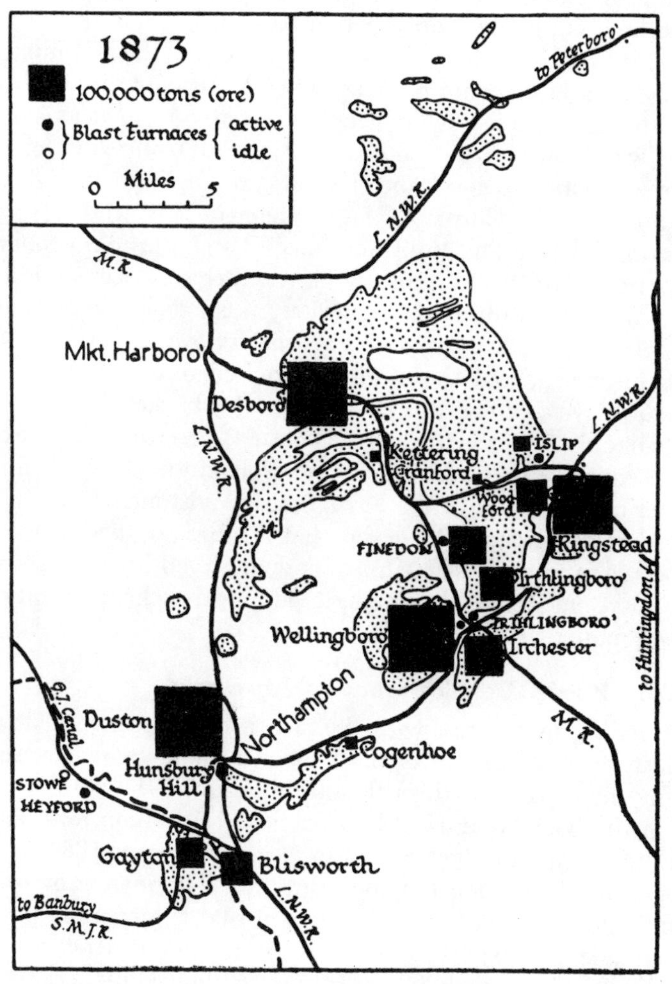

FIG. 6.—Quarries and blast furnaces in relation to railways and canals, 1873.

commenced the erection of a furnace near Finedon, alongside the Midland railway, and within easy reach of extensive ore deposits on the eastern side of the Ise valley; and their plant was increased to three furnaces by 1869. This company is especially interesting in that about 1871 it began to sink a colliery at Alfreton, Derbyshire, for the supply of coal to its furnaces,—a premature example of that vertical integration which is one of the most important features of the modern iron industry. The year 1866 also witnessed an extension of the business of Thos. Butlin & Co. Three years previously, the pioneer furnace at Wellingborough had been added to, and now an entirely new works was planned, on the Irthlingborough side of the Midland railway at Wellingborough (hence known as the Irthlingborough works). The third smelting works to be constructed during this period was in the nature of a white elephant. At Stowe, near Weedon, a furnace was blown in in 1866, by the Stowe Iron Ore Co., but the venture soon came to grief, and the works remained derelict for several years until Mr. McClure purchased them in 1873 and reconstructed the existing furnace and built another. To continue the story—after three years of fairly continuous work the furnaces became idle again, and remained so until another company was formed in 1889 to revive them. Another year or two of smelting followed, but by 1892 they had again ceased, and the plant was pulled down some years later. The trouble with Stowe was that it was not backed by extensive ore deposits like the other works. It lay near the south-western extremity of the productive ore-field, where the Northampton Sands formation has deteriorated into a mere ferruginous sandstone.

It is mere coincidence, however, that all three of these works should have been abandoned. Finedon was demolished in the early years of this century, and Irthlingborough, a victim of a scheme of rationalization, was pulled down in 1932.

1874–80. Rapid Development of Smelting

This period, in contrast with the last, is chiefly concerned with the early history of a number of blast furnace plants which were set up following the rapid growth in the output of ore. The ore production, which in the boom year 1873 had reached 1·4 million tons, fell somewhat in the succeeding years, and except for the year 1880, stood at a level of about 1¼ million tons per annum. The increasing quantities used in the local furnaces therefore meant that a decreasing proportion was available for export to other iron-making districts; and this proportion fell from nearly 90 per cent in the preceding period to about 70 per cent.

The destination of the Northamptonshire ore remained substantially the same as in the preceding years, but one or two additional features are noteworthy. The South Wales Coal Measures ores were being rapidly depleted, the production falling from 1¼ million tons in 1872 to ½ million tons in 1875, and there was consequently a much greater demand for imported ores—a demand which was partly satisfied by the Northamptonshire area, and especially by the Blisworth-Gayton district, which was by this time connected to South Wales by a much shorter route than heretofore—via Stratford-upon-Avon and Broom Junction.[21] The demands of Derbyshire were also increasing, and during this period that region was importing from Northamptonshire more than it produced in its own mines; and moreover the Northamptonshire ore could be unloaded at the furnaces at less cost than the local claybands, which were becoming more and more difficult to work.[22] These facts began to induce Derbyshire ironmasters to acquire leases of the Northamptonshire stone for themselves. In 1871 the Stanton Ironworks Co. began to work the ore near Wellingborough; in 1873 the Wingerworth Iron Co. commenced quarrying near Stamford; and in 1879 the Staveley Coal and Iron Co. took over the quarries at Cranford. By 1879, in fact, the Stanton Company was receiving close on 200,000 tons of ore a year from its quarries at Wellingborough and Desborough. A certain amount of ore seems to have found its way, strangely enough, into the Cleveland district,[23] where it was employed for mixing with the Marlstone ores in order to reduce the percentage of phosphorus in the pig-iron.

About the same time the Scunthorpe district, which by 1872 had four works in operation, began to use the siliceous Northamptonshire ore with eminent success; [24] and so great was the need for a siliceous ore to mix with the calcareous Lower Lias ores that the working of the Northampton Ironstone was commenced near the city of Lincoln, where in 1877 the Greetwell mines [25] and quarries were opened. The second phase of blast furnace development in Northamptonshire began in 1871, when Mr. C. H. Plevins, a Derbyshire coal-owner, who in the previous year had acquired the Heyford plant, proposed to lay out a site for four furnaces near Islip, on the north side of the Kettering and Huntingdon branch of the Midland railway. The first one was blown in in 1873, and the plant was completed in 1879, when the concern became known as the Islip Iron Co. This company, like the Glendon Co., had the advantage of controlling its own coal supplies

[21] In 1872 the Northampton and Banbury Junction railway from Blisworth via Towcester and Banbury was opened, tapping the Blisworth quarries, and giving an alternative route to the Midlands via Banbury and the Great Western line; in 1879 the connection from Towcester via Stratford to Broom Junction, near Evesham, was completed, thus giving a much more direct route to South Wales.

[22] *Mining Journal*, Sept. 13, 1879.

[23] Ibid., Apr. 11, 1874.

[24] Ibid., Mar. 11, 1871.

[25] Shallow mining was necessary here because the ironstone is directly overlain by the Lincolnshire limestone.

in Derbyshire; it also had an extensive area of good ore within easy reach of the furnaces. Then in 1873 the erection of a new two-furnace plant at Hunsbury Hill, about a mile south-west of Northampton, alongside the Northampton and Blisworth line of railway, was commenced, both furnaces being blown in towards the end of 1874. In 1875 the Cransley works were projected, on a site about a mile south-west of Kettering, and connected with the Midland railway by a private branch line. The plant was completed in 1878. Finally, in 1877, the works of the Kettering Iron and Coal Co., at Warren Hill, on the north-west outskirts of the town, and adjoining the Midland railway, were commenced, and two furnaces were blown in during the summer of 1878.

Such is a catalogue of the development of the early blast-furnace works in Northamptonshire. The first twenty-five years of the life of the iron industry were marked, in addition, by a number of developments, some of which proved abortive, in the working of the local pig-iron. This was the great age of wrought iron, and it was only natural, even allowing for the absence of coal, that some efforts should have been made to increase still further the prosperity of Northamptonshire as an industrial region, by building plant to work up the pig-iron into finished products. Although several attempts were made to set up puddling works, however, none of them succeeded, and the finished iron industry of the county, such as it is, has been confined to the foundry business.

In 1868 the "Wellingborough Bar Iron Company" was formed, and the erection of puddling furnaces was commenced on a site adjoining the Midland railway quite close to the East End smelting works. Two mills were to be built, one for rolling iron rails and the other for boiler plates and sheets; [26] but only a small portion of the plant was assembled when financial difficulties caused a cessation. In 1871 the company was revived, and the construction recommenced; [27] but the work was never completed, and the site and plant were disposed of in 1874. [28]

At Holt, on a small outlier of the ironstone just within Leicestershire, an ambitious programme for four blast furnaces, with "ample and suitable space for rolling mills in their vicinity," was prepared about 1867 or 1868, [29] and the project was revived some years later. Nothing came of it.

In the official statistics for 1885 and 1886 we find a record of a "Patent Iron Scrap Forge Works," owned by Mrs. Whitworth, at

[26] *Mining Journal*, June 5, 1869.
[27] Ibid., Mar. 11, 1871.
[28] Ibid., Sept. 5, 1874.
[29] Ibid., Sept. 19, 1868.

Northampton. In 1885 there were five puddling furnaces here, only one of which was working. Presumably the concern was a failure, for it is only recorded for these two years, and no other reference to it has been discovered.

A more successful venture was that of Messrs. Stenson & Co. at Northampton ; here the local pig-iron was successfully puddled, but the industry cannot be said to have been based entirely upon local product, for large quantities of scrap were mixed in with the pigs. Most of this iron went (as considerable amounts do at the present time) to Bedford and other places for the manufacture of agricultural implements.[30]

An interesting experiment with Northamptonshire ores was conducted at Towcester by Siemens, the inventor of the open-hearth steel process. About 1875 two " rotatory " furnaces were erected, which were designed to produce wrought iron direct from the ore, without the intervention of the blast furnace and pig-iron.[31] This patent process was experimented with for several years, and when the results were good, the iron was considered to be equal to Swedish bars in toughness and quality ; [32] but the low grade ores of Northamptonshire were not really suited to this type of furnace, and the production of wrought-iron was abandoned in 1878, though the furnaces continued to be used for producing spongy iron for filtering purposes until 1882.[33]

The production of cast-iron was established on a firmer basis than the puddling industry, for the amount of fuel necessary for reheating pigs is very much less than that required for the long puddling process and the driving of forge hammers and rolling mills. Consequently, it alone has survived. Messrs. Williamson, at Wellingborough, set up their foundries in the 'sixties, and produced railway chairs, kitchen grates, and material for iron bridges—and the production of this type of material is still carried on in Wellingborough and Northampton.

The period 1851–80 has been treated in such detail because it demonstrates the factors which were at work in modelling the Northamptonshire iron industry : the development of the various parts of the ore-field and their dependence on geographical conditions ; the results of an increased production of pig-iron upon the output and export of ore ; and the failure of all the attempts at establishing puddling plant in the district. By 1880 the industry was well set, and its prospects were most favourable. A certain amount of reshuffling of the population must have accompanied the rise of iron-making,

[30] *Mining Journal*, Sept. 19, 1868.
[31] Described in *Mining Journal*, Sept. 13, 1879.
[32] *Mining Journal*, Feb. 21, 1880. [33] *Journ. Iron and Steel Inst.*, 1883, p. 654.

for although the quarries were scattered, and did not need a large labour supply, the blast furnace works at Heyford, Finedon and Islip especially must have caused the expansion of the villages near to which they were situated. Even Wellingborough, already the seat of an important leather industry, increased its population by some 35 per cent between 1860 and 1870,[34] to the detriment of Market Harborough and others of the neighbouring small towns which still remained as market centres for farm produce. The next period, 1880–1900, would perhaps have been productive of far more sweeping changes, had not the whole aspect of the iron industry of Britain been changed by the substitution of steel for wrought iron. As it was, the iron industry in Northamptonshire came near to stagnation for a while, and we can pass over the period with a few brief comments only upon the changes in the geographical distribution of the quarrying industry.

1880–1900. The Effects of the Great Depression

Fig. 1 shows that the output of ore, apart from the years 1886–7 and 1892, of which the former was marked by a deep industrial depression and the latter by a series of coal strikes, remained fairly stationary around the 1¼ million tons level until the end of the century, when the recovery from the great depression was beginning to be evident. Fig. 3 shows that this 1¼ million tons continued to represent approximately the same proportion (about 7 or 8 per cent) of the whole British output of ore ; in other words, the British production also remained roughly stationary during this period, the expansion of the steel industry being fed principally by imported ores. Whilst the output of ore did not increase, however, the increased efficiency of the furnace plant and the substitution of larger and faster-driven furnaces for the old original ones, considerably augmented the production of pig-iron in Northamptonshire, with the result that less ore was available for export to the other iron-making districts (Fig. 2). The output of iron per furnace in 1879 was under 10,000 tons per annum ; by 1896 this had increased to over 20,000 tons (Fig. 4).

In Fig. 7 the relative importance of the various quarries, as judged by their output, is shown for 1881, the beginning of this period. A comparison with Fig. 6, for 1873, reveals several interesting facts. In the first place, a number of those workings which had been begun during the boom period of 1870–3 had but a short life, the result either of financial difficulties or of geological circumstances. The early workings at Islip, Slipton and Burton Latimer, the quarries at Ringstead, Towcester, Shutlanger, Burleigh Park, and certain of

[34] *Mining Journal*, Sept. 30, 1871.

the Desborough workings may be cited as examples of undertakings which only lasted a few years. Secondly, the ore-field became more scattered, and the production less concentrated upon a few large quarries. The decline of the south-western area (Duston and Blisworth), and of the eastern area (Newbridge and Ringstead) was both relative and absolute, whilst the rise of the Kettering, Cranford and Brixworth districts made the distribution of quarries much more even. The only district which changed little was that of Wellingborough, where the production went on steadily, to supply the constant demand of the local furnaces. Thirdly, by 1881, the Mid-Lincolnshire region had commenced the mining and quarrying of the ore.

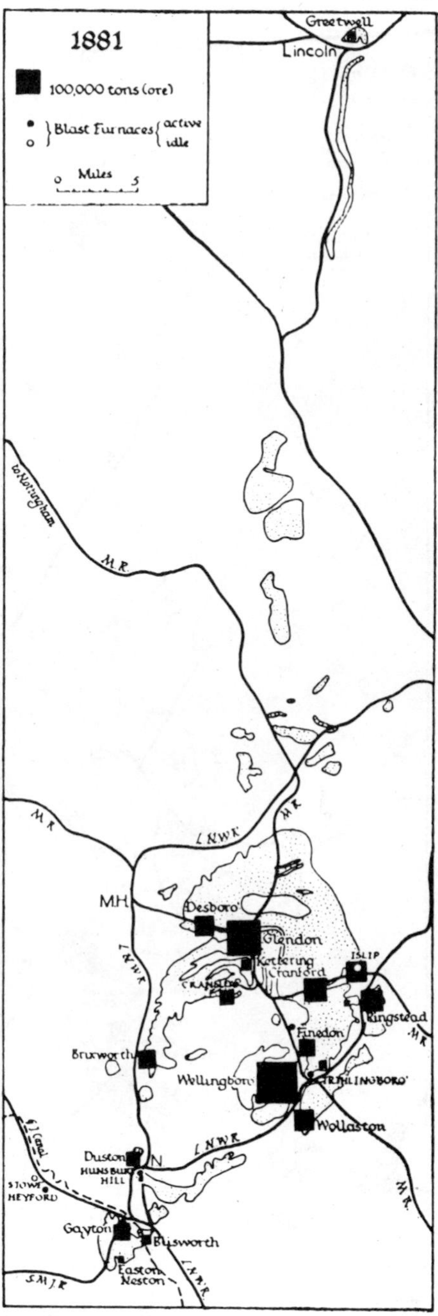

Fig. 7.—Quarries and blast furnaces in relation to railways, 1881.

The map for 1895 (Fig. 8) shows some even more striking changes; but before this is considered it will be as well to review briefly the progress of quarrying during the 'eighties. The construction of a special branch line of the Midland railway, from Ashwell to Cottesmore in 1882, opened up an entirely new field for the development of quarrying, and introduced industry into the county of Rutland,—an extensive area of outcrop ore which had hitherto been unavailable owing to the lack of rail

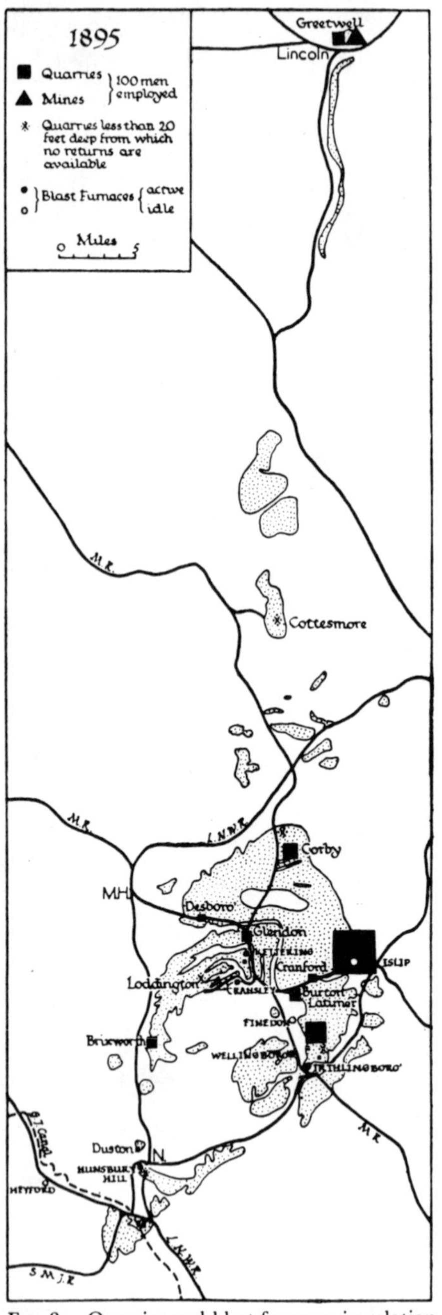

Fig. 8.—Quarries and blast furnaces in relation to railways, 1895.

transport facilities. Another new line of railway, the Midland connection between Kettering and Manton, opened in 1879, brought within the possibility of development the ore-bearing region around the Corby and Weldon inliers, and the outcrop ore along the scarp overlooking the Welland valley near Gretton,—areas in which quarrying was commenced in 1882 and 1888 respectively, by Lloyds Ironstone Co., a subsidiary of Alfred Hickman's of Bilston in the Black Country. The decline and final extinction of the Cransley ore output, about 1890, led that company to seek new ore-fields, and in 1891 the Loddington workings were opened, to be connected two years later by a standard gauge branch line with the Midland railway at Kettering. In addition to the opening of new areas, the period witnessed the decline of a number of older ones which had been

Note on statistics: From 1854 to 1881 the annual volumes compiled by Robert Hunt, Keeper of Mining Records at the Royal School of Mines, give the output and value of ironstone from every quarry and mine, and county totals; also lists of blast furnaces, and occasional data on railway and canal transport of ironstone and iron. In 1882, publication of statistics was taken over by the Home Office, Mineral Statistics Branch. Publication of individual pit outputs prohibited under Coal Mines Act, but allowed under Metalliferous Mines Act (and stratified ironstone comes under Coal Mines Act). Publication commenced of annual Lists of Mines and triennial Lists of Quarries, containing employment figures for each quarry and mine; but for shallow quarries under 20 feet deep (including numerous shallow ironstone pits) no statistical information is available at all. County output figures are given—but difficulty arises because the output of Rutland is included with the Oxfordshire Marlstone ores, and the Northampton Sands output of Lincolnshire is not distinguishable from the large output from the Middle and Lower Lias in that county.

working continuously for 30 or 40 years. By 1895 (Fig. 8) the south-western area had very greatly declined, and the immediate neighbourhood of Wellingborough had also almost ceased to produce, whilst the workings east of the Nene valley had become extinct. The Islip area had expanded rapidly owing to the demands of the three Islip furnaces; whilst the Corby, Gretton and Cottesmore areas are new features on the map. The importance of the Greetwell mines and quarries had also increased materially. On the whole, however, the output from the new areas only just balanced the reduction due to the decline of the older workings, and so there was no great absolute increase in the total production.

A noteworthy feature of this period is the increasing extent to which Derbyshire ironworks were acquiring leases of the Northamptonshire ore, so as to assure themselves of a regular supply at the cheapest possible rates. The Stanton company, which was the first to enter the county, increased its supplies and reserves by commencing the quarrying of the ore at Finedon (1883), Wellingborough (Harrowden) (1886), and Irchester (1892); [35] the Sheepbridge company opened up the Cottesmore area in 1882 and started a quarry at Brixworth in 1884; the Staveley company acquired interests in the Duston workings in 1885, and worked the short-lived Waltham area from 1882–5; and the Bennerley company began at Gretton in 1888 and at Geddington in 1898; whilst in the Park Gate company, of Rotherham, which became associated in the Duston workings in 1886, we have the first Yorkshire firm to enter the field.

Finally, an event occurred in 1895 which was destined to have very far-reaching results, particularly on the relation of the ironstone industry to the landscape. This was the introduction by Lloyds Ironstone Co. at Corby of the first "steam navvy" for excavating the iron ore.[36] When this machine was combined with a belt conveyor for the stripping of overburden, in 1897, the first areas of "hill-and-dale" were produced. Hitherto all the worked-out land had been restored to its agricultural use, but now in several parts of the orefield the increasing overburden began to be removed by mechanical means and restoration of the irregular, soil-less ridges thus produced was much more difficult. Another twenty years were to elapse, however, before mechanical excavation became at all common, and some examples of hand labour were still to be found in 1930.

[35] H. B. Hewlett, *The Quarries* (a series of articles in *The Stantonian*, reprinted, 1935), pp. 25–40.

[36] W. Barnes, "Excavating Machinery in the Ironstone Fields," *The Engineer*, Aug. 1942.

The history of the blast furnaces during the last two decades of the nineteenth century, like that of the quarries, shows two trends. First, there is the decline and extinction of one or two of the older works, and secondly, the opening of new works and the building of new furnaces by those already in existence. The original East End works had closed down in 1876, the firm of Butlin then concentrating upon the Irthlingborough plant. The Heyford works, having exhausted its local ore supplies, closed down in 1891, and was demolished some years later, and the Finedon works, which in 1883 had five furnaces out of six in blast, closed down in the same year, and was likewise subsequently demolished ; and the Stowe furnace, which had sprung into activity again in 1890, blew out only a year or so later, and was never again worked. This sudden extinction of works about 1890–2 is to be associated with a very definite period of depression in the iron industry as a whole which caused those firms which were financially unstable to go out of business. The reduction in the pig-iron ouput which followed this drop in the number of furnaces was very marked in 1893 (Fig. 4).

As regards the improvements, the most noteworthy is the erection of an entirely new works, at Wellingborough, on the North side of the town, alongside the Midland railway, in 1885. Three years later this establishment became the Wellingborough Iron Co., with two furnaces, to which a third was added in 1898. At the other works, the improvements and reconstructions were largely the result of the great depression of 1886, and of the trade boom which developed shortly after, about 1889. The Cransley furnaces were idle from 1886–9 as a result of the financial disaster caused by the depression, but in 1890 a new company, the New Cransley Iron and Steel Co., was formed and the plant largely reorganized. At Hunsbury Hill the furnaces were out of blast in 1888 and 1889, until they were bought up in 1890 by the executors of P. Phipps, the local brewing magnate. At Kettering and Irthlingborough, new furnaces were erected in 1889 and 1884 respectively. The result of all these changes is shown in Fig. 8, which shows the furnaces in and out of blast towards the end of the period, in 1895.

1900–30

Whilst this period, taken as a whole, witnessed a great and far-reaching expansion in the output of the Northampton Ironstone and to a lesser extent of the pig-iron produced in Northamptonshire, the 1914–18 war and the post-war years of depression had such a disturbing effect upon both industries, that not until about 1928–30 was the pre-war level of production attained. Moreover, after about 1912, Rutland and South Lincolnshire appeared as important ore-producing regions, and in consequence our field of study must be widened, for Northamptonshire, though still dominant, was now no longer the only area to be reckoned with. Fig. 3 shows that the proportion of the British output of ore produced in Northamptonshire rose rapidly in the early years of this century, and this rapid rise in the relative importance of the Inferior Oolite ore was continued by the advent of the other two counties as large producers, so that this particular formation was producing by 1930 well over 30 per cent of the total British output, with Northamptonshire alone responsible for about 20 per cent.

The period 1900–30 divides itself naturally into three parts :

(*a*) *1900–13*

Speaking generally, the decade preceding the War was one of extensive development in the iron and steel industry as a whole, and the boom periods of 1900, 1907 and 1913 did much to restore to Britain some of the prosperity which she had lost during the critical years when steel was being substituted for wrought iron. But as this development was mainly on the side of steel production, Northamptonshire did not share in it to such an extent as the great steel districts of Teesmouth, South Wales and Scotland. Nevertheless, the rapidity with which the ore output expanded rivalled the phenomenal increase which characterized the period 1867–73 (Fig. 1) ; whilst the pig-iron output increased at a rate comparable with that of the 1875–80 period (Fig. 4).

The 1900 trade boom resulted, like that of 1873, in the opening of a number of new workings for the ore, especially in the Wellingborough and Islip districts. In 1901 the first mines in Northamptonshire were opened in the latter region, where the rising ground between Slipton and Islip, and the down-faulting of the ore-bed, rendered quarrying a difficult process. Subsequently, the expansion in the output of ore seems to have been obtained more by an increase in the output of each quarry rather than by an increase in the total number of quarries. This was undoubtedly due in part to the use of mechanical excavators. There were nearly a dozen " steam navvies " in the various Corby quarries by 1913, and others were in use at Glendon, Kettering, Loddington, Desborough and Wellingborough. The chief alterations in the distribution of the quarries during this period were concerned with the further decline of the south-western area (Gayton ceased in 1900 and Duston in 1908), and the development of the Corby area, the region north of Kettering (Rothwell-Geddington), and the Islip-

Cranford area. The proportion of the ore moving out of the county rose again to 70 per cent during the first few years of the century. Two factors combined to produce this result. In the first place, the development of the furnaces did not keep pace with the increased ore output, until 1910 when the building of a new works at Corby resulted in an abrupt drop in the proportion of ore exported ; secondly, more and more " outside " firms began to work the ore on their own account, for the supply of their works in Derbyshire and elsewhere. Of the Derbyshire iron companies, Stanton began working at Orton in 1902, Clay Cross at the Cranford mine in 1908, and Staveley at Lamport in 1912 ; a Leicestershire firm, the Holwell Iron Co., of Melton Mowbray, began working at Colsterworth (Lincs) in 1912 ; a Scunthorpe works, the North Lincolnshire Iron Co., acquired the highly siliceous ore at Easton (near Stamford) in 1913 ; and even a Middlesbrough firm began to feel the necessity of supplementing its local ore supplies, Bell Bros. commencing at Wakerley in 1913.

The furnace plant which was working in 1900 underwent few changes of importance, other than a general increase in efficiency, and consequently in productive capacity, between that year and the War. In 1908, however, the construction of an entirely new smelting works was commenced by Lloyds Ironstone Co. of Corby, on a site alongside one of the main lines of the Midland railway just north of Corby village. Two large furnaces, each capable of producing about 1,000 tons of pig-iron per week (twice as much as the average of the rest of the Northamptonshire furnaces) were erected, and the whole of the plant was planned on the most up-to-date lines, with mechanical charging and the use of the waste gases for the production of electricity. The necessity for large supplies of ore to feed these furnaces meant a rapid increase in the productivity of the Corby area, which quickly became the largest single producing district in the whole field. Partly as a result of the introduction of these two large furnaces, the first of which was blown in in 1910 and the second in 1911, the average production per furnace rose in 1911 to over 27,000 tons per annum (Fig. 4).

(b) 1914–18

For various reasons the Northampton Ironstone field, and the industry based upon it, did not undergo such remarkable developments as were witnessed in other parts of the country. The output of ore in Northamptonshire fell considerably from the high level of 1913 during the first two years of the war, owing mainly to the reduction in the available supply of labour ; and though, with the opening up of new areas, it rose again in 1917–18, it was still below the level of the pre-war period (Fig. 1). The inferior geographical position of Northamptonshire, combined with the absence of coal, which had already prevented the county from participating in the expansion of the steel industry, again precluded any development in this direction, for the existing steel centres were more favourably situated for the erection of new plant. Moreover, even the production of basic pig-iron, for use in the new steel furnaces, was hindered by the fact that the Northampton Ironstone is very siliceous and is deficient in manganese, so that the addition of manganese ore, and in some cases of a small proportion of calcareous Marlstone ore from Oxfordshire, was necessary for this branch of the smelting industry.[37] However, certain important developments did take place, in both quarrying and smelting.

The need for more and more iron-ore, whatever its quality, soon made itself apparent, and numerous new workings were opened up. The county of Rutland especially underwent a great development, and quarrying was commenced on a large scale, in 1916, at Market Overton and Uppingham, the former locality having access to the Saxby and Bourne line of the M. & G.N. Joint Railway, and the latter to the Seaton and Uppingham branch of the L.N.W.R. A new portion of the field, previously almost untouched, was developed on the sides of the Nene valley between Northampton and Wellingborough, Whiston commencing in 1916 and Earls Barton in 1918. An especially interesting feature is the opening, in 1916, of the Irthlingborough mine. The Ebbw Vale Steel, Iron and Coal Co. had in 1915 been seeking new ore deposits to supplement their imports from Spain, which might at any moment become difficult or impossible to obtain. After extensive trials, they acquired a large tract of land (nearly 3,000 acres) and commenced work upon a mine and calcining plant, which was designed to produce some 12,000 tons of ore per week for feeding the South Wales furnaces. Within a year or so they were employing over 200 men, and the mine was the largest single producing unit in the whole of the ore-field.

Fig. 9 shows the location of the mines and quarries working in 1918. Comparing with Fig. 8, the decline of the south-western district is still further evident, whilst the dominance of the central area of the field (bounded by Corby, Brixworth, Irthlingborough and Islip) is beginning to diminish, with the opening up of Rutland and South Lincolnshire. The number of individual quarries is also a feature which had developed during the first two decades of the century.

[37] *Mem. Geol. Surv.*, "Mineral Resources," Vol. XII, 1920, p. 161.

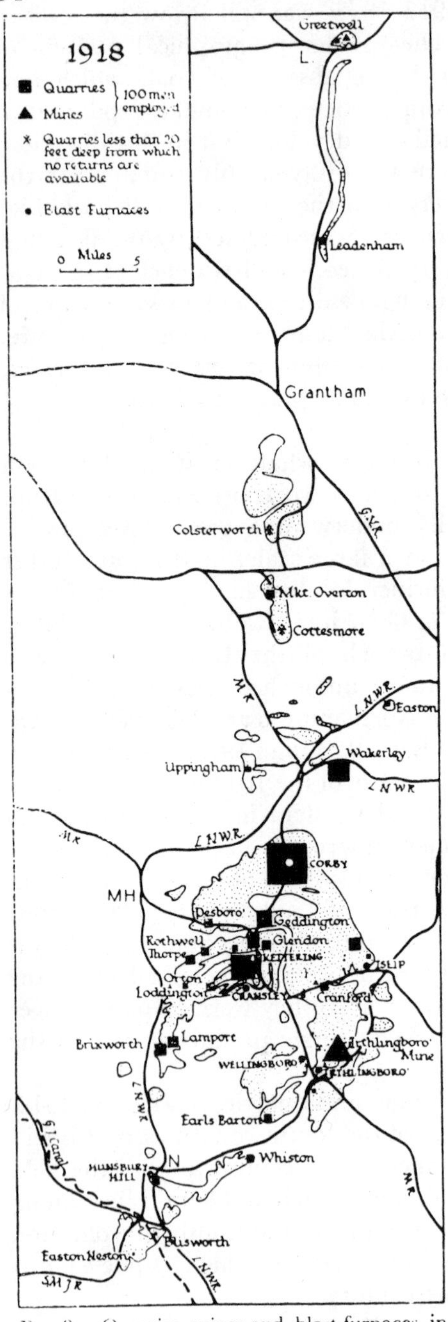

Fig. 9.—Quarries, mines and blast furnaces in relation to railways, 1918.

The vital need for steel, and the uncertain prospects of foreign hæmatite ore supplies resulted in the formation of the "Basic Plan" for the reorganization of the British steel industry and its adaptation to the use of phosphoric Jurassic ores.[38] The only Northamptonshire plant, however, which was definitely enlarged as a result of this plan was Corby, where a new furnace was completed in 1918. All the other furnaces were too small and out-of-date to permit of the addition of new and improved plant without upsetting the whole economy of the works. Whilst there was only one real addition, several existing furnaces, as at Islip and Hunsbury Hill, were turned over to the production of basic pig-iron.

(c) *1919–30*

The expansion of the quarrying and smelting industries was just making itself felt towards the end of the war, and the post-war boom of 1920 gave a further stimulus. Had this expansion continued uninterrupted, production, aided now by the employment of large excavating machines in many of the quarries, would quickly have surpassed the 1913 level.

[38] F. H. Hatch, *The Iron and Steel Industry of the United Kingdom under War Conditions*, 1919, especially Chapter VII.

Disaster overtook the whole of the iron and steel industry in 1921, however, and gave a most serious set-back to that part of it which was concerned with the Northampton Ironstone, especially in Northamptonshire. The 1921 depression and coal strike resulted, in fact, in the closing of the following ironstone quarries: Easton; Wakerley; Orton; Finedon Park; Burton Latimer; Wellingborough (Ditchford); Whiston; Earls Barton; Wootton; Blisworth; Hunsbury Hill; Easton Neston (Towcester). Four of the quarries (Hunsbury Hill, Wootton, Whiston and Blisworth) ceased work as a result of the blowing out of the Hunsbury Hill furnaces, which, casting their last pigs on January 28, 1921, never worked again and have since been demolished. Two noteworthy facts emerge from a study of these abandoned workings. First, amongst the victims of the crash were several of those wartime quarries which had started in areas where the geological conditions and the nature of the ore were unsatisfactory; second, the whole of the south-western district of the field, which had been steadily declining in its relative importance for fifty years, was put out of action at one blow. Once the Hunsbury Hill furnaces ceased, the adverse geographical circumstances proved fatal to the continuance of the quarrying. The siliceous deterioration of the ore, the inferior rail transport facilities (Blisworth having to rely on the Grand Junction Canal and Easton Neston on the S.M.J. railway), and the distance from the coalfields, all combined to render the working of the ore unprofitable.

The extinction of the south-western region was followed by the rapid rise of the extensive area of outcrop ore lying in North Rutland and South Lincolnshire. The Stainby and High Dyke mineral line, joining the main line of the G.N.R. south of Grantham (Fig. 9), which was sanctioned in 1912 and completed in December, 1917, had opened up this vast tract of easily-worked ore. Owing to the post-war depression and to the great amount of preliminary cutting and constructional work which was necessary, the extraction of ore did not really commence on a large scale until about 1924–5, since when the area between Sproxton and Colsterworth has come to occupy a very high place in the list of producing districts. In 1918 the Ashwell and Cottesmore mineral line was extended to Burley, thus opening up a further section of this area of surface ore, which began to be worked about 1921–2. Other new workings have been opened, at Pilton, south Rutland, in 1922, quite close to the Luffenham and Manton line of the Midland railway; in 1925 at Pitsford, connected by a short narrow-gauge line to the Northampton and Market Harborough branch of the L.N.W.R.; and in the same year at Harringworth, on the flanks of the Welland valley and adjoining the Midland main line.

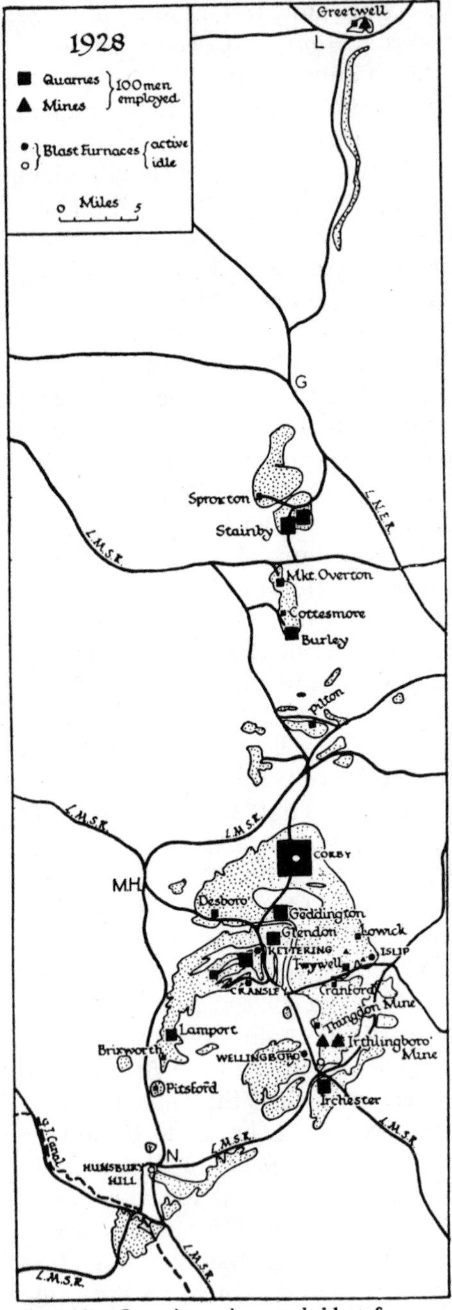

Fig. 10.—Quarries, mines and blast furnaces in relation to railways, 1928.

Fig. 10 shows the location of the mines and quarries in 1928. The reduction in the number of separate workings, compared with 1918, is at once obvious It is a feature partly due to the increasing dominance of the ore output by big iron and steel combines, which, with a plan of rationalization in view, closed down some of the smaller and less profitable quarries and concentrated on machine-aided mass-production from a few large ones. This is especially the case with the quarries owned by the Stanton and Staveley companies, Staveley having closed down Burton Latimer, Lamport and part of the Cranford group, whilst Stanton ceased work in several of their Glendon pits. The control of the ore, apart from that used locally, by the "foreign" ironworks was now almost complete. During the nineteen-twenties the Frodingham group and Stanton developed the South Lincolnshire area, Cargo Fleet acquired Irchester, Dorman Long opened Burley, Staveley commenced at

Note on statistics: A further change took place in 1921, when the functions of the Home Office Mineral Statistics Branch were transferred to the Mines Department. Output statistics by counties and geological formations are available in the *Annual Report of the Secretary for Mines*, with Mines and Quarries Lists as before. But statistics of blast furnaces, etc., cease to be available in such detail; the data published by the National Federation of Iron and Steel Manufacturers (from 1934, British Iron and Steel Federation) group works regionally, and Northamptonshire is lumped in with Derbyshire, Notts, Leics, and Essex; lists of furnaces in operation are, however, published regularly in the *Iron and Coal Trades Review*.

Pilton and Pitsford, whilst the Stanton company acquired the whole of the extensive workings of Messrs. Pain in the Rothwell and Weekley area and at Market Overton.

Although Fig. 10 is constructed on the same basis as Fig. 9, it does not form nearly such a reliable index of output, the reason being the discrepancy which exists between the output from quarries worked by hand labour and quarries worked by machinery. Thus the size of the blocks representing the South Lincolnshire quarries in Fig. 10 gives an inadequate idea of the actual output of this area, where mechanization is complete ; whilst the importance of the mines, with their large man-power per ton of output, is somewhat exaggerated.

The post-war period, then, witnessed an increase in the output of the Northampton Ironstone, responsibility for which rested largely with the new producing area of North Rutland and South Lincolnshire. Northamptonshire itself somewhat declined from its 1913 production, owing to the closing down of the whole south-western region and of a number of other quarries ; but its position relative to the rest of Britain remained, with the decline of the Cleveland area, about the same (about 17 or 18 per cent) (Fig. 3).

This same post-war period, however, witnessed the rise of a new problem in the ironstone fields, that of the devastation resulting from quarrying, and the conflict of quarrying with other forms of land-use and with amenity. Before the 1914 war there was very little mechanization in the quarrying industry (*vide supra*, p. 51) ; very few quarries worked ore from beneath more than 20 ft. of overburden, and from the beginning of the industry in 1852 until the war almost every acre of land from which ironstone has been extracted had been restored to agricultural use.[39, 40]

The advent of the mechanical shovel and the dragline for removing and dumping overburden was accompanied by the appearance of extensive areas of " hill-and-dale," and the producers, faced with the considerable cost of restoring a level surface, began increasingly to make payments to the landowners in lieu of restoration. In the shallower quarries—in South Lincolnshire for example—restoration to agriculture was relatively simple, even with mechanization ; but with overburdens in central Northamptonshire reaching forty to fifty feet, by 1930, the position was becoming serious, and the industry was becoming very much more obvious to the local inhabitants and

[39] S. H. Beaver, *The Land of Britain*, Part 58, Northamptonshire, pp. 372–4.
[40] See also the interesting figures given in *Memorandum dealing with the working of stratified ironstone in the Jurassic System in England* (National Council of Associated Iron Ore Producers, Kettering, 1944).

to those travelling by rail or road through the county than ever before.

Conclusion

The year 1930 really marks the end of a major episode in the history of the Northamptonshire iron industry. The great depression which followed the boom of 1929 placed the entire British iron and steel industry under a cloud, and forced considerable sections of it to make drastic, even revolutionary plans for rationalization and reconstruction. During the depression were born many great schemes which came to fruition, fortunately, before the outbreak of another war in 1939, and one of the greatest of these was the plan for a steel works at Corby. Stewarts and Lloyds, who had taken over Lloyds Ironstone Company with its quarries and furnaces at Corby in 1920, and had acquired the Islip Iron Company in 1930, planned a great tube manufacturing plant, supported by basic Bessemer steel produced through the medium of a large new battery of blast furnaces from the local ironstone. The capital was made available in 1932, building commenced in 1933, and the first steel was melted in 1936. Northamptonshire had at long last come into its own, and its geographical parallelism with Lorraine —in respect of the age and nature of its ores, its lack of local fuel, and its development of the "Thomas" (basic Bessemer) process—was rendered complete. Further, the Wellingborough works was acquired by the Stanton organization, and completely reconstructed in 1932. And the progress of mechanization in the quarries—shovels and drag-lines operated by diesel and electric power and no longer by steam —was rapid all over the orefield but especially in those areas controlled by Stewarts and Lloyds and Stanton. This materially increased the depth at which ironstone could be worked—an 85-ft. overburden was removed at Corby—and correspondingly the amount of unrestored land. But all this, culminating in the phenomenal output figures achieved during the 1939–45 war, is another story, and it is appropriate to end this detailed study at 1930.

IV

AIR TRANSPORT

Some Preliminary Considerations

By R. OGILVIE BUCHANAN, M.A., B.Sc.(Econ.), Ph.D.

A study of the geography of air transport, like that of any other form of transport, is a study of the fitting of the medium into the pattern of world economy. It is concerned with air routes, air bases, nature and amount of air traffic, and their reflex effects on trends in production and population patterns. This essay deals directly with none of these elements. It is rather an attempt to distinguish something of the economic character and problems of the aircraft and the ground installations that are the working tools of commercial air transport, in the hope that it may serve as useful groundwork for the study of the actual and potential geography of air communications.

It was not till 1903 that the first heavier-than-air machine, that of Wilbur and Orville Wright, made a successful flight, and the aeroplane was still in the hit or miss stage when war broke out in 1914. From 1914 to 1918 developments were almost wholly conditioned by military requirements, civil air transport did not begin until 1919, and the second world war saw again a practically complete concentration of design, production and operation on military types. Now military requirements (particularly in the stress of total war) differ from civil requirements in that within the limits of technical possibility they must be satisfied regardless of costs. The military machine, therefore, however good it is for the purpose for which it was designed, is normally a wildly extravagant machine to use for civil purposes,[1] and there are limits to the losses that even a nationalized air transport organization can face. But, if the military machine is not directly appropriate to civil air transport, the advances made during the war in the design of air-frames and of power units and in the evolution of ground control devices for military air operations can be adapted for civil purposes, and a period of rapid development in design and operation of civil aircraft and in the general organization of air transport services is a natural corollary of the end of World War II.

[1] This is not wholly of necessity a matter of direct running costs. The effect may be indirect. The Lancaster bomber, for instance, designed to carry a very concentrated load close to its axis of balance, would give a much poorer performance with the same weight of load distributed throughout the fuselage.

Among the characteristics of the aircraft, those most commonly mentioned are speed, small size, independence of surface obstacles and costliness. Only the first three of these are " prime " characteristics. The fourth, costliness, is derivative from other features, and as much of the rest of this essay will be concerned with cost elements no more need be said on the point here. The small size of the vehicle is of course obvious. The air giants of the present have about equalled the tonnage of the mediæval ship, and even that statement does not adequately indicate the disparity in size between the air vehicle and the ocean vehicle of today. More significant than over-all size is the pay-load capacity. In the aircraft that capacity is to be measured in weight, rather than in volume, and the proportion it bears to all-up weight of the loaded vehicle is much smaller than for any surface transport vehicle.

Speed of vehicle, in miles per hour while flying, is already high, both absolutely and relatively, and will continue to increase. It is a function partly of the aero-dynamic qualities of the airframe and partly of the power provided. " Aero-dynamic qualities " in this context includes the reduction of headward air resistance (drag) to the lowest possible point by the best possible streamlining of the air-frame as a whole and of all exposed parts, and by the refinement of smoothing of all outside surface. Both lines of attack have necessitated long-sustained, expensive research, and the attainment of the most exacting standards of workmanship. Power is still provided mainly by the orthodox internal combustion engine, and so remarkable have been the advances in design that 3,000 horse-power can be developed from a unit occupying only a few cubic feet. Nevertheless the ratio of engine weight to all-up weight is very high, and is a prime factor in reducing the ratio of pay-load weight to all-up weight as compared with surface vehicles. The other chief element in limiting pay-load weight is the amount of fuel that must be carried—and even the most economical aero-engine is greedy of fuel. A rough average fuel consumption rate may be taken to be rather less than half a pound weight of fuel per horse-power hour. An aircraft, therefore, powered with four 2,000 horse-power engines, uses about a ton and three-quarters of fuel per hour, or about fourteen tons on an eight-hour journey. It must of course carry enough additional fuel to afford a safe margin over and above estimated consumption. The longer the non-stop flight, the more drastically does fuel-load curtail pay-load, and the long distance and endurance records of the past have all been set up by craft that were in effect flying petrol tanks and nothing more. It is equally obvious that high-grade aviation spirit is a very expensive fuel. At half a crown a gallon the fuel cost of the fourteen tons mentioned above is £560. Compared with the orthodox type of aero-engine the newer "jet" engine gives two significant economies : the weight/power ratio of the engine itself is decidedly less, and the disposable pay-load correspondingly greater, for the same all-up weight and power output ; and, secondly, the paraffin fuel, though it is consumed at a comparable rate, is much cheaper than high-octane petrol. The commercial promise of these two sources of reduction of aircraft operating costs need not be laboured, but, important as they are, they are only part, ultimately perhaps only a minor part, of the contribution the "jet" engine has to offer. Its most distinctive characteristic is that its efficiency increases as atmospheric density decreases. It gives its most efficient output of power at high altitudes, where the bugbear of drag is markedly reduced, where atmospheric turbulence and frequency and severity of icing conditions are noticeably less, but where the petrol aero-engine loses much of the efficiency it has at low levels.

The significance of this power position lies in the fact that it must be costed against a relatively small weight of pay-load, with the result of high power cost per unit. Given full-loads, some progressive economy is gained with increasing size of aircraft : in general, disposable pay-load weight increases somewhat more rapidly than total weight of loaded aircraft. The advantage gained by the large aircraft in power cost per unit of pay-load may be offset, however, to a greater or smaller degree, at any rate for passenger airliners, by another significant element in direct operating costs. This is the cost of the operating crew. Like the power cost it is large as compared with surface transport, but unlike the power cost it tends to increase with increasing size of aircraft. A passenger train carrying 600 passengers requires no bigger crew than a passenger airliner carrying thirty passengers. In general, while power requirements per ton-mile in air transport may be very roughly averaged as about forty times the requirements in railway transport, crew requirements may be tentatively suggested to be about ten to fifteen times as much. These power and crew costs are the outstanding elements of the direct running costs incurred in the machine itself, most of the others partaking to a greater or smaller degree of the nature of overhead costs.

Among the overhead costs first mention should be made of the initial price of the aircraft. The amount of that price is related on the one hand to the nature of the aircraft itself and of its intricate equipment, and on the other to the conditions of its production. The new aircraft is a miracle of design and construction, but a miracle behind which lies an immense amount of research, experiment and development work, some of it directed specifically to the requirements of the new type, some of it not. The new type must bear its share of these basal overhead costs as well as its own specific costs if the aircraft manufacturing industry is to remain solvent. Nor has the manufacturer of civil aircraft anywhere in the world, except possibly in the United States, been able to avail himself fully of the economics of mass production methods. Apart from the difficulty of maintaining standards of workmanship required to approach closely to perfection, a difficulty that could probably be got over, the limited demand for any particular type has always kept production volume low. In addition obsolescence was extremely rapid before the recent war, and is very unlikely to be any less rapid during the next decade or so. In the nineteen-thirties a new type aircraft on first going into service already had its successor on the drawing-board, and its expectation of profitable working life was only about five years, which meant that the operator had to write off as depreciation some 20 per cent per annum of the initial cost of the machine. For a large airliner that could amount to more than £500 a week, whether the aircraft was employed or not.[2] If costs of this order of magnitude are to be covered by the operation of machines of which only 15 to 25 per cent of the all-up weight can be given to pay-load, then clearly air transport must be able to offer service for which the consuming public is prepared to pay a high price.

The specific service offered by the aircraft is usually said to be speed. More accurately it is saving of time, and this depends on other things as well as on the rate of movement of the aircraft in the air. It depends, for example, on the fact that the aircraft has a large measure of freedom from the effects of surface obstacles to straight line, or great circle, movement. It can save time by saving distance, and just how much distance can, in theory at any rate, be saved in an extreme case may be illustrated by route distances from London to Tokio. In rounded figures they are as follows: by sea via Suez and Singapore, 13,000 miles; via North Atlantic, Canada and North Pacific, 12,500 miles; by the trans-Siberian railway route, 8,600 miles; by great circle via the Arctic, 6,000 miles. It should be noted, however, that flying of regular services along great circle routes of this kind is still a matter for the somewhat distant future. The prime purpose of commercial air services is to move profitable pay-load, and for air services as for surface services those routes that provide the greatest supply of and demand for such pay-load movement will be the major world routes.

[2] This extreme rapidity of depreciation of the aircraft implies also rapid depreciation of manufacturer's specialized plant with further addition to overhead costs.

It is not without relevance to add that such routes are precisely those on which the provision of airports, fuel, servicing facilities and so on are most easily and cheaply provided. It might be urged that since the aircraft is a small vehicle there should be no difficulty in filling it with through traffic,[3] so that the availability of point to point traffic would become a matter of indifference, and regular, if not frequent, express services could be flown over " empty " routes. There is some force in this as an expression of faith, and the view may possibly be justified at some future time, but for the present and the more foreseeable future world-spanning great circle routes do not come into the picture—absence of bases and restricted economical range of aircraft alike forbid them.

Nor is the great circle route necessarily available even for the shorter distances. One of the major calamities of commercial aviation in pre-war Europe was the prevalence of government restriction of permissible routes over national territory. The beginnings of this policy can be seen in the acceptance by the Paris Convention in 1919 of the complete and exclusive sovereignty of any power over the whole atmosphere above its national territory and territorial waters, with, as a corollary, the decision that any power might proclaim " prohibited " areas and take any necessary steps to exclude foreign aircraft from such areas. It was left to Germany with its clandestine rearmament programme after 1933 to carry the idea to its logical conclusion, the prohibition of all foreign aircraft from using the German air except along a few specified narrow corridors of entry and of passage. Germany was by no means the only sinner, and the great majority of international routes in Europe was thus arbitrarily penalized. In the post-war world there is little indication that Russian treatment of foreign aircraft over U.S.S.R. territory will be any more generous than that of pre-war Germany. Complete freedom from inter-state restrictions of this kind over the continental territory of the U.S.A. was a major advantage enjoyed by American lines.[4]

[3] Cf. E. P. Warner, Wilbur Wright Memorial Lecture to the R.A.S., 1943 : " In the United States before the war, of the passengers using domestic airlines :

(i) half travelled 250 miles or less
(ii) three-quarters travelled 400 miles or less.
(iii) nine-tenths travelled 750 miles or less

so that even in air-minded U.S.A., through passenger traffic from terminal to terminal probably amounted to less than one-tenth of total passengers."

[4] It is not wholly without significance that the flying-boat, when it used foreign seaports, inherited something of the age-old tolerance extended to foreign shipping in the commercial harbours of the world, and escaped some at any rate of the new suspicion that met the land 'plane.

A further point that is relevant to the amount of time-saving that air services can provide is the number and length of halts a vessel must make on its journey. This is a function of a number of variables, including, for example, the range and capacity of the aircraft type, the nature of the route and the organization and equipment of the airway and the airports. Much of the potential advantage of the early Imperial Airways route to India, and later to the Far East and to Australia, was still-born through the limitation of flying to daylight hours only. Real success on long trunk routes demands uninterrupted flight by night as well as by day. Up till 1939 commercial night flying was limited to routes provided with " air lights," visual beacons whose distance from each other depended on visibility and varied from an average of about fifteen miles over Western Europe and North-eastern U.S.A. to about forty miles in the clearer atmosphere of the Middle East. By the end of the war radar installations had displaced visual beacons as the appropriate markers for the airways, but off the war routes little progress had been made, or has since been made, in providing the necessary equipment.

Apart from night flying the number of halts on a trunk route is largely a question of finding the optimum length of stage for the type of aircraft in use. This is a mean somewhere between the extremes of having a stage so long that fuel-load eats disastrously into the weight available for pay-load and a stage so short that even though relatively large pay-load be carried all the way landing fees, servicing and total time taken become excessive. Technical developments imply in general a longer optimum stage for large aircraft and a progressive advance towards eliminating more and more intermediate halts by express services on the long trunk routes. It will be obvious that this will apply with particular force to aircraft designed to give their most economical performance at very high altitudes. For a really effective " strato-line " it will cease to be a question of what is the longest stage that can be flown economically and become a question of what is the shortest stage that will justify the long climb upwards to the effective operating altitude.

At present, length of flying stage is not always a matter of choice. Up to 1939 it was still true that there was no world net of air routes comparable to the world net of sea routes. The Atlantic Ocean still formed a major gap, though two routes already crossed the Pacific Ocean. The problem of the 1,800 mile gap between West Africa and Brazil had been practically solved, and an experimental attack on the North Atlantic by flying-boat was in progress when war broke out. Here Greenland and Iceland seem to provide stepping stones conveniently spaced to reduce to manageable proportions the distances of open ocean to be crossed, and this route approximates fairly closely to a great circle route between eastern North America and North-Western Europe, but possible airfield sites on these stepping stones have such a high proportion of bad flying weather that it was considered preferable to accept the penalty of longer total journey and longer ocean crossing for the sake of less difficult weather conditions at airports. Up to the end of the war, while the civil services were still run by flying-boat, two routes were used according to season. The summer route, London–Foynes–Gander–New York, had an ocean stage of 2,000 miles from Foynes to Gander in the total of 3,600 miles. The winter route, London–Foynes–Azores–Bermuda–New York, had its longest ocean stage, 2,100 miles between Azores and Bermuda, and a total length of 4,700 miles. With the land 'plane services inaugurated at the end of the war the Foynes–Gander route has become the all-seasons route. Even so the ocean crossing causes fuel-load to cut significantly into pay-load, and is probably twice as long a stage as operators would choose if their choice were free. The trans-Pacific routes from the United States to the Philippines and East Asia and to New Zealand are tenuous links making use of island stepping stones too widely spaced for maximum economy, and they diverge very widely from the great circles joining their terminals.

The day, then, of the non-stop express, or even "limited," service, flown in very long stages along strict great circle routes, is not yet. In the more foreseeable future the major contribution to be made lies in the provision of more and better services in and between the closer areas of dense population and heavy production. On the shorter routes these services can be flown point to point great circle, so far as governmental prohibitions permit. Ultimately the sphere of air transport is surely to provide a world network of routes with a range of specialized services, attuned to the varied requirements of different regions and different kinds of traffic, and analogous to the variety of service offered by ocean, by rail and by road transport each in its own domain. The general picture no doubt will include a backbone of express services of large, fast, high-flying aircraft, using first-class airports along intercontinental routes and supplemented by slow services and branch services catering for intermediate airports along the trunk routes and for airports not on them. But an essential prerequisite is that air services shall, in addition to speed, provide standards of safety and regularity of the same order as those already achieved by surface transport.

For safety the airliner is already a more efficient instrument than is commonly realized. Progressive improvement of the ability of the air-frame to withstand the severest stresses, the increase of the power and reliability of engines and the development of instrument flying have made the airworthy aircraft an effective answer to the general meteorological hazards confronting it, and considerable progress has been made in dealing with such special problems as icing. The high degree of safety already achieved may be illustrated by the American civil aviation record for 1943 of 51,300,000 miles flown per accident,[5] a record very little inferior to that of American railways and superior to that of American roads. In general the meteorological hazards diminish with height above ground, but even in the stratosphere they do not disappear, and the atmospheric turbulence experienced there, though much less pronounced than in the troposphere, presents nevertheless a considerable problem at the very high speeds that are necessary to justify stratosphere flight.[6] The greatest risk of accident, however, occurs and will probably continue to occur during take-off and landing. Here fog is the chief enemy, more particularly in landing, and, despite the advances made with devices for blind landing, diversion of aircraft from fog-bound to fog-free airfields is still preferred.

Regularity has been achieved to a much less satisfactory degree than safety. Indeed the good safety record has been gained to a considerable extent by the sacrifice of regularity. Apart from daily aberrations such as keeping outgoing aircraft grounded or diverting incoming aircraft in bad flying weather, both winter services and night services have shown up to the present relatively poor results in number and dependability. In 1939, for example, European winter routes had barely 40 per cent of the mileage of summer routes, and were subject to much more frequent interruption. The performance on night routes was even less good. Much of the reason is to be found in the fact that dependability of flying services is as much a matter of ground facilities and organization as it is of aircraft and aircrew, and only in limited areas have the necessary ground developments kept pace with the evolution of the aircraft and its various instrumental aids to flying.

The hub of an airways system is the airport, the counterpart to the seaport, and, like the ocean port, increasing in magnitude and complexity with the increase in size and speed of the vehicles concerned. A modern first-class airport makes heavy demands on space, equipment and surface communications. It must provide taking-off, landing, parking, fuelling, repair and overhaul facilities for aircraft ; lighthouses

[5] Civil Aeronautics Administration.
[6] Cf. Air Commodore L. G. S. Payne, in the *Daily Telegraph*, April 25, 1947.

and beacons; flying control accommodation and equipment; an efficient meteorological service; radio and radar installations; customs office; post and telegraph office. It should also, and in the future will certainly have to, provide waiting rooms, shops, hotels, cinemas. Finally it must have fast services by surface transport, as frequent as air time-tables require, to the heart of the region served. Of these items—none of them distinguished by cheapness—the most expensive are those directly concerned with the landing ground itself.

The first point to notice is the area of land occupied. This is conditioned primarily by the number, length and arrangement of the runways, and the essential fact here is that modern large airliners, with heavy wing loading and high stalling speeds, touch down at not less than 80 miles an hour, many of them at more than 100 miles an hour, and some at more than 120 miles an hour. Runways, then, must be of great length and strength. Even main runways have not, indeed, yet reached the 5,000 yards' length foreseen in 1944 by the Air Ministry,[7] but rapid increase in length has marked the course of events in recent years. A width of 200 yards is already standard practice for runways in the larger airports. The orthodox lay-out of runways is basically X with the chief leg running in the direction of the prevailing wind, and this design lends itself to expansion and elaboration to handle increasing traffic at a busy terminal.[8] From the point of view of cost the significant feature of this land area is that it is situated within the range of surburban, if not indeed urban, land values. Cost of runways, which are built in solid concrete, preferably with a middle course of a material that will give a degree of resilience to the structure, and with a surface coating to damp down somewhat the initial landing shock, added to the site cost, goes far to explain the estimate, for example, of over £31,000,000 for the London Airport (Heathrow), a capital expenditure which at 3 per cent would require an annual revenue of £600,000 to cover airport overhead costs alone.

To the air transport operator the airfield costs are direct charges, since such share of them as he pays is chiefly in the form of landing fees. Clearly the real incidence of airport costs is related to the density of the traffic, and it would seem that unless traffic reaches the maximum capacity of the airport for a large part of every day landing fees adequate to cover airport costs would be crippling to the operators

[7] Pamphlet, "Technical Characteristics of Aerodromes," 1944.

[8] The more recent conception of a tangential lay-out of runways would appear to have the merits of reducing the risk of collisions on the ground and of economizing airport area. It is said, however, to have introduced difficulties of ground control of airborne traffic, and has not yet found general acceptance.

and would ensure the defeat of the purpose of the airport. Airport owners are commonly public authorities, either local or national, and able therefore at the expense of the ratepayer or the taxpayer to subsidize the airlines by charging landing fees at less than full cost rate. Whatever the view that may be held of the immediate or the ultimate desirability of such a practice, it must be recognized that part of the real cost of air transport is being paid by the community at large, and that until the practice ceases air transport remains in the nurseling stage, unable to stand on its own feet in level competition with other transport media.

On the subject of airport sites it might be mentioned in parenthesis that while they must be as close as possible to the centres of the great cities they serve, they are still somewhat distant as compared with the stations serving land transport, and the time taken in getting to and from the airports (" dead " time) may considerably affect the total time of journey. Just how significant the dead time is depends on the length of the flight and on the speed and frequency of ground services. In the extreme case, as, for example, between London and Birmingham before the war, it could neutralize the whole advantage of the greater travelling speed of the aircraft. More generally, it was the reason why, so long as aircraft speeds did not much exceed 100 miles per hour, development of internal air services was slow in Great Britain, with its short distances and close networks of road and railway carrying fast and frequent services. The contrast is striking on the one hand with the trunk Imperial routes to West Africa, to South Africa, to the Far East and to Australia and New Zealand, and on the other hand with the internal air routes of the U.S.A., U.S.S.R., and Australia, with their continental magnitude and more open or even rudimentary networks of surface communications. Every increase of aircraft speeds and every addition to the frequency of air services reduces the effect of " dead " time even in such countries as Great Britain or Belgium, and the advent of efficient helicopter service at the airports might well dispel the last vestige of such effect.

Other aspects of ground organization necessary to enable civil air transport to translate its potential contribution into actual performance include standardization of classification and control of airways, of traffic control procedure at and between airports, of radar equipment and service (including signals) not only at airports but at all necessary intermediate points, of meteorological information (actual and forecast), of rule of the road, of design of air navigation maps, and so on. These are obviously matters for international action, and international agreement in air matters is notoriously difficult to achieve. The inter-war organization of I.C.A.N. (International Commission for Air Navigation) achieved much, but there was much that it failed to achieve, and the very success of war-time technical developments has added new fields in which lack of uniformity in international practice will hamper the efficiency of air service operation. The Chicago Conference, 1944, gave small hope that post-war differences of national opinion would be noticeably less acute than pre-war.

These are technical and diplomatic matters, important not so much for their own sake as for the assistance they give, or do not give, to air service operators in finding a solution to their central commercial problem. This problem has already been indicated in part, but may be elaborated a little here. Since a very high proportion of total costs is represented by overhead costs, economy can be achieved only by spreading the costs over the maximum possible number of freight units. This implies a traffic flow sufficient to keep airports fully engaged, to eliminate empty cargo space from the individual aircraft, and to reduce idle time of aircraft to the essential minimum required for adequate maintenance. Such a traffic flow would have spectacular effects in the reduction of costs per ton mile, and would permit profitable working at low freight charges for suitable freight. But until some such flow is achieved low freight charges would merely increase losses and hasten bankruptcy. That is the dilemma which it was the purpose of government assistance to resolve, but government assistance before 1939, influenced by non-commercial considerations of national prestige and military security, for the most part degenerated into a bitter subsidies race in which the prime purpose of making civil air transport an independent, self-supporting element in the transport equipment of the world was completely lost.

V

RAINFALL REGIME IN THE SOUTH-EASTERN UNITED STATES

By P. R. CROWE, B.Sc.(Econ.), Ph.D.

The following paper is purely descriptive in character. Its object is to establish a regional classification of rainfall types on an objective basis. Yet it is the work of a geographer seeking valid generalizations and not of a meteorologist for whom the study of rainfall is an end in itself.

The problem of evolving a disciplined method of describing the seasonal incidence of rainfall has already been discussed in earlier papers.[1] Its importance seems undeniable in view of the significance of precipitation in human affairs; its difficulty is implicit in the baffling variability of the rainfall record. Yet objective description is a necessary first stage in a scientific analysis. If description is at fault, classification is likely to be in error, and if classification is illogical, the search for mechanisms and correlations will be launched upon false premises.

The graphical method employed is designed to throw emphasis on the seasons of most rapid increase or decrease of rainfall rather than on the months which appear to record " maximum " or " minimum " values. Classification is thus based on the " nodes " rather than on the " turning points " of the curves of annual régime. This appears to be particularly desirable in a humid region where the annual curve often lacks boldness of character and where the range of variation of monthly figures is very remarkable. Under such conditions, however, significant contrasts between adjacent months (" sharp " changes) are likely to be rare and, even when they occur, they may not be shared by any considerable group of stations. Emphasis must shift, therefore, to contrasts between *alternate* months (" graded " changes) and it is with respect to these that many of our distinctions will be drawn. It may also prove worth while to introduce the expression " slow change " to cover all cases where valid contrasts can be drawn only over a still longer time-interval. The point to be borne in mind throughout is that the very characteristics which compel us to think in the rather unfamiliar terms of May–July or October–December

[1] See References.

contrasts are just those which make the time of occurrence of arithmetically derived maximum or minimum values almost entirely fortuitous.

The present study has been confined to the area extending southwards from the latitude of St. Louis and Washington towards the Gulf of Mexico. Since most of the plotting was done before the War,

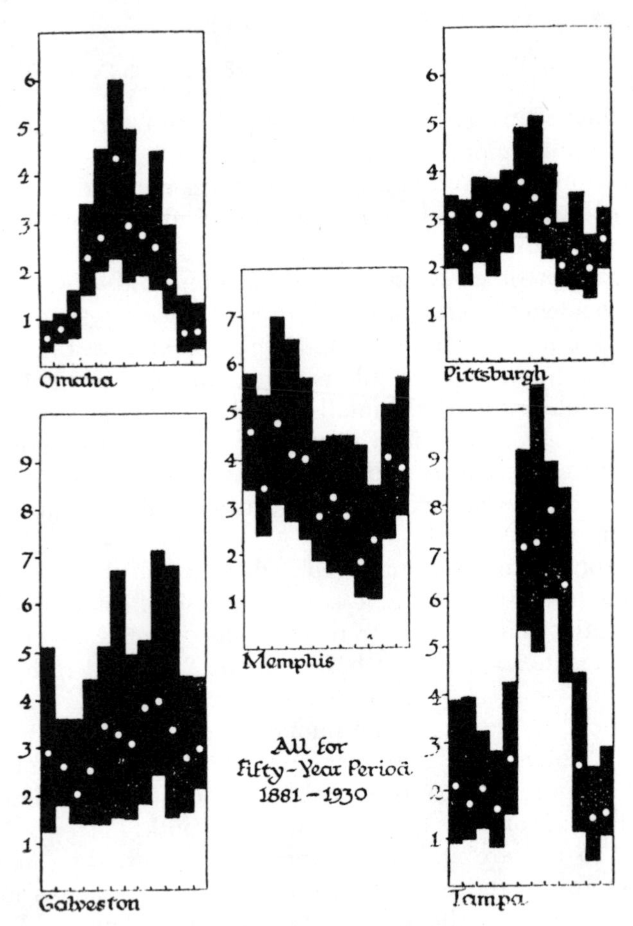

FIG. 11.—Selected rainfall graphs.

the fifty-year period 1881–1930 has been taken as standard. Half of the stations employed have records covering the whole of this period, the remainder have records covering at least forty years.

The interest of the rainfall of the South from the climatological point of view is due in large measure to its unusual wealth of transitions. This point is well illustrated by the five diagrams shown on Fig. 11.

Omaha, Nebr., Pittsburgh, Pa., Tampa, Fla., and Galveston, Tex., stand at the four corners of a square with a side of approximately 800 miles, whilst Memphis, Tenn., is not far from its centre. At Omaha and Tampa there is marked concentration of precipitation in the summer months but at Galveston and Pittsburgh, although the summer is still the wetter half of the year, the seasonal contrast is much less emphatic. The summer maximum thus weakens eastward in the north and westward in the south. In the centre, however, over a wide area around Memphis, it is the *winter* half of the year that has the heaviest rainfall !

The following table presents these facts in numerical form. Figures derived from the summation of monthly medians and of the monthly averages given by the Weather Bureau are given for comparison but it will be noted that the percentages remain virtually unchanged.

		Summer (*Apr.–Oct.*)	*Winter* (*Oct.–Mar.*)	*Summer as Percentage of Year*
		in.	in.	%
Omaha	Medians	17·7	5·7	76
	Averages	21·3	7·2	75
Tampa	Medians	32·5	11·4	74
	Averages	35·8	14·7	71
Pittsburgh	Medians	18·3	15·5	54
	Averages	19·6	16·2	55
Galveston	Medians	20·2	16·7	55
	Averages	24·4	20·4	55
Memphis	Medians	18·7	23·0	45
	Averages	22·4	25·9	46

There are thus three main rainfall provinces in this area. In the west, the *Plains Borderlands* (E) have most rainfall in summer, especially in early summer. In the east, the *Atlantic Coastlands* (J) have a mid-summer maximum which locally reaches a very high degree of prominence. Between these lie the *Mississippi Lowlands* (G) where rather more than half of the annual precipitation falls during the winter half of the year. These broad contrasts correspond closely to R. DeC. Ward's distinction between " Missouri," " Atlantic " and " Tennessee " types of régime,[2] but our interest is in their more precise definition and in the examination of the nature of the transitions, which he ignores.

The Plains Borderlands (E)

It was shown in " The Rainfall Régime of the Western Plains " [3] that the most noteworthy features of the régime of the eastern sections of the Great Plains are a significant increase from March to April

[2] *Climates of the United States*, pp. 186–92. [3] *Geog. Rev.*, 1936, p. 483.

and a significant decrease from October to November (Omaha, Crosbyton). A rainy season of seven months' duration (April to October) is thus succeeded by a drier season five months long

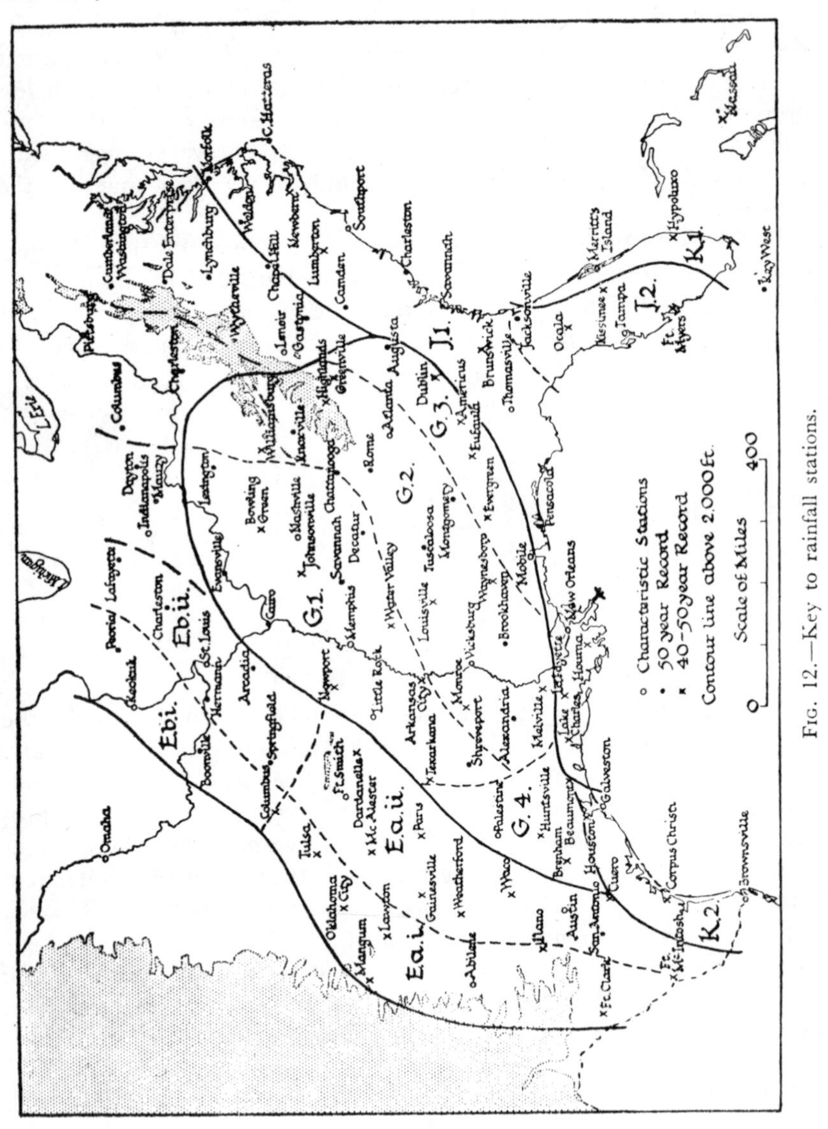

Fig. 12.—Key to rainfall stations.

(November to March) and the transitions from one season to the other are remarkably abrupt. They are, of course, related in part to wide temperature contrasts between the seasons.

It is to be anticipated that, as we approach the Mississippi Lowlands with their higher winter temperatures and much heavier winter precipitation, this clear-cut distinction between the seasons will progressively weaken. The zone where this occurs and where the essential features of the Mississippi or "Tennessee" type are still undeveloped may be conveniently described as the Plains Borderlands. It occupies a zone about 250 miles wide extending diagonally across the country from central Texas to the shores of Lake Michigan.

Subdivision of this extensive area is possible both longitudinally according to the stages by which the Plains features are gradually eliminated, and transversely in the general neighbourhood of the Osark Plateau. It is not easy to decide which of these distinctions is the more fundamental but, in view of the range of latitude involved, it seemed wise to make the primary subdivision on a transverse basis. Using the names of the river basins broadly involved, the zone has thus been divided into a southern or Red River region, and a northern or Illinois River region. It is convenient to name further sub-divisions after characteristic stations (Fig. 12).

The Red River Region (Ea.)

Abilene area (Ea.i). In this area, which forms a fringe about 100 miles wide along the south-eastern border of the Plains Province as previously defined, the sharp changes in April and November, though not entirely absent, are usually replaced by successive graded changes between February–April and March–May and again between September–November and October–December. The transition from summer to winter is thus only slightly less abrupt than farther west and the five winter months still form a period of remarkably light precipitation. The chief characteristic of this zone, however, is the significant decrease of rainfall from May to July (recalling conditions in Montana and Wyoming). This feature increases in intensity southwards until at stations beyond Llano, Tex., July and August, with median rainfalls of about an inch, are scarcely wetter than the midwinter months.

Austin–Fort Smith area (Ea.ii). As we move eastward across the Red River region, early winter precipitation rapidly increases so that the contrasts between September–November and October–December disappear. Indeed at several stations it is no longer possible to draw valid comparisons between the mean rainfall of any of the months from July to January or February (Fort Smith). The increases from February to April and from March to May are still well-marked, however, as also is the significant decrease from May to July already recognized to the west. The net result of these developments is to

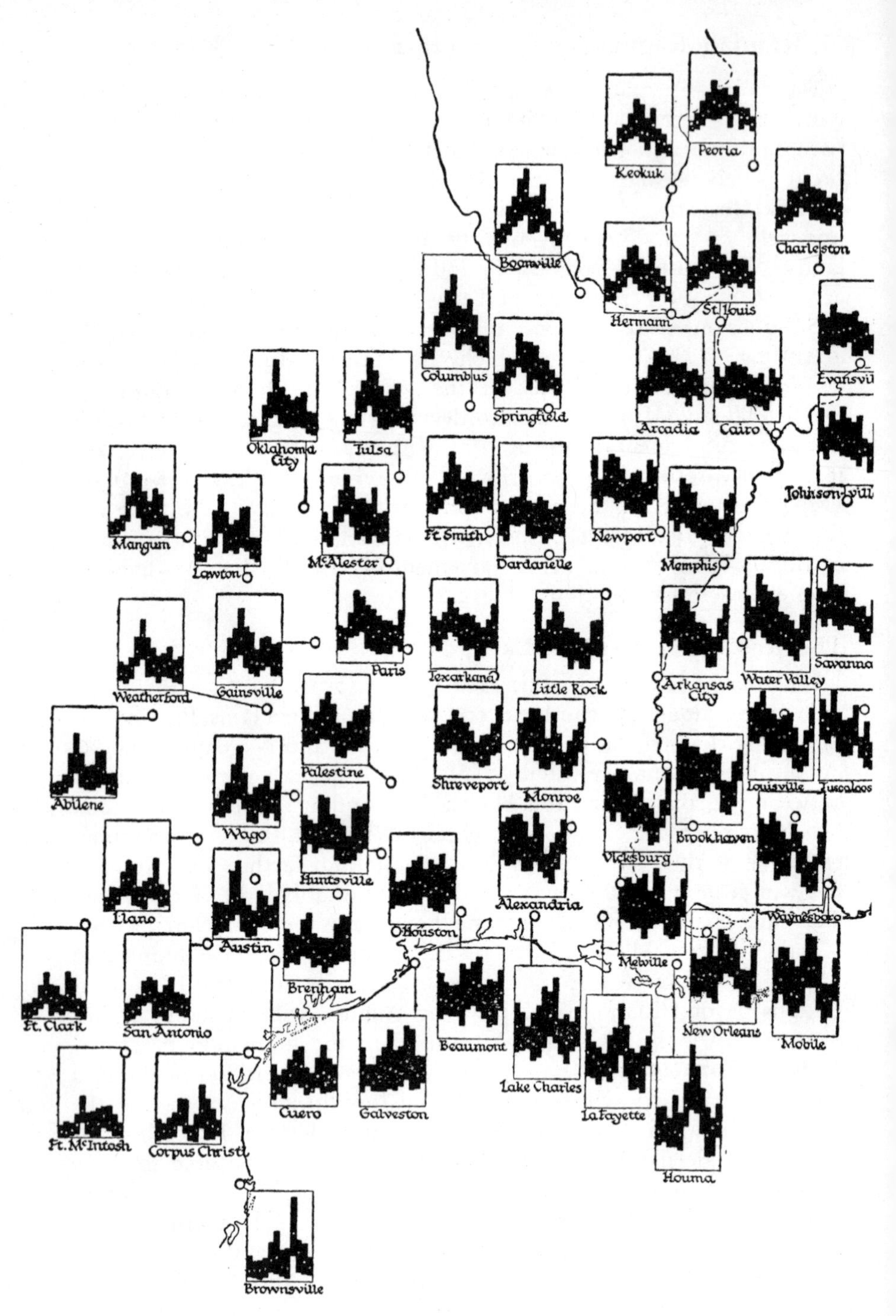
Peoria
Keokuk
Charleston
Boonville
Hermann
St Louis
Columbus
Springfield
Arcadia
Cairo
Oklahoma City
Tulsa
Mangum
Lawton
McAlester
Ft. Smith
Dardanelle
Newport
Memphis
Weatherford
Gainsville
Paris
Texarkana
Little Rock
Arkansas City
Water Valley
Abilene
Palestine
Shreveport
Monroe
Louisville
Waco
Huntsville
Vicksburg
Brookhaven
Llano
Austin
Houston
Alexandria
Waynesboro
Ft. Clark
San Antonio
Brenham
Beaumont
Melville
New Orleans
Mobile
Ft. McIntosh
Corpus Christi
Cuero
Galveston
Lake Charles
La Fayette
Houma
Brownsville

Fig. 1

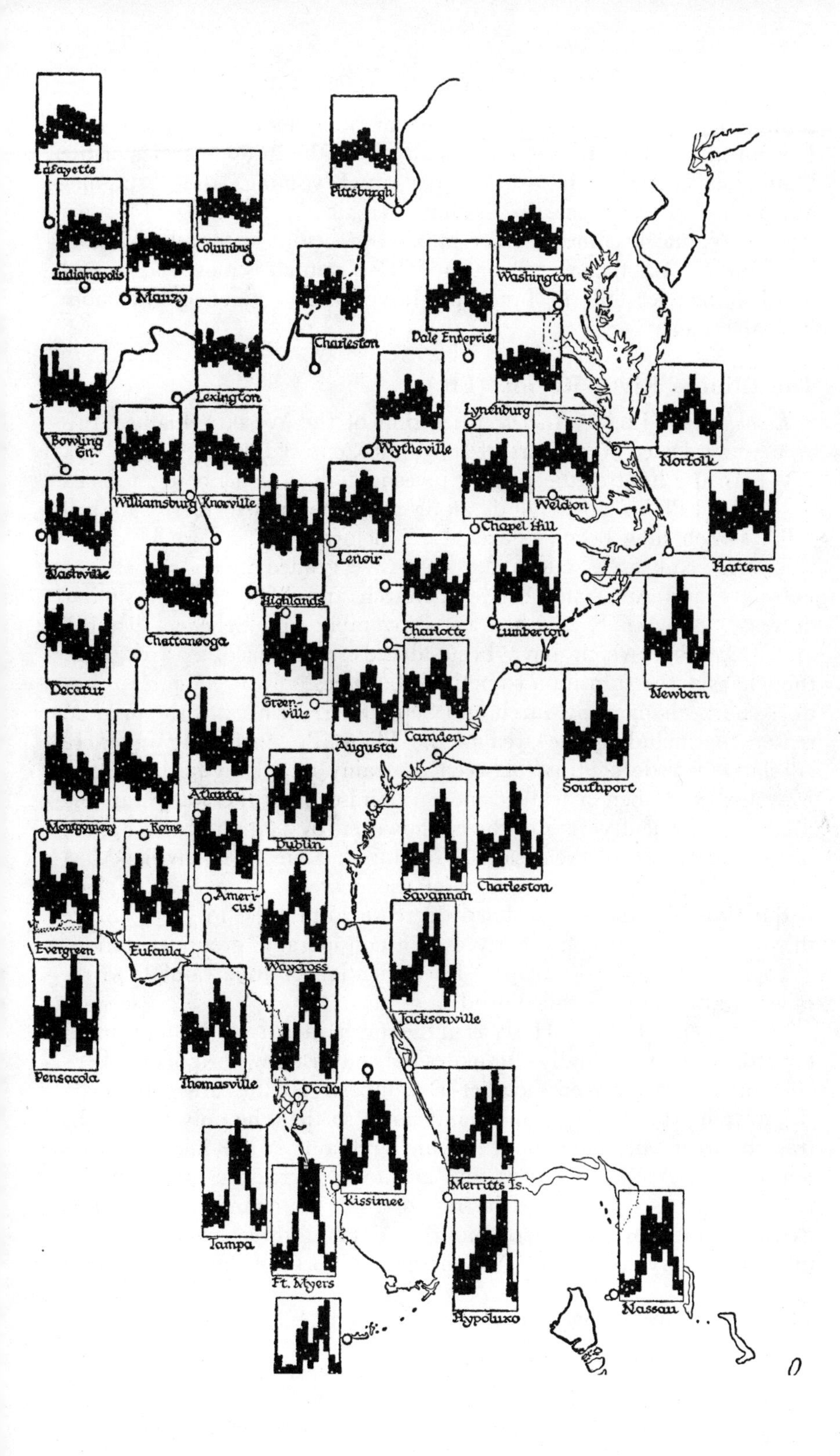

Lafayette
Pittsburgh
Columbus
Indianapolis
Mauzy
Washington
Charleston
Dale Enterprise
Lexington
Bowling Gn.
Lynchburg
Wytheville
Norfolk
Williamsburg
Knoxville
Weldon
Chapel Hill
Hatteras
Nashville
Lenoir
Highlands
Charlotte
Lumberton
Chattanooga
Decatur
Newbern
Greenville
Augusta
Camden
Southport
Atlanta
Montgomery
Rome
Dublin
Savannah
Charleston
Americus
Evergreen
Eufaula
Waycross
Jacksonville
Pensacola
Thomasville
Ocala
Merritts Is.
Kissimee
Tampa
Ft. Myers
Hypoluxo
Nassau

confine the really wet season to the two months, *April and May* (Austin). It is an interesting fact therefore, that along the humid border of the Plains, as along their arid border in Wyoming, the "summer maximum" usually regarded as characteristic of the Plains or Missouri type of régime becomes a May maximum before it completely loses its identity. In other words, on the Plains, rainfall is most widespread in May, heavier falls in June and July being confined to the more central regions.

The Illinois River Region (Eb.)

Keokuk area (Eb.i). In the discussion of the Western Plains [4] this area was referred to as part of the West-Central Lowland Zone (i.e. D3). In the light of the present investigation it seems better to class it with the Plains Borderlands along with the area formerly marked "F" which now becomes area Eb.ii below.

The Keokuk area retains the successive graded increases between February and April and between March and May and the similar decreases between September and November and between October and December which have been already recognized in area Ea.i, though here the transition sometimes becomes abrupt enough to produce sharp changes in March and September. This was the original reason for including the area in the Plains Province. Again, there is relatively little contrast between the rainy months, especially from May to September, a feature more reminiscent of the west than the south. A difficulty is introduced however by stations immediately to the north-west of the Osarks (Columbus, Kans.; Boonville, Mo.) where, although the significance of the decrease from May to July is questionable, there is a sharp decrease from June to July. Since this does not occur consistently over a considerable group of stations it appears to be due to a local intensification of June rainfall which might repay more detailed study.

St. Louis area (Eb.ii). Here, as in the south, increasing winter rainfall towards the east gradually eliminates the contrasts between September–November and between October–December. At the same time even the March–May change loses significance so that the only valid contrast between alternate months which remains is the increase from February to April. There is thus a notable increase of precipitation early in spring but during the rest of the year changes are so slow as to require comparison of monthly data at three- to four-month intervals (Lafayette, Ind.). Yet the general form of the annual curve

[4] Op. cit., p. 483, and map, p. 478.

is still reminiscent of the Plains and *May and June* are usually significantly wetter than the midwinter months from December to February. The season of maximum rainfall is thus a month later than in area

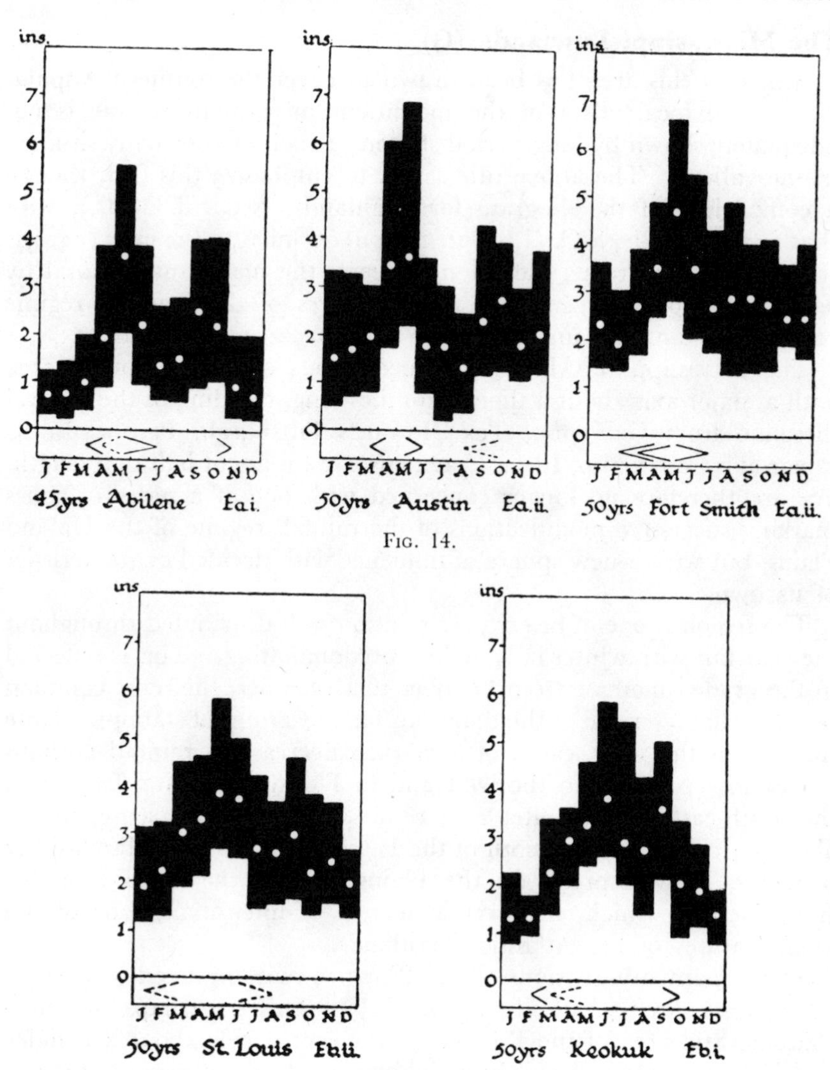

Fig. 14.

Fig. 15.

Ea.ii to the south but a distant relationship to the Plains Province is still maintained.

The presence of the famous " oak openings " of Illinois and Indiana in area Eb.ii and of the Black Prairies of Central Texas in area

Ea.ii, although unquestionably partly related to edaphic factors, certainly suggests that the title " Plains Borderlands " is not entirely inappropriate.

The Mississippi Lowlands (G)

Although this area has been drawn to cover the southern Appalachians, the local effect of the mountains on rainfall régime is not adequately shown by long-period stations, which are normally situated in the valleys. The above title serves to emphasize this fact, though a comparison of the diagram for Highlands, N.C. (3,350 ft.), with that for Greenville, S.C. (1,039 ft.), about 50 miles to the east, suggests that even when greater elevation increases the mean annual total by as much as 60 per cent, the main features of the regional régime remain substantially unchanged.

The Mississippi Lowlands Province covers a great elliptical area with a major axis about a thousand miles long, trending in the general direction from Galveston, Tex., towards Pittsburgh, Pa., whilst its minor axis from Cairo, Ill., to Americus, Ga., is about half that length. We are therefore no longer concerned with one of a series of zones marking successive modifications of the rainfall régime of the Upland Plains, but with a new sphere of influence with decided characteristics of its own.

The region is one of heavy precipitation well-distributed throughout the year but with winter falls slightly predominating. This is reflected in the graded increase from October to December, the most common positive characteristic of the diagrams for this group of stations. Note that this is the very season of a marked decrease in rainfall both on the Plains (Abilene) to the west and in Florida (Merritts Island) to the south-east. The winter rain results, in a negative sense, in the absence from this region both of the February–April and March–May increases, so widespread on the Plains, and of the April–June (or July) increase which, in varying degrees of intensity, is one of the main features of the Atlantic Coastlands.

Three main sub-divisions of the Province may be noted.

The Little Rock—Nashville Region (G.1). This is perhaps the most characteristic area of the Province. It forms a belt about 250 miles in width, extending from the neighbourhood of Shreveport, La., to Lexington, Ky. The increase from October to December is at once introduced and at nearly all the stations it forms the most abrupt transition experienced throughout the year. This feature may well result from the clash of Gulf and Continental air masses as winter spreads over the continental interior.

During the rest of the year significant contrasts, even between alternate months, are very rare and, indeed, from December to May, and at some stations even from November to July, it is virtually im-

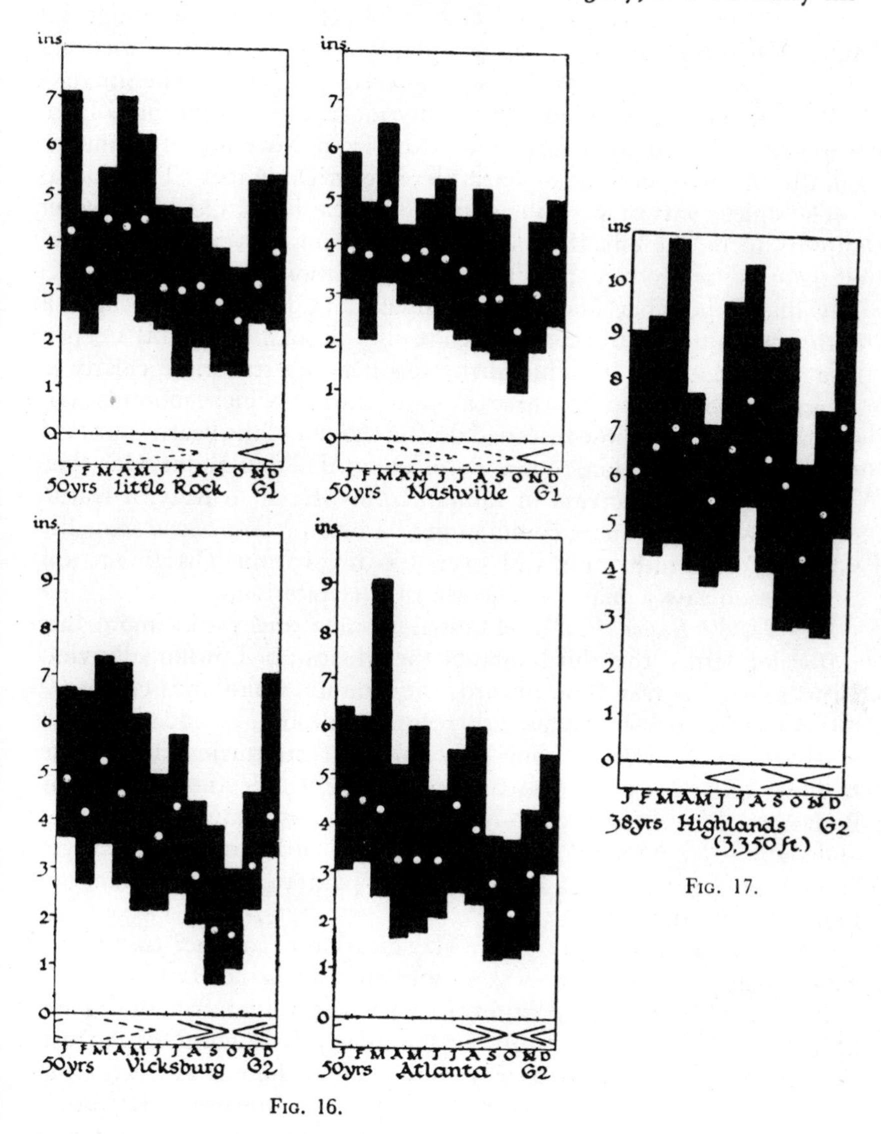

Fig. 16.

Fig. 17.

possible, in view of the wide range of variation, to distinguish one month from another. A slow decrease of rainfall from spring to autumn is nevertheless apparent and *March and April* are usually significantly wetter than September and October. Thus is introduced the relative drought of the Fall, a feature contributing to the noted charm of that season over most of the Eastern States.

The Vicksburg–Atlanta Region (G.2). This forms a zone about 150 miles in width on the south-eastern flank of the region just discussed, from Alexandria, La., to beyond Knoxville, Tenn. The increase from October to December is generally more abrupt than in G.1 and is usually followed by a further increase from November to January. Locally this may become a sharp increase in December (Tuscaloosa).

The chief contrast with the Little Rock–Nashville region, however, appears to be the effect of increasing midsummer rains. One result of these is to produce a significant diminution of rainfall between July and September or between August and October. Locally, at both extremities of the region (Knoxville, Brookhaven), this steepens to a sharp decrease in September, thus defining still more clearly an autumn spell of two to three months during which monthly falls are only about half those recorded during the rest of the year. Towards the centre of the region, round Tuscaloosa, Ala., and Rome, Ga., there is also a marked decrease of rainfall from March to May, a feature which becomes still more common in G.3 below. It is worth recalling that at Fort Smith, only just over 400 miles from Tuscaloosa, this same season saw a marked increase in precipitation.

The Mobile Region (G.3). Limited in area and rarely more than 100 miles across, this third zone of the Mississippi Lowland Province forms a genuine transition towards the Atlantic Coastlands type. All the main features of G.2 are still recognizable but, in addition, there is a significant increase from May to July, introducing midsummer rains which are just as heavy as those of late winter and early spring. Both April–May and October–November are thus relatively dry periods and there is a genuine " double maximum " in February–March and July–August, even when due regard is paid to the wide range of monthly variation.

The Palestine Area (G.4). The classification of another transitional zone at the south-western extremity of the Mississippi Lowlands presents greater difficulty. Within a small area round Palestine, Tex., although April–May is usually wetter than July–August (or August–September) as in the Austin-Fort Smith area, there is an increase of rainfall, with the approach of winter, as in the Mississippi Lowlands. Precipitation is very variable, however, and only slow changes can be recognized. Indeed, at Houston where, in addition, August and September are affected by rains from the Gulf Coast, the threefold transition results in a diagram from which it is impossible to draw valid distinctions between the mean falls of any two months out of the twelve. There is thus " rain at all seasons " in the most complete sense of the term.

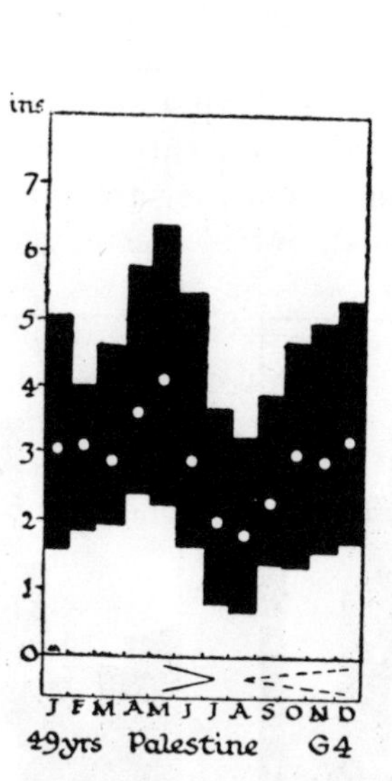

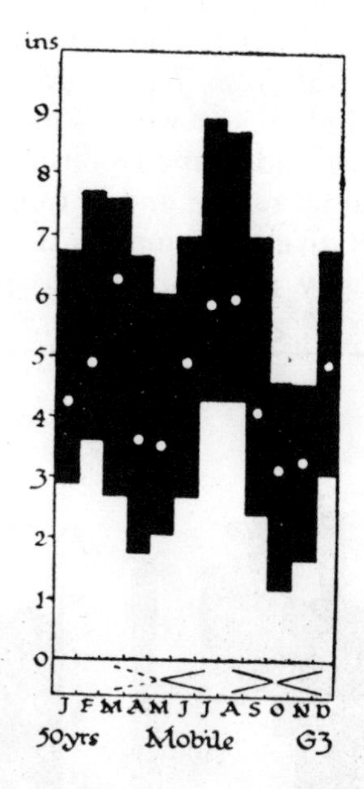

FIG. 18.

It will be seen later that at the opposite extremity of the province, round Mauzy, Ind., where the Plains, Great Lakes and Mississippi regions approach each other to form a similar threefold transition, very much the same effect is produced.

The Atlantic Coastlands (J.)

As its name implies, this province is essentially littoral. Its typical régime is best developed in Florida where very abrupt changes in rainfall (" major breaks ") are beautifully exemplified. Related types are found in a zone about 100 miles wide which extends along the Atlantic coast to North Carolina, and, with a number of complex transitions, in a much narrower fringe along the northern coast of the Gulf of Mexico.

The essential feature here is the reappearance of a genuine mid-summer maximum though the source of the humidity and the mechanism by which it is precipitated must obviously be quite independent of conditions on the Great Plains. Superficial resemblances between graphs for the two areas may possibly occur but the dispersion diagrams show clearly enough that the characteristic rainfall " breaks " of the Plains, particularly those in April and November, are certainly not repeated in this area. Taking the province as a whole, the most violent contrasts in rainfall amount occur between May and June, on the one hand, and September and October, on the other. The four months, *June, July, August and September*, thus clearly form a distinct season, indeed the word "monsoon" would seem to be appropriate. This view finds some confirmation in recent analyses of the frequency of torrential rains,[5] and of thunderstorms.[6] Not only are these shown to be relatively common in the Deep South but they are most frequent in precisely the four months just named. It is even suggested that

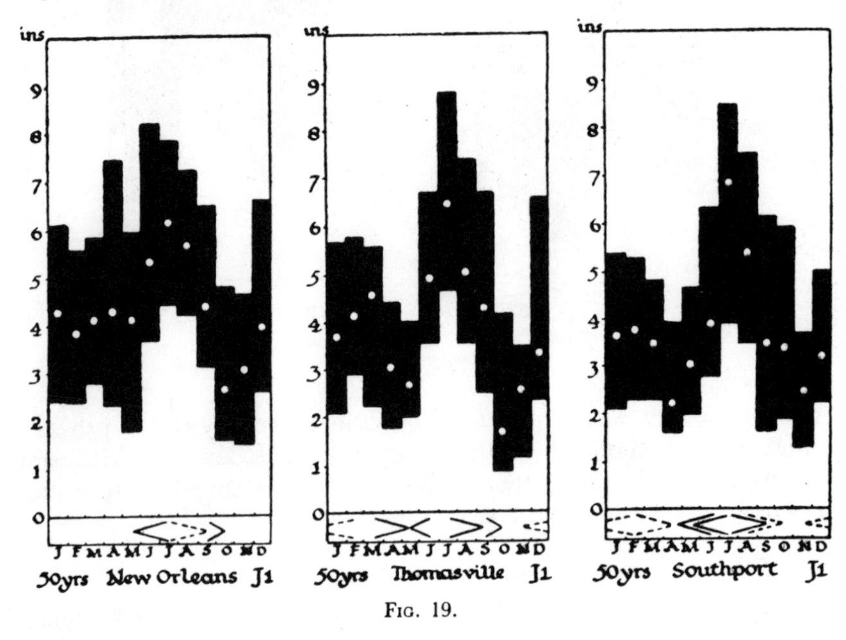

Fig. 19.

the nature of the rainfall in this area may partly explain the prevalence of very sandy soils there.

The New-Orleans-Thomasville-Southport Region (J.1). This region extends over about 1,200 miles of coastland from the eastern borders of Texas to the northern border of North Carolina. Rainfall is heavy (from 45 to 50 inches) and even the drier months yield mean records of from two to three inches. The chief feature, however, is the period of summer rain introduced and concluded, in the central area between

[5] S. S. Visher, "Torrential Rains as a Serious Handicap in the South," *Geog. Rev.*, 1941, pp. 644–52.

[6] W. H. Alexander, "The Distribution of Thunderstorms in the United States," *Monthly Weather Review*, 1935, pp. 157–8.

Thomasville and Savannah, by the sharp changes in June and October which have been referred to above. Yet the onset of the monsoon rains is rather more abrupt than their cessation and even at Thomasville, September is appreciably drier than July. As the extremities of the region are approached, both northwards through the Carolinas and westwards towards the Mississippi delta, the sharpness of the transition gradually weakens and the summer rains are introduced by successive increases between April–June and May–July and concluded by similar decreases between August–October and September–November. This has the effect of producing an approximation to a July or a July–August "maximum."

Over a small area in southern Alabama (Pensacola) June is no wetter than the first five months of the year and the monsoon rains are introduced by a sharp increase in July. In other respects this area falls within the present region. Ward's "North Gulf Coast Type" thus appears to be in error since the features he emphasizes are those of region G.3 which are not experienced at truly coastal stations.

At the northern and western extremities of this region the monsoon feature becomes so weak that it is a little difficult to say where precisely its influence comes to an end.

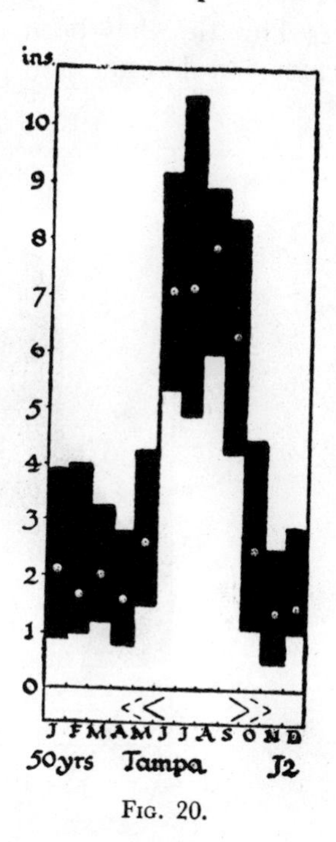

Fig. 20.

The Tampa Region (J.2). Covering the eastern portion of peninsular Florida, this region provides diagrams which are ideal for demonstration purposes. The monsoon season of four months' duration is marked off from the rest of the year by breaks reaching "major" dimensions at three stations. Within that season there is no significant contrast between the component months despite minor differences in their arithmetic means. Again, at Tampa and Fort Myers, for instance, there is a dry season lasting six months and no contrast can be drawn between any pair of these. May and October are thus left as transitional periods, clearly distinguishable from both the wet and the dry seasons in the south (Fort Myers) but merging gradually with the latter as we travel north. The violence of the rainfall changes in this region is exceeded only by those of Monsoon India. Thus, over the fifty-year period 1881–1930 at Tampa, June recorded more rain than the preceding May on forty-five occasions whilst September recorded more than the following October on forty-one occasions.[7] To those accustomed to think in terms of monthly means the number of failures may be, perhaps, more impressive than the number of successes! Yet a little experience would soon show these to be rainfall changes of a very high degree of dependability!

The Hurricane Coasts (K.)

This title has been chosen, rather tentatively, to cover two small areas in Florida and Texas whose only feature in common is the

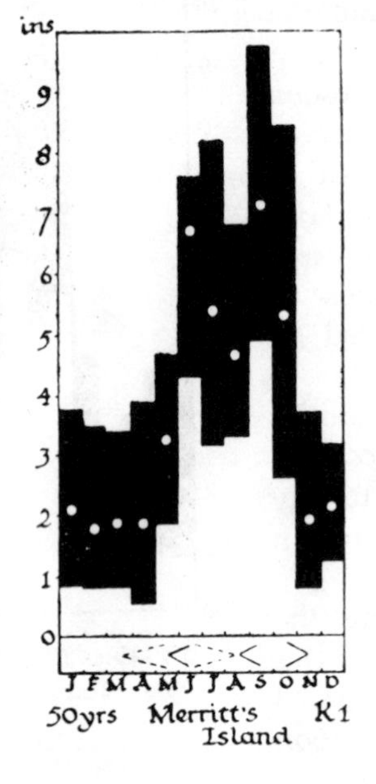

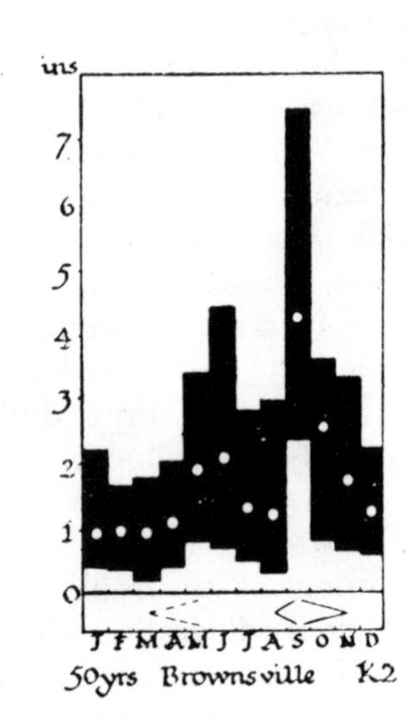

Fig. 21.

predominance of precipitation during September and October, the hurricane months. Both are low, eastern coasts with a north-south trend and the association of this ground plan with exceptional precipita-

[7] See P. R. Crowe: A Note on H. A. Matthews, "A New View of Some Familiar Indian Rainfalls (*Scot. Geog. Mag.*, 1936, pp. 84–97) in *Scot. Geog. Mag.*, 1936, pp. 187–8; and also P. R. Crowe, "A New Approach to the Study of Seasonal Incidence of British Rainfall," *Quart. Journ. Roy. Met. Soc.*, 1940, pp. 285–316, for further elaboration of this method. A "major" break is theoretically equivalent to a majority of at least seven-eighths, i.e. of 43 to 44 out of 50 in favour of the wetter month.

tion during the hurricane months probably deserves more careful analysis. Almost certainly much of the rain must be brought by storms which do not, in fact, attain hurricane intensity although they are probably weaker members of the same family.

The Merritts Island area (K.1). At first glance one is tempted to classify the régime of this small area as a further variation of the Atlantic Coastlands type. In view, however, of its equally close relationship with the régime experienced at Nassau, Bahamas and Havana, Cuba, a separate category appears to be advisable.

As in eastern Florida, the initial break of the rains occurs in May–June but rainfall then declines sufficiently to make comparison between August and June at least almost justifiable. The striking features, however, are the sharp increase from August to September and the even more abrupt decrease between October and November.

The diagrams suggest, indeed, that September and October rains differ from those earlier in the year not only in amount but also in character. At Key West the onset of the rains is a little earlier but their cessation in November is just as abrupt as at the more northerly stations.

The reader may be interested in having these features checked against frequency counts like those quoted above. The fifty-year period is the same as at Tampa and data are given for each of the four chief periods of rapid change.

	Number of Occasions when :			
	May > Apr. (*years*)	*June > May* (*years*)	*Sept. > Aug.* (*years*)	*Oct. > Nov.* (*years*)
Merritts Island . .	36	37	37	41
Key West . . .	35	(31) [8]	36	40

Direct comparison of pairs of months, year by year, in this fashion is sensitive to any correlation which may exist but ignores the amount of the difference ; hence the verdict may differ in detail from that derived from the dispersion diagrams. The figures fully confirm the broad conclusion that these are periods of unusually vigorous change in rainfall conditions, though at the same time, they place all generalization regarding rainfall régime into proper perspective in relation to the actual facts.

[8] Less than a two-thirds majority, hence not a significant change. The dispersion diagram is still more emphatic on this point.

The Brownsville area (K.2). This included a restricted portion of the coast of Texas which it is difficult to fit into any of the categories already given. A not very clearly developed increase in rainfall from March to May recalls the Plains Borderlands to the west but the dominant feature at Brownsville and Corpus Christi is the sharp increase between August and September, immediately followed by a more gradual decline. At Brownsville, over the period of fifty years, September recorded more rain than August on forty-one-and-a-half occasions, the fraction being due to one case of equality.

The North-Eastern Province

Our interest in the area from Indianapolis to the Atlantic must be confined, in the present discussion, mainly to the determination of the northern limits of the regions already recognized in the South. Without plotting the data for a considerable number of stations towards the Great Lakes and beyond, it is impossible to determine a logical grouping with any confidence. No index letters or numbers have been given therefore in this area though three subdivisions have been tentatively noted.

The main feature of this region is the relatively even distribution of rainfall throughout the year and such distinctions as can be recognized are unquestionably only of minor importance. From the point of view of the present paper, " even " distribution of rainfall is approached when only " slow " changes can be recognized and it is completely attained when no valid contrast can be drawn between the mean fall of any pair of months in the year. This latter condition is not quite fulfilled by any of the stations plotted but the changes are often slow enough to make classification difficult.

Around *Indianapolis* a slow increase from February to May is followed by an even slower decrease from May to October. The curve is thus a very weak reminder of those of Eb.ii to the west.

Around *Pittsburgh*, and apparently over a considerable portion of the Cumberland Plateau, contrasts in the rainfall of the first eight months of the year are slight but a decrease between July and September, sometimes appearing as a sharp change in September, introduces a drier spell during the last four months. The evident affinities of this type with G.2 to the south scarcely supports Ward's view that " the Ohio type is essentially the Missouri type with somewhat more cyclonic rain added during the winter."

Finally, east of the Appalachians from *Lenoir*, N.C., to *Washington*, D.C., a weakened form of J.1 appears to be characteristic, slow changes occurring between April and July and between August and November.

The existence of a true summer maximum, in the sense that June or July are really significantly wetter than December, January and February is, however, rather doubtful at most stations on the Piedmont though it is better developed in the Shenandoah Valley (Dale Enterprise).

This brief review of the North-east makes it clear that the most striking features of the rainfall régime of the South are rapidly lost as the upper Ohio basin and the Middle Atlantic states are approached.

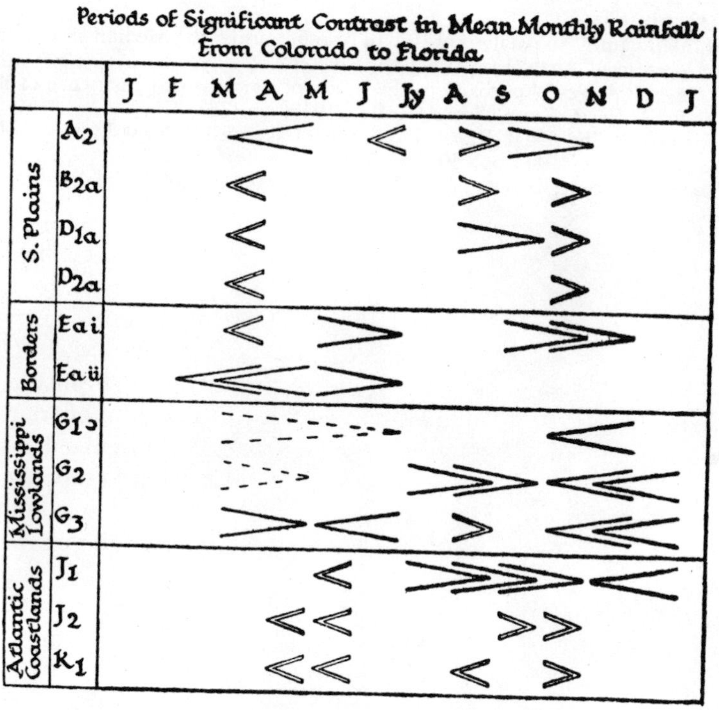

Periods of Significant Contrast in Mean Monthly Rainfall
From Colorado to Florida

Fig. 22.

Fig. 22, on the other hand, expresses, in summary form, the nature of the transition between the South and the Plains which it was the task of the first part of this paper to examine. Defined as the area included under G, J, and K in Fig. 12, the South thus forms an intelligible whole from the point of view of rainfall distribution, though falling into at least two distinct provinces, the one coastal and the other continental. In the interior, winter and spring rains, introduced in November or December and continuing into early summer, form the chief feature of a not very clearly defined annual cycle. The coastlands however, experience a much more clearly-marked régime with June, July, August and September closely approximating to a wet monsoon.

REFERENCES

P. R. Crowe, "The Analysis of Rainfall Probability," *Scot. Geog. Mag.*, 1933, pp. 73–91.

S. R. Savur, "The Use of the Median in Tests of Significance," *Proc. of Indian Acad. of Sc.*, 1937, pp. 564–76.

E. E. Lackey, "Annual Variability Rainfall Maps of the Great Plains," *Geog. Rev.*, 1937, pp. 665–70.

L. S. Mahalingam "An Analysis of Indian Rainfalls using the Median as a Statistic," *Sc. Notes India Met. Dept.*, Vol. VII, No. 82.

P. R. Crowe, "A New Approach to the Study of the Seasonal Incidence of British Rainfall," *Quart. Journ. Roy. Met. Soc.*, 1940, pp. 285–316.

P. R. Crowe, "The Dual Rainfall Régime of Roswell," New Mexico, *Monthly Weather Review*, 1941, pp. 40–7.

E. E. Lackey, "The Reliability of the Median as a Measure of Rainfall," *Geog. Rev.*, 1942, pp. 323–5.

VI.

THE SEVERN WATERWAY IN THE EIGHTEENTH AND NINETEENTH CENTURIES

By W. GORDON EAST, M.A.

Less has been written about the river Severn in the related fields of geomorphology, geography and history than the interest of the subject demands, and further research remains to be undertaken before a balanced picture can be drawn of its genesis and the part which it has played in the human history of southern Britain. On the geomorphological side many competent papers bear on the problem, not yet wholly solved, of the origin of the Severn by the union of two separate drainage systems, the one above and the other below the Ironbridge gorge, which appear to have had quite divergent histories.[1] An economic historian in recent years has reviewed one important phase of the river's history in the light of historical documentation.[2] It is attempted here to show that, from the standpoint of historical geography, yet another approach may be made to the study of the Severn's many-sided history.

Ranked alongside the Thames, Trent and Yorkshire Ouse as one of the chief rivers of medieval England and as one of the king's highways, the Severn offered, at least in law, free passage to all men with their goods and chattels. "Time out of mind," recorded an Act of Parliament of 1532,[3] "the public have had, and used, a pathway one foot and a half wide on either side of the Severn for drawing their trows, barges and other boats up the river by ropes or lines without paying any toll whatever." There is in fact plenty of evidence of the river's use both during the Middle Ages and later, as Leland

[1] The Montgomeryshire Severn was probably, as C. Lapworth argued, at first tributary to the Dee but its headstream may once have drained to the Thames near Oxford. The lower Severn, or rather the Severn-Avon, is believed, following W. M. Davis, to have developed as a subsequent stream after the mid-Tertiary elevation. The union of the Montgomeryshire Severn and the Severn-Avon below Ironbridge occurred after the last glaciation, the Ironbridge gorge having been cut in pre-Cambrian rocks by the overflowing waters of glacial lake Lapworth which occupied the Cheshire–North Shropshire Plain. On these problems, see L. J. Wills, "The Pleistocene Development of the Severn from Bridgnorth to the Sea," *Quarterly Journal of the Geological Society*, Vol. XCIV, 1938, p. 161, and *The Palæogeography of the Midlands*, 1948.

[2] T. S. Willan, "The River Navigation and Trade of the Severn Valley, 1600–1750," *The Economic History Review*, Vol. VIII, 1937, no. 1.

[3] Cited by H. J. Marten, *Report to the Severn Commission*, 1892, pp. 3–4.

relates, by sea-going vessels up to Gloucester and even Tewkesbury as well as by smaller craft up to Shrewsbury and even Welshpool. It is with the eighteenth and nineteenth centuries that this paper is concerned, a stage in the river's human history of particular interest in that it then achieved its greatest importance as a waterway. For the very reason that the transport needs of developing heavy industries in the western Midlands and the new enthusiasm for canal construction made greater claims on the Severn, there are available a number of competent surveys which allow the reconstruction with some precision of the physical conditions of navigation from its estuary up to its navigable head.

Contemporary writers gave contrasted verdicts on the usefulness of the Severn as a waterway in these two centuries. One enthusiast for water transport, who did the river more than justice, wrote [4] *c.* 1795 :

> There is no river that has such a length of navigation (namely, in England) as the Severn : you may navigate a vessel of fifty tons, and not a lock the whole way up to Welsh Pool, except in excessive drought, which does not happen every year, and, when it does, not above a month, seldom two.

Joseph Plymley of Shropshire, too, boasted [5] in 1802 about the absence of locks and weirs along the whole stretch of 155 miles from the mouth of the Bristol Avon to Welshpool, implying the adequacy of the waterway without such artificial aids. In contrast, the engineer Telford, who had actually surveyed the river, found its navigation " very imperfect " and wrote [6] in 1797 : " It has been suffered to remain in its natural and imperfect state, not one obstacle has been removed, nor has one improvement been yet introduced." It is easy to show from the experiences of the actual users of the river above Gloucester that the many imperfections long persisted. The waterway exponent cited above admitted the presence of a number of shallows between Tern-Bridge (near Wroxeter) and Redstone Ferry (below Bewdley), and that water might sink to as low as eighteen inches though only in the driest summer. A witness before a Commission in 1841 [7] described how cargo had to be re-shipped nine times between Worcester and Bristol. Another reported delays of three weeks at Deerhurst shoal (below Tewkesbury) and then again at Upton shoal. Owing to the varying conditions of the river, barges had frequently to work light without a paying load. Telford calculated that for an average of nearly two months a year navigation above Worcester was " quite

[4] Cited by J. Phillips, *A General View of Inland Navigation*, 1795, pp. 152–3.
[5] J. Plymley, *General View of the Agriculture of Shropshire*, 1803, p. 95.
[6] Telford's account of the Severn is given in J. Plymley, op. cit., pp. 317–33.
[7] Severn Commission, *Parliamentary Papers*, Vol. IX, 1841.

stopped" and, for the greater part of the year, "very irregular." Nor were the conditions of the Severn estuary below Gloucester much better. Richard Thomas of Falmouth, who carefully charted the estuary in 1815,[8] whilst emphasizing the dangers of its navigation, noted his failure to find that anything material had ever been done for the improvement and protection of the Severn waterway. In short, it was less than a half-truth to suggest that Nature had done so much for the river Severn as to leave comparatively little for art to accomplish. It had in its natural state some usefulness as a waterway, however irregular and precarious, but the complaints of the time underlined its many and continual difficulties. Most of these were of a physical character : shoals, shifting channels, floods and even freezing ; in addition certain bridges, notably that at Upton above Tewkesbury, were obstacles, and towing paths were inadequate, to which was due the continuance until after 1800, especially above Worcester, of the old practice of haulage by "higging men." In order to make clear and to localize these shortcomings of the Severn, it is necessary to study closely and illustrate with sketch-maps the seasonal régime of the river, the profile of its bed and, lastly, its tidal phenomena. For this purpose the three stretches of the navigable river should be distinguished : the estuary up to Gloucester, the middle course thence to Stourport, and the "upland" river between Stourport and Welshpool.

The Severn receives from its upper basin, because this is not only hilly but also mainly composed of impermeable rocks, proportionately more of the water which falls on it than either the Thames or the Trent receives from theirs.[9] On the basis of calculations made for the years 1882–9, it was found [10] that the discharge of the river at Worcester averaged as much as 46·2 per cent of the rainfall received in the basin above the town and that two days elapsed before this rainfall affected the level at Worcester. Little more than half of the rainfall was thus lost to this part of the river by evaporation and percolation. One difficulty about navigation above Gloucester was the variability in the rainfall both annually and seasonally which reacted on the depth of water in the river. But the low water which was characteristic during the summer six months was not necessarily due to lower rainfall in that period but rather to the increased rate of evaporation and percolation. During the years 1882–9, for example, the percentage of rainfall reaching the river was consistently less during these months even when the

[8] R. Thomas, *Hints for the Improvement of the Navigation of the Severn*, 1816.

[9] W. Parkes, "On the Estuary of the Severn," *Minutes of the Proceedings of the Institution of Civil Engineers*, Vol. V, 1846.

[10] *The Investigation of Rivers : Final Report*, Royal Geographical Society, 1916, pp. 12–13.

rainfall exceeded that of the winter six months, which it did in four out of the eight years. Generally, then, rainfall and temperature conditions within the basin explain the better depths of water above Gloucester between October and March, although the chances that flood and even freezing might interrupt navigation were greater during this season. Average figures obviously conceal the considerable variability of the weather and its effects on the river. For instance, it is clear how important were the rainy spells following droughts, which produced a river "fresh" and thus provided a "flash" of water sufficient in depth to allow barges and trows to cut short their delays at shoals and, more than that, sufficient in velocity to propel them effectively—if their journey lay downstream. That the average conditions differed strikingly from those of particular years is shown by the record of 1796 in Shropshire, when there were not two months (as Telford noted) in which barges could be navigated, even downstream, with a paying load. This referred to barges of more than 20 tons.

Thanks again to Telford's survey,[11] detailed figures are available to show the daily depths of water in the river at Coalport in the Ironbridge gorge during the years 1790–1800. At this time Coalport was springing up as a new river port to serve the many busy collieries, blast furnaces and pottery works of the Shropshire coalfield. Telford's figures are summarized in the table below and their bearing on the navigability of the river is shown. The barges employed between

State of the Severn at Coalport

	1790	*1791*	*1792*	*1793*	*1794*	*1795*	*1796*	*1797*	*1798*	*1799*	*1800*	*Yearly Averages*
No. of days recorded	312	301	305	307	314	333	366	342	348	365	356	332
No. of days with less than 2′ 6″	179	149	74	146	136	136	234	118	171	119	148	146
% of days with less than 2′ 6″	57%	50%	24%	48%	43%	41%	64%	35%	49%	33%	42%	44%
No. of days frozen over	0	0	0	0	0	(Jan.) 32	(Dec.) 19	0	(Dec.) 4	(Dec./ Jan.) 37	3	8·6
No. of days of flood (i.e. over 12′ of water)	(Jan.) 3	(Dec.) 3	(Apr., May, Nov.) 5	0	(Feb.) 4	(Feb.) 5	0	(Dec.) more than 4	0	0	4	2·5

Shrewsbury and Gloucester ranged between 20 and 80 tons; a 30–40 ton barge laden drew 3 ft. 8 in. to 4 ft. of water, and a fully laden 20 ton barge (the smallest in use) drew about 2 ft. 6 in. Since there

[11] J. Plymley, op. cit., pp. 317–33.

are gaps in Telford's figures—there were few readings for Sundays between 1790 and 1795—his recordings of lowest water levels, below 2 ft. 6 in., are shown also as percentages of the total recordings for each year. The table shows how variable was the usefulness of the "upland" river from year to year—compare the best years 1792 and 1797 with the worst year 1796—and that, on the average, it was useful for the *smallest* barges only about half the year.

The graph (Fig. 23) gives a more detailed picture. It shows the inconstancy of water depths, the tendency to summer shallows and to deeper water with occasional floods and freezing in winter, and the many sudden "freshes" which occurred. In 1790 there were floods in January and low levels during spring and summer. In 1799 the levels were high for August and September—each year had its peculiarities. In 1795 a sudden thaw after a long spell of freezing produced severe floods for nearly a week, one result of which was the destruction of the stone bridge at Buildwas which had long been an obstruction to navigation because of its narrow arches.[12] It was replaced a year later by the famous cast-iron bridge at the new industrial settlement of Ironbridge a few miles downstream. In the light of the physical difficulties of navigating the Shropshire Severn it is not surprising that during the American War of Independence the Government had to wait three months before there was sufficient water in the Severn to carry downstream even one of the cannon cast at Coalbrookdale.[13]

The river floods were aggravated towards the end of the eighteenth century by the enclosure and embanking for the benefit of agriculture of lands adjacent to the upper Severn and Vyrnwy in Montgomeryshire and north Shopshire. Embankments were built from a little below Welshpool as far as the junction with the Vyrnwy and some miles up the latter stream as far as Llanymyneich.[14] The effect of the new embankments, while they held, was to impound and send rapidly downstream waters which would previously have overflowed and then slowly and usefully seeped back into the river. Actually these early efforts at embanking were defeated by the assaults of the winter torrents. Although valves had been let into the banks to allow the escape of water for irrigation purposes, the river breached its new banks continually—in 1794, in 1795 after a thaw which swept away bridges and mills as well as embankments, and again, after repairs had been made,

[12] Buildwas had suffered badly from one of the much feared summer floods in 1773: J. Fletcher, *Sudden Stoppage and Desolation between Coalbrookdale and Buildwas Bridge, May, 1773*, 1773.

[13] *Telford MSS.*, Library of the Institution of Civil Engineers, pp. 72-3.

[14] W. Davies, *General View of the Agriculture of North Wales*, 1813, p. 263; J. Plymley, op. cit., p. 287.

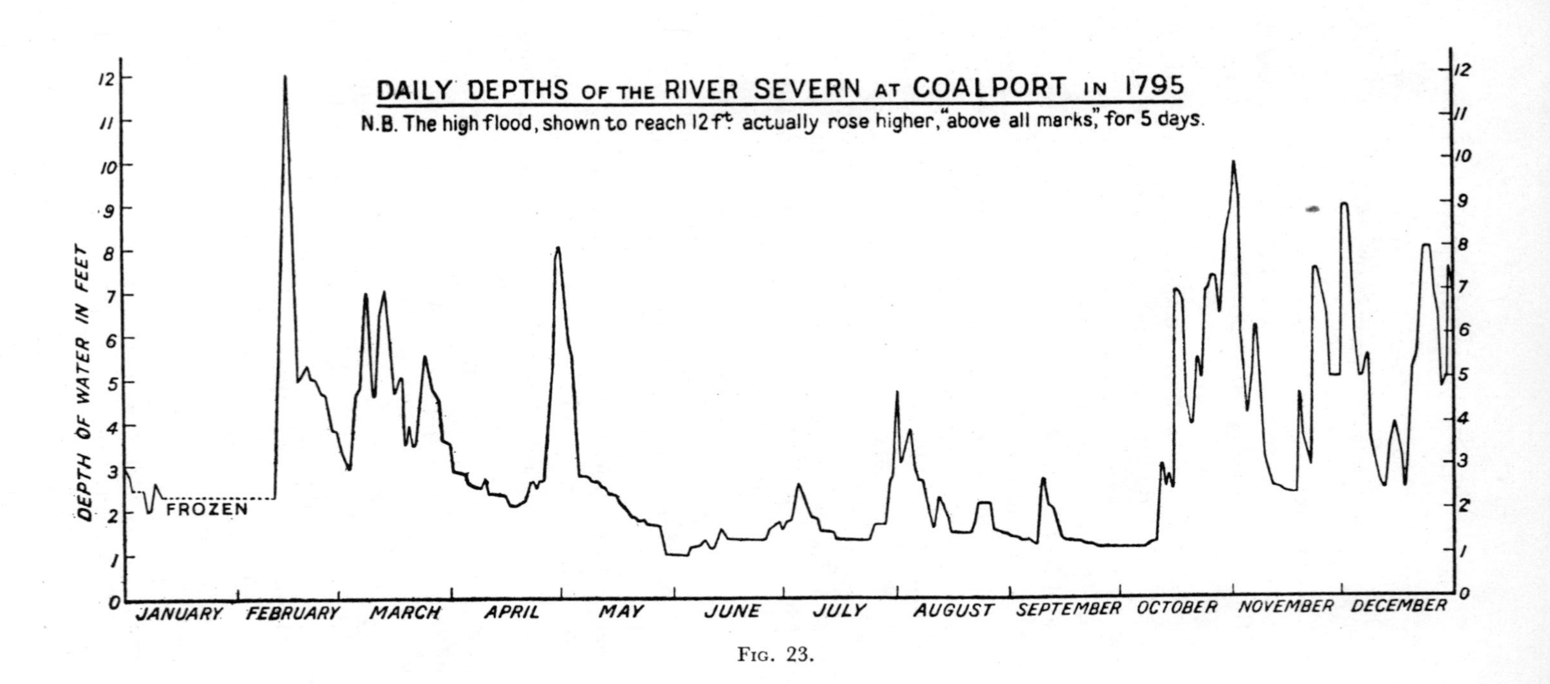
DAILY DEPTHS OF THE RIVER SEVERN AT COALPORT IN 1795
N.B. The high flood, shown to reach 12 ft. actually rose higher, "above all marks," for 5 days.
DEPTH OF WATER IN FEET
FROZEN
JANUARY
FEBRUARY
MARCH
APRIL
MAY
JUNE
JULY
AUGUST
SEPTEMBER
OCTOBER
NOVEMBER
DECEMBER
0
1
2
3
4
5
6
7
8
9
10
11
12

FIG. 23.

in 1797, 1806 and 1808. Until nearly a century later, when the construction of the Vyrnwy reservoir removed one major cause of flooding, flooding especially in winter remained a continual nuisance as far down as Gloucester. In 1848 " ordinary high floods " so-called were known to raise the river above its summer low level as much as 20 ft. at Worcester, 19 ft. at Tewkesbury and 15 ft. 6 in. at Gloucester.[15] On this stretch of the river the recurrent floods were all the worse if they happened to coincide with spring tides. If they occurred in early spring or autumn, they were welcomed by farmers, for they fertilized the meadows, but summer floods were particularly harmful.

Next in importance to the seasonal régime of the river among the physical conditions of navigation were the gradients and unevenness of the river's bed which conditioned the strength of the current and the distribution of shoals. If, as was asserted, the Severn obeyed the ordinary law of gradual diminution of slope except between Longney and Sharpness below Gloucester, that is not to suggest that its longitudinal profile was evenly graded. The variations of slope between Gloucester and Buildwas can be readily seen in the following table drawn up in 1797.[16]

Gradients of the River Bed

Section of River	*Distance*	*Total Rise*	*Rise per Mile*
	miles	ft. in.	ft. in.
Gloucester–Worcester	30	10 0	0 4
Worcester–Stourport	13	23 0	1 9
Stourport–Bridgnorth	18	41 9	2 4
Bridgnorth–Buildwas	11	29 6	2 8

This table shows the gradualness of the ascent of the tidal river from Gloucester to Worcester, in contrast to the increasing steepness of what came to be called the " improved " river between Worcester and Stourport and of the " upland " river above. Fig. 24 illustrates the profile of the river below Coalbrookdale. The uneven surface of the river bed between Stourport and Tewkesbury was due to " hard shoals," beds of marl rock and compact gravel intersecting the channel, which caused shallow rapids at low water but, acting as natural dams, held up the water in the intermediate reaches. The shoals between Tewkesbury and Gloucester, in contrast, were " shoals of deposit " which consisted of banks of quicksand and mud deposited by winter floods whenever a sudden increase in the width of the river caused

[15] Report to the Admiralty upon the Improvement of the Severn Navigation, *Parliamentary Papers*, Vol. XXXI, 1847–8, p. 429.
[16] By Telford, see J. Plymley, op. cit., p. 288.

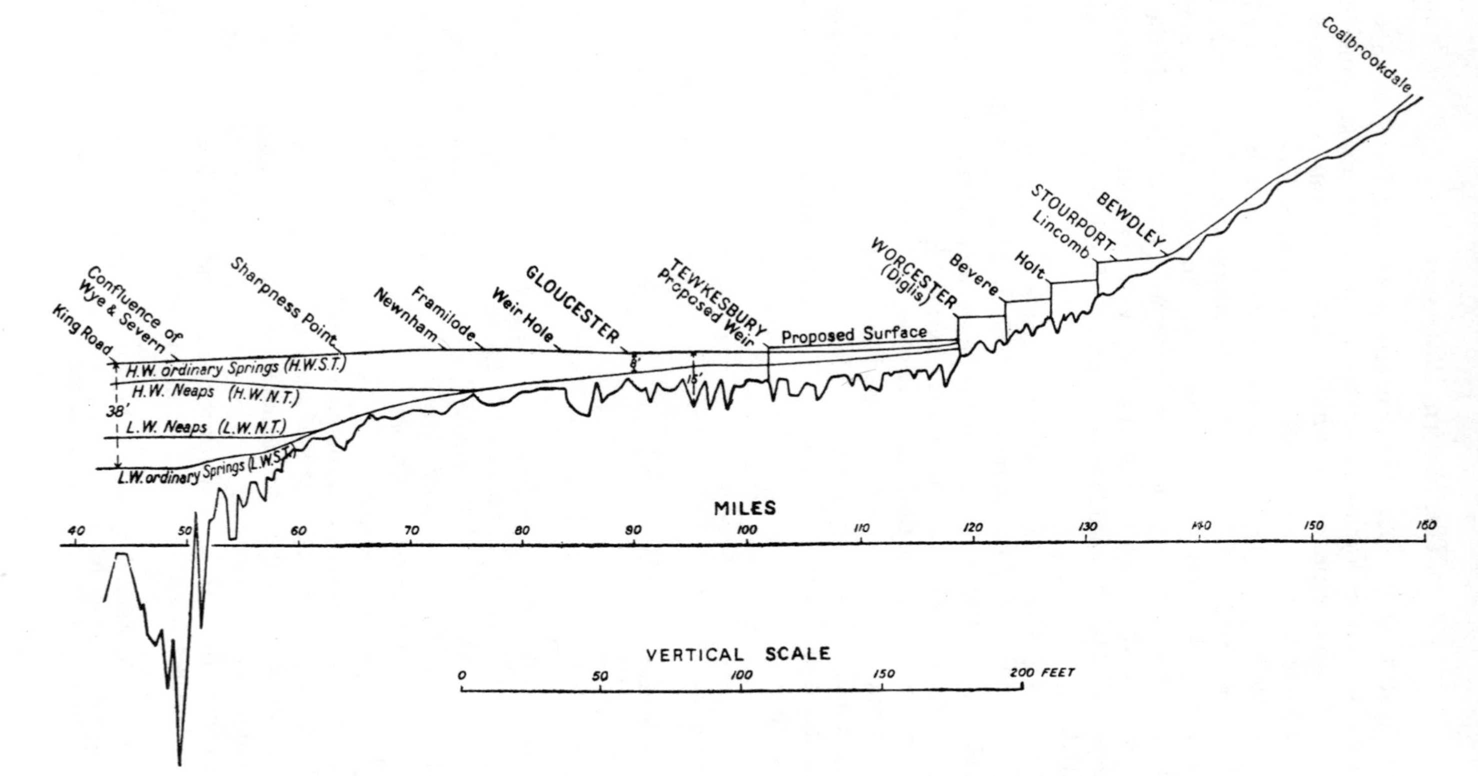

FIG. 24.—Longitudinal profile of the river Severn, showing the effects of the tide on water depths below Worcester and the sites of weirs built by the year 1844.

a decrease in the velocity of the current.[17] The sharp break of slope and the resultant change in the strength of the current just above Worcester, which may have had some bearing on the original siting of this town, was well recognized by the users of the river. Only half a man per ton was needed—and then not at all times—to haul vessels from Upton up to Worcester, whereas from Worcester up to Stourport one man per ton was required. William Jessop's survey of the river in 1784 [18] recorded twenty-eight shallows in the 32-mile stretch between Stourport, the new river port which grew up where the Staffordshire and Worcestershire canal joined the Severn (Fig. 25), and Deerhurst which gave its name to a most troublesome shoal just below Tewkesbury. Between Stourport and Bewdley, which were only four miles apart, there were five more shoals.

How much better were the navigable conditions between Worcester and Gloucester than in the stretch above? A rocky ledge in the estuary above Sharpness served as a natural weir to pen up water below Worcester. This stretch of the river was gently graded and its level was raised by spring tides 1 ft. 9 in. at Upton and 1 ft. 6 in. at Worcester. Moreover, spring tides supplied motive power up to Worcester and were more useful than the ebb and river currents combined for the downstream trade. But the larger barges and trows which sailed along this stretch—they ranged between 30 and 140 tons and were equipped with sails and lowering masts—suffered many hindrances. There were bad shoals at Upton and Deerhurst, which at their worst offered only 2 ft. of water ; occasional floods ; and before dredging was undertaken seriously in 1842, at best only a 4-ft. channel could be hoped for. Actually a 4 ft.–4 ft. 6 in. depth was adequate between Worcester and Gloucester, if it could be relied on, but even after improvements had been made by dredging and by locks a dry summer might give only 3 ft.–4 ft. at the shoals which formed (after 1844) above and below Diglis lock. The Severn Commission reported [19] in 1846 that the river had been made navigable to vessels of 6 ft. draught from Gloucester right up to Stourport, but there was an element of wishful thinking in this report as the drought in 1847 proved. The stone bridge at Upton,[19] to pass under which masts

[17] E. Leader Williams, " Account of the works recently constructed on the River Severn . . . near Tewkesbury," *Minutes of the Proceedings of the Institution of Civil Engineers*, Vol. XIX, 1859–60, pp. 527–8.

[18] W. Jessop, *Two reports on the navigable state of the River Severn and the means for improving navigation from surveys May and October 1784* (Telford MSS. at the Library of the Institution of Civil Engineers).

[19] Report to the Lords Commissioners of the Admiralty, *Parliamentary Papers*, Vol. XXVII, 1849, pp. 177 and 199.

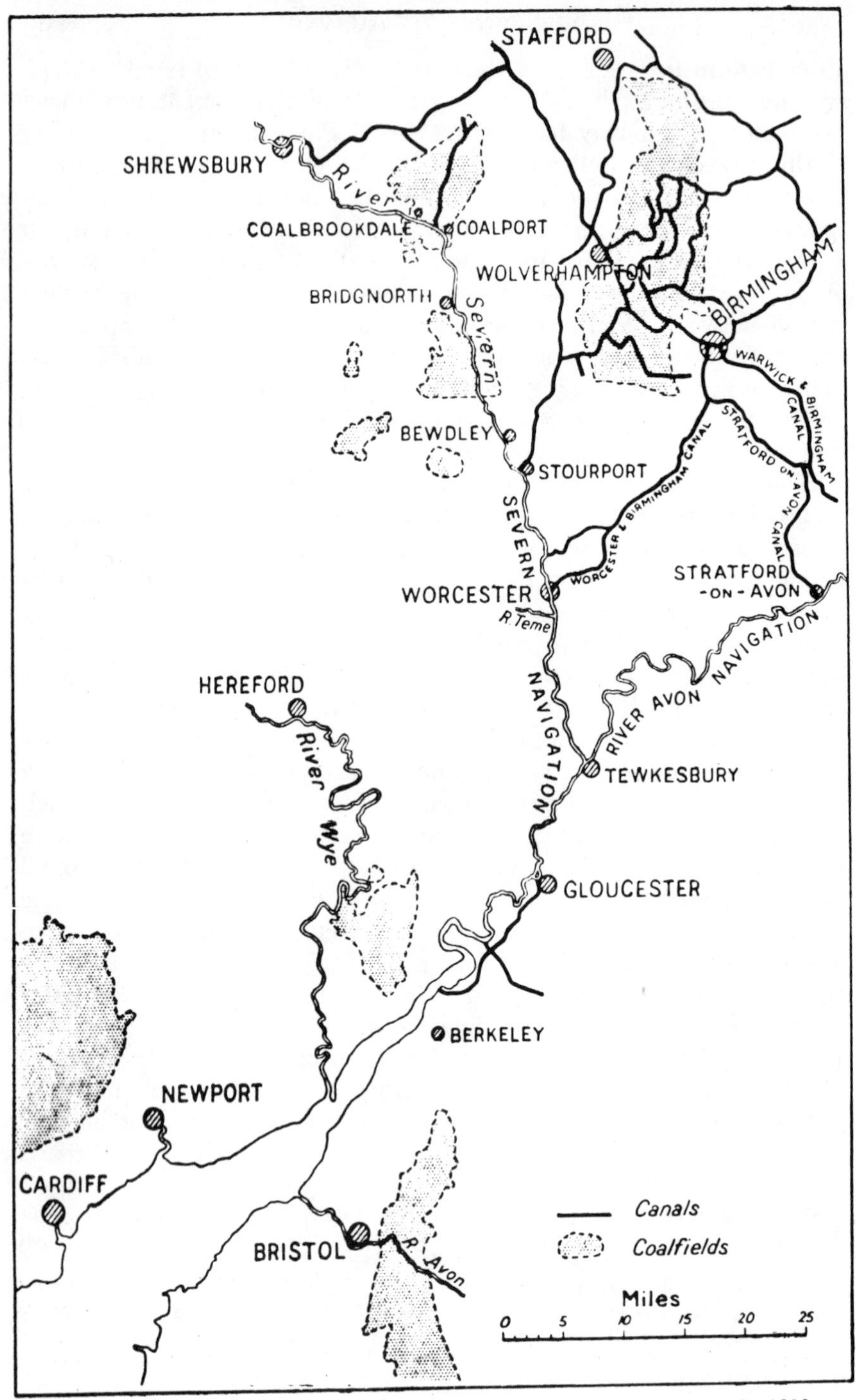

Fig. 25.—The Severn Navigation and waterway connections with it, as in 1892.

had to be lowered, was a considerable obstacle: with 100 feet of piers and abutments in a river only 300 ft. wide, it acted, like old London Bridge, as a dam to both tide and flood and caused dangerous currents and troublesome shoals.

Before turning to the estuary of the Severn, where the tidal factor obtrudes in importance, let us note briefly what efforts were made and measures taken to improve the waterway between Gloucester and Stourport. At the instance of the Staffordshire and Worcestershire Canal Company, the river was surveyed by Jessop between Coalbrookdale and Gloucester in 1784 and application was made to Parliament for powers to improve it by dredging the channel below Worcester and by building four locks together with dams or weirs at Diglis, Barbourne shoal, Holt and Larford (Lincomb). This application was made jointly by this canal company and the ironmasters of Coalbrookdale, but it met sturdy opposition from the landowners of Worcestershire and Gloucestershire, who argued that locks and weirs would aggravate floods. Actually all that the Act of 1790 allowed the Canal Company to do was to remove shoals and thus to deepen the channel between Stourport and Diglis (near Worcester) "by all necessary and proper works, except locks and weirs." In the course of this work the Company tried to contract the stream and increase the scour by erecting above the shoals timber-framed jetties filled with stones, but owing to strong opposition of bargemen, landowners, and others these were successfully indicted as a nuisance in 1793 and removed.[20]

Nothing further was done until 1836, when the newly-formed Severn Navigation Company ordered Thomas Rhodes to make a fresh survey of the river below Stourport with a view to large-scale improvements.[21] He advised in his reports the construction of locks and weirs and hoped somewhat optimistically to make Worcester a seaport for ships of 200 to 300 tons and Stourport accessible by trows and barges of from 70 to 100 tons by securing a minimum depth of 12 ft. between Gloucester and Worcester and of 6 ft. 6 in. thence to Stourport. But the Bill introduced into Parliament in 1837, which sought powers for the Severn Navigation Company to carry out Rhodes's recommendations, was strenuously opposed by a variety of interests and rejected. After further attempts to secure powers had failed, the Severn Navigation Act in 1842 allowed for the first time construction of locks and weirs, but only between Worcester and Stourport. As a precautionary measure the Act stipulated that three clear months, including January,

[20] H. J. Marten, op. cit., p. 5.

[21] E. Leader Williams, *Facts and Suggestions. The River Severn as it was, is, and ought to be,* 1863, pp. 8–12.

should elapse between the completion of the first and the building of the second lock and weir. The first lock and weir were built at Lincomb and since no flooding occurred above the weir during the stipulated period, three other locks and weirs were completed in 1844 at Holt, Bevere and Diglis respectively. Below Worcester improvement was to remain limited to dredging, by which means, however, a depth of 6 ft. could not be maintained. The project for a weir and lock near Tewkesbury was designed to remedy this deficiency but was delayed owing to opposition of the Admiralty which held conservancy and jurisdiction rights over the tidal portion of the river and refused to obstruct tidal action by these works : they were not constructed until 1858. In 1853, too, a new bridge with a widened span was built at Upton, whilst a lock and weir at both Maismore and Llanthony near Gloucester were erected in 1871 so as to maintain a 6-ft. depth at all times between Gloucester and Tewkesbury. Already a series of Acts passed between the years 1772 and 1811 had authorized the making of " roads or passages " for horses along the banks of the Severn from Bewdley Bridge down to Lower Parting below Gloucester.[22] There is a certain irony in the fact that when eventually in the eighteen-forties the river was being improved by locks and weirs a sturdy competitor had appeared in the railway. Even so, it did not lose its utility for heavy traffic and improvements continued. Thus the Severn Navigation Act of 1890 authorized alterations in the locks and the removal of shoals by dredging so as to obtain a minimum depth of 10 ft. between Worcester and Gloucester and 7 ft. between Worcester and Stourport.

For seamen using the river, as for Richard Thomas who charted it, the Severn estuary effectively began at King Road, the anchorage at which vessels awaited the tide before continuing either up the Severn or up the Avon to Bristol. Below King Road stretched east-westwards the broadening sea estuary of the Bristol Channel ; above it lay the ascent of the Severn in a north-easterly direction. At King Road vessels ascending the river embarked pilots, for the fifty-mile stretch to Gloucester entailed difficulties and hazards. Special care and local knowledge were necessary owing to the meandering of the river, the rocks and sands exposed at low water, the twisting and variability of the channel, inadequate depths of water in the upper part of the estuary, and lastly the remarkable velocity of the tidal currents. The head of the estuary was reached at Gloucester where, since Roman times, stood the first bridge across the Severn.

Figs. 26 and 27 are based on Thomas's survey of 1815, which gives

[22] H. J. Marten, op. cit., p. 4.

detailed soundings and was drawn on a scale of two inches to the mile. [They are generalized to show in black the low-water channel.] Upstream from King Road vessels sailed through the Shoots where the tidal waters, fast-moving and attaining at times as much as 10 knots, have scoured the deep passage between the rocks of Lady Bench and of English Stones. A more precarious channel was to be found at low water across English Lake. It is at this narrows, where the Lower Lias limestones have resisted with some effect the forces of erosion, that the Severn tunnel has been bored and the erection of the Severn barrage is contemplated. Here too was the New Passage ferry (Fig. 28) of the eighteenth century, by means of which mails passed between London, Bristol and Southern Ireland.[23] It was called " New " since it replaced the Old Passage, an " ugly, dangerous and very inconvenient ferry "—as Defoe described it—which used the much shorter crossing between Beachley and Aust Head. Aust Head, and on the opposite bank Dod Rock, jut out into the river, where at low water several rocks were exposed. Above Aust Head the valley widens again and the main channel passes first north of Oldbury Sands and then between Lydney Sand and some rocky projections on the left bank. The estuary then tapers to another narrows, half a mile wide, between Sharpness and Lydney. Because of the narrowing and consequently increased scour, the water was deep here, although at low water depths sank above and below the narrows to as little as one or two feet in places. Above Sharpness insufficient depths of water, except at high water springs, virtually prohibited passage to sea-going craft. The reason for this was the sharp increase in the gradient of the river bed between Sharpness and Longney. Here was formed a natural weir of Lias limestone which usefully held up the water in the reaches above Longney but offered little water between Longney and Sharpness, little more than a foot in places at low water. Hock Crib,[24] where the river meanders northwestwards in the shape of a horse-shoe, was reckoned the most difficult part of the whole navigable river, owing to its tidal current, rocks, the shifting channel, and the quicksands of the Nouze (Fig. 27). The increased fall of the river below Longney is shown by the following gradients : 9 in. per mile from Longney to Framilode ; 1 ft. 2 in. from Framilode to Hock Crib ; and 2 ft. per mile thence to Sharpness. These figures may be contrasted with that of the almost level bed from Gloucester to Longney, falling only one inch per mile.[25]

[23] R. Thomas, op. cit., p. 3.

[24] W. Parkes, " On the estuary of the Severn," *Minutes of the Proceedings of the Institution of Civil Engineers*, Vol. V, 1846, pp. 300–9.

[25] Report to the Admiralty upon the Improvement of the Severn Navigation, *Parliamentary Papers*, Vol. XXXI, 1847–8.

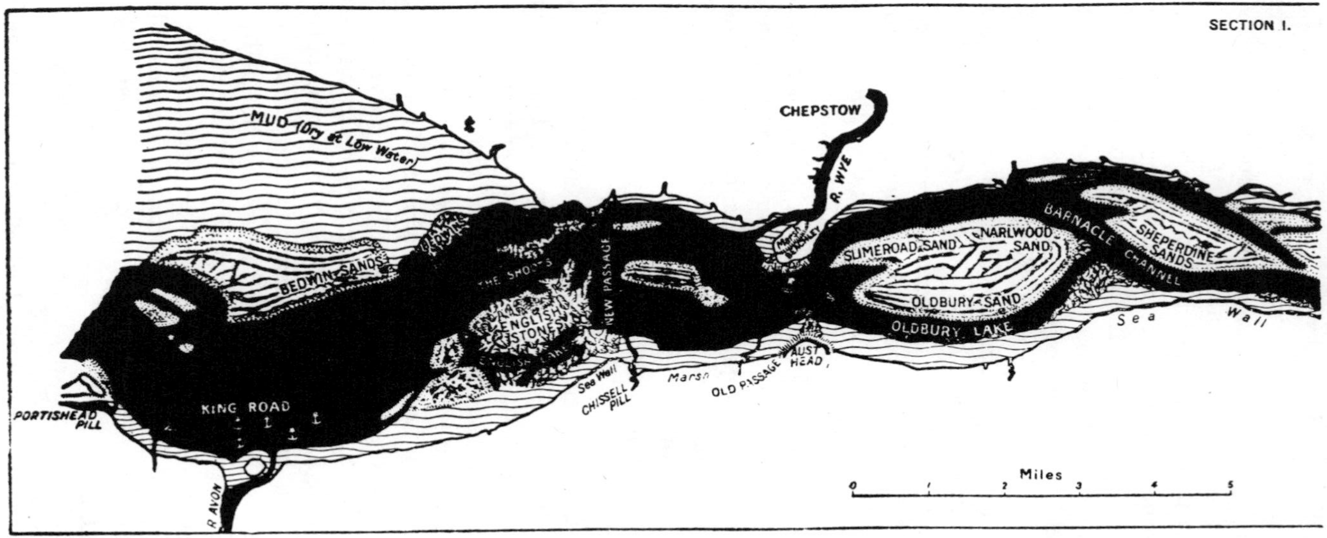

SECTION I.
MUD (Dry at Low Water)
CHEPSTOW
R. WYE
BEDWIN SANDS
THE SHOOTS
ENGLISH STONES
NEW PASSAGE
SLIMEROAD SAND
NARLWOOD SAND
OLDBURY SAND
OLDBURY LAKE
BARNACLE CHANNEL
SHEPERDINE SANDS
Sea Wall
Sea Wall
CHISSELL PILL
Marsh
OLD PASSAGE
AUST HEAD
KING ROAD
PORTISHEAD PILL
R AVON
Miles
0 1 2 3 4 5

Fig. 26.

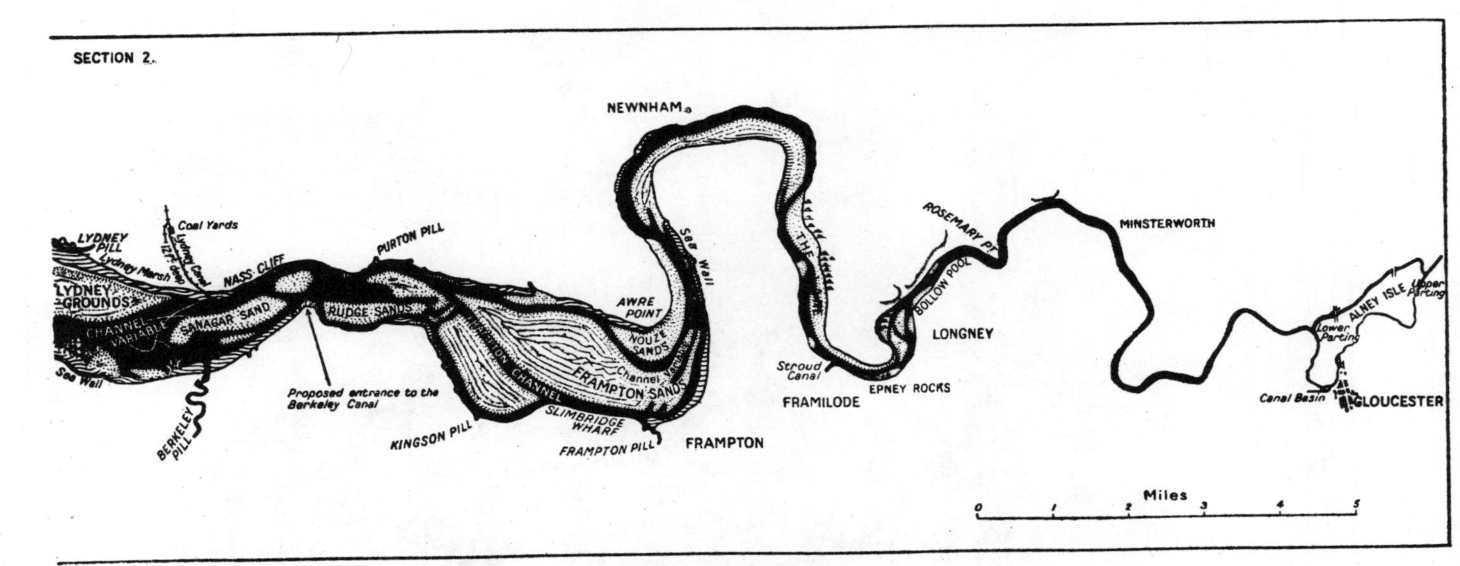
SECTION 2.
NEWNHAM
LYDNEY PILL
Coal Yards
Lydney Canal
12 ft deep
Lydney Marsh
LYDNEY GROUNDS
NASS CLIFF
PURTON PILL
CHANNEL VARIABLE
SANAGAR SAND
RUDGE SANDS
Sea Wall
Proposed entrance to the Berkeley Canal
BERKELEY PILL
KINGSON PILL
CHANNEL
SLIMBRIDGE WHARF
FRAMPTON PILL
FRAMPTON
FRAMPTON SANDS
Channel
NOUZE SANDS
AWRE POINT
THE
Stroud Canal
FRAMILODE
EPNEY ROCKS
LONGNEY
BOLLOW POOL
ROSEMARY PT
MINSTERWORTH
ALNEY ISLE
Upper Parting
Lower Parting
Canal Basin
GLOUCESTER
Miles
0
1
2
3
4
5

FIG. 27.

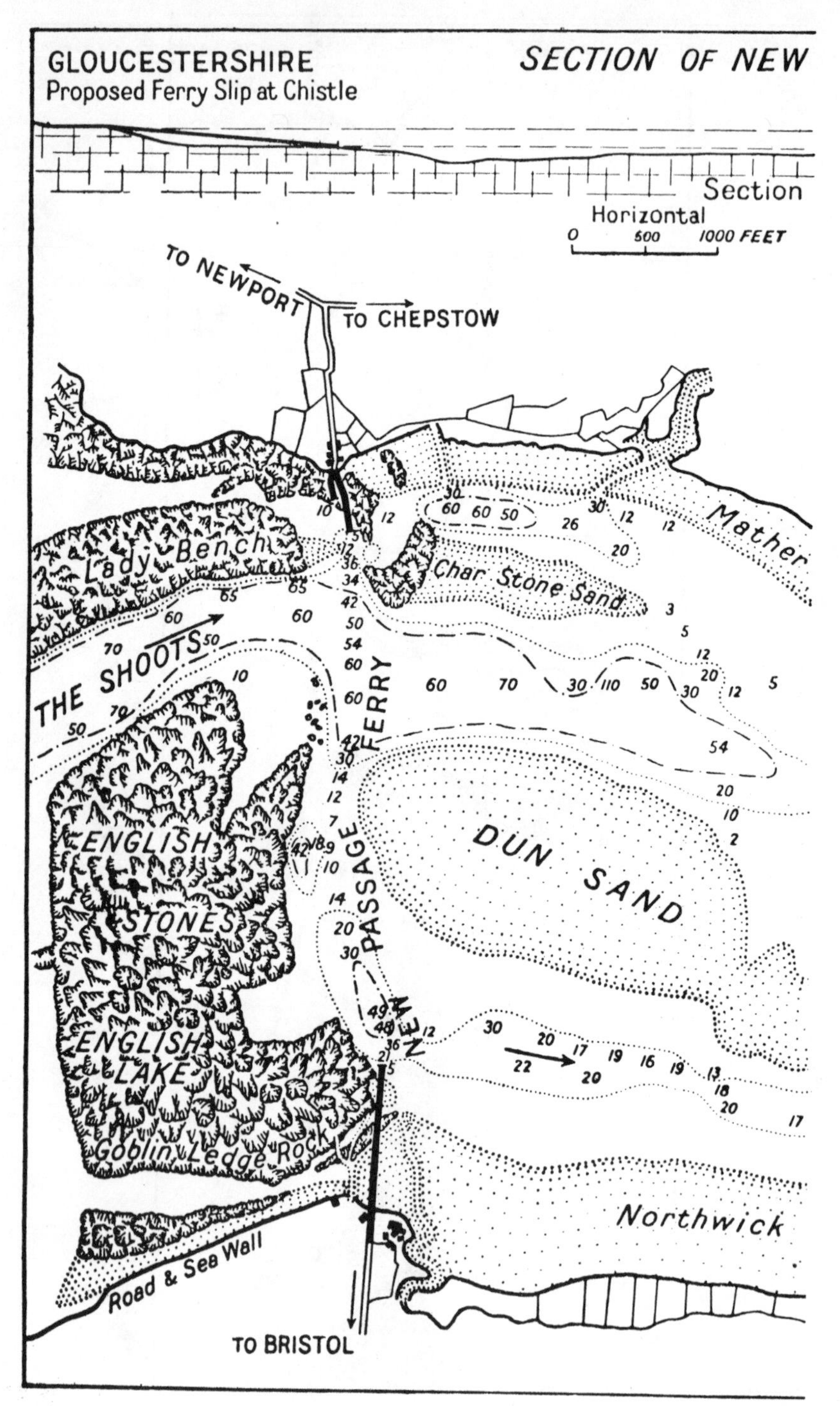

FIG. 28.—The Old and New Passage Ferries (based on

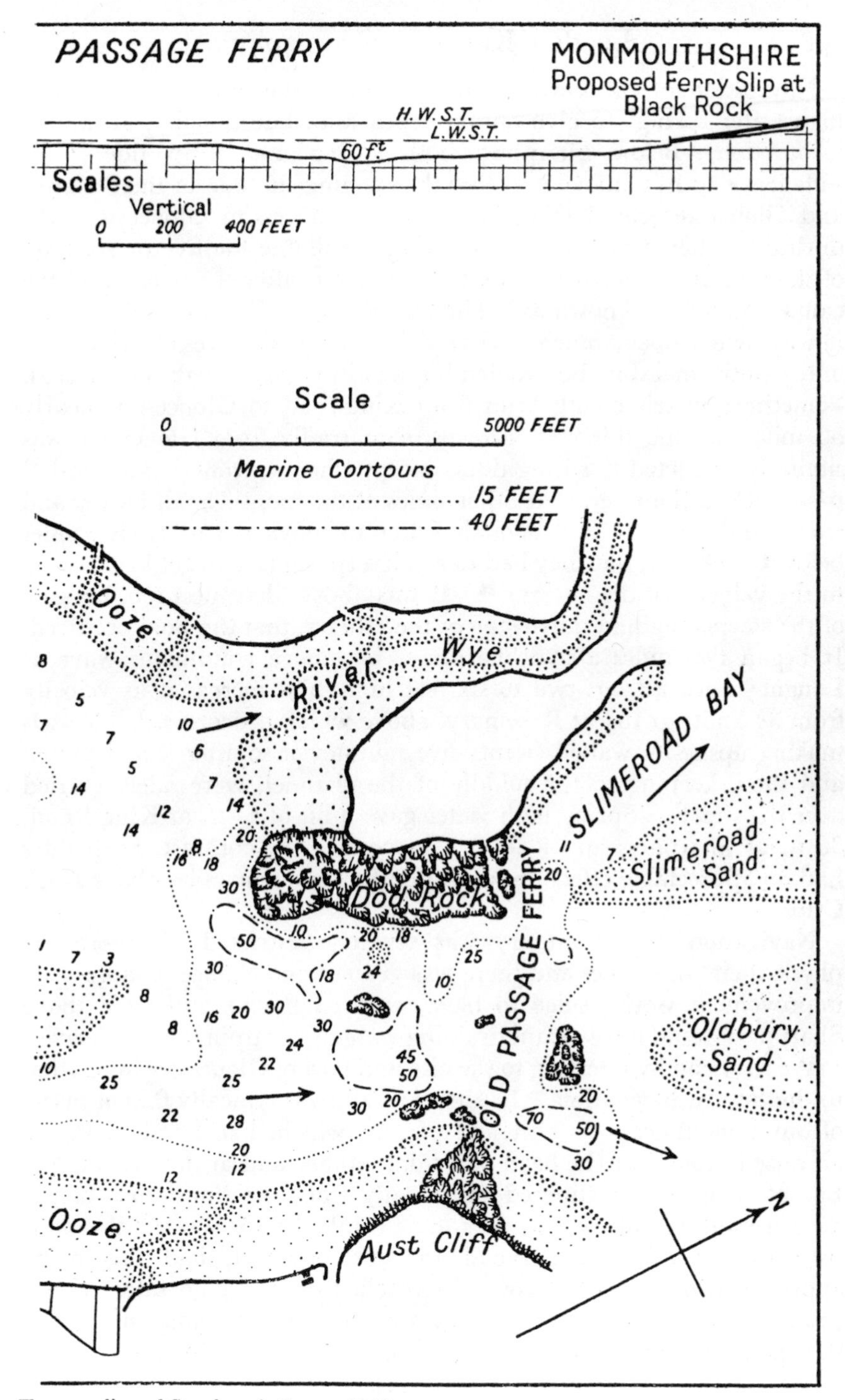

Townsend's and Steadman's Survey 1822).

Navigation of the Severn estuary up to Gloucester depended on high-water springs to give ample depth and, additionally, an aid to locomotion. Below Sharpness both spring and neap tides gave sufficient depths ; above it, before the opening in 1827 of the Berkeley and Gloucester canal (Fig. 25), ships had access to Gloucester only during the hours of high water springs available for five or six days of the month. The river route to Gloucester, after the opening of the canal, came to be known as "The Old Road." Despite its difficulties it was, when open, much preferred by the smaller vessels, since not only could canal dues be avoided but a swift passage might be effected. Sometimes vessels could "run from King-road to Gloucester, nearly 50 miles, in one tide." [26] Downstream traffic from Gloucester was similarly restricted to spring tides, except when a strong "land fresh" passed down the river. At other times of the month small barges and trows could usually find enough water to move ten or twelve miles below Gloucester, but they had to await a spring tide to get back owing to the velocity of the current.[26] It was above Sharpness too, because of the steepening and narrowing of the estuary, that the Bore occurred. It began two miles above Sharpness, became a continuous wave at Longney with a front two to six feet high, and increased its velocity from 3¾ knots to 13¾ at Rosemary, above which it decreased. Vessels making upstream waited twenty-five minutes or so after it had passed and then, keeping to the middle of the channel, were safely carried over the shoals. Spring high water gave a lift of 47 ft. at King Road, 30 ft. at Sharpness and 8 ft. at Gloucester. In contrast, neap tides had no appreciable effect on water depths some four miles above Hock Crib.

Navigation in the estuary was severely restricted, therefore, by physical circumstances and there was virtually no chance of large-scale improvement works, since to have removed the natural weir above Sharpness would have meant draining water away from the upper river.

We may end by referring to the city and port of Gloucester in relation to the Severn navigation. Its site (Fig. 29) was typically that of many of our famous seaports past and present which, like London, stands on raised ground at the head of a tidal estuary and at the first bridge. But as a seaport it suffered historically from the limitations of the Severn estuary : already in the later Middle Ages it was outdistanced by Bristol, relatively a parvenu, the port of which had also to be approached by river, but could be reached on any high tide. Gloucester's chance came, though belatedly, with the opening in 1827 of the Berkeley and Gloucester Ship Canal which, by giving the city a

[26] W. Parkes, loc. cit., p. 3.

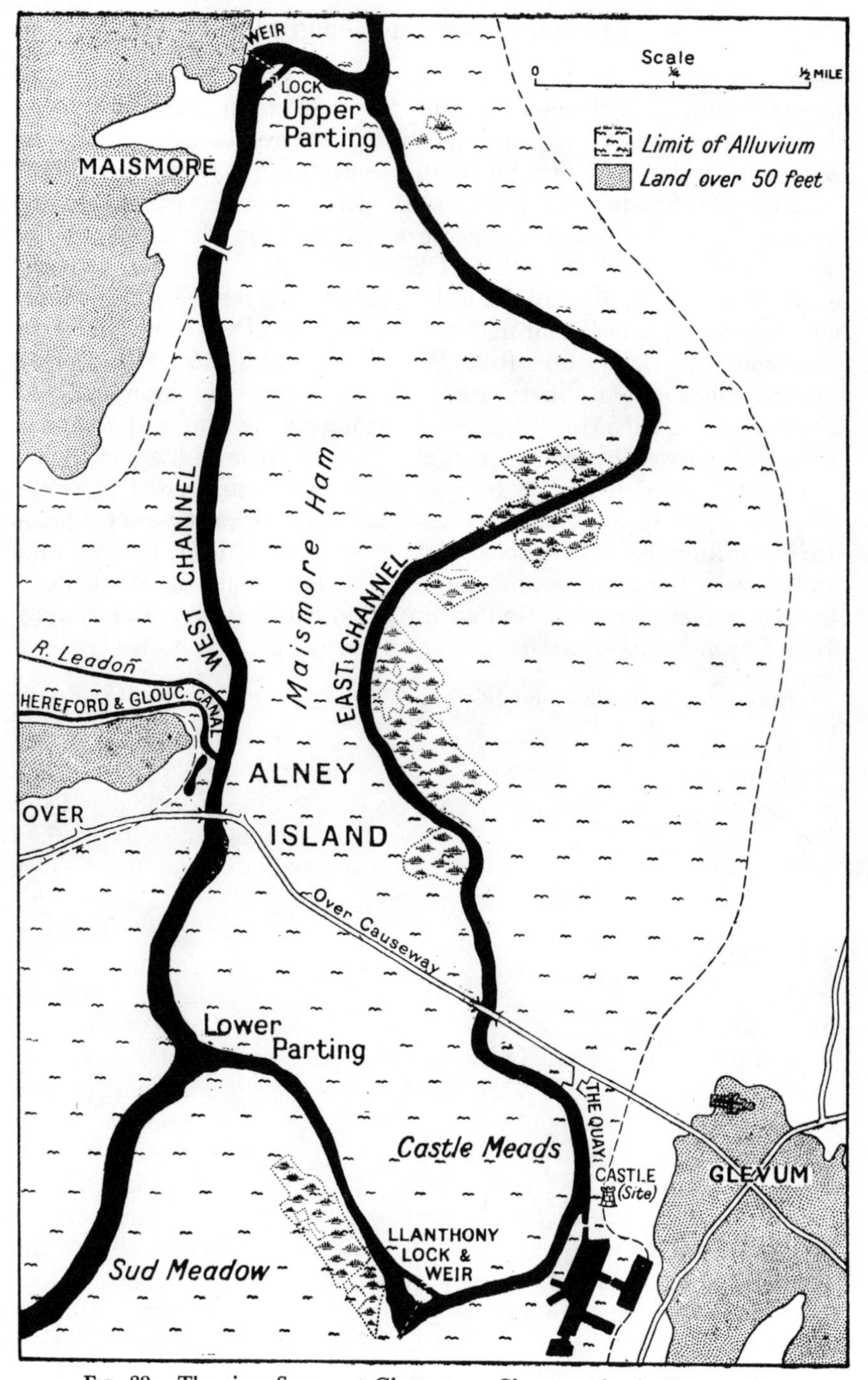

FIG. 29.—The river Severn at Gloucester. *Glevum* marks the Roman site.

short controlled waterway to the estuary at Sharpness, greatly improved its accessibility. This canal admitted vessels of up to 600–700 tons drawing up to 14½ ft., on any high tide : as a result, Gloucester was, according to Parkes, writing in 1846, " within a few years . . . raised . . . from an unimportant place to considerable eminence as a mercantile city." Any attempt to measure the scale of the trade of the Severn and, in particular, that of Gloucester, the chief link between up-river and sea-going shipping, lies outside the scope of this paper, although we may note that in 1845 364,000 tons of shipping passed up the estuary by " The Old Road," compared with 205,000 tons that entered the canal at Sharpness, and that upstream shipping from Gloucester totalled 551,000 tons.[27] We may glance, too, at the variety of cargoes carried by the river in the days of sail and haulage, some of which reached it at first by pack-horse routes and later by canal (Fig. 25) : above all, coal, pig- and bar-iron, ironware, pottery, glassware and lime from Coalbrookdale ; grain, cheese, lead, flannels and timber from Shrewsbury and above ; Droitwich salt via Worcester ; whilst upstream were distributed imported goods (wines, sugar, etc.) from Bristol and Gloucester.

[27] Report to the Lords Commissioners of the Admiralty, *Parliamentary Papers*, Vol. XXVII, 1849, p. 10.

VII

CLIMATIC CLASSIFICATION

By F. KENNETH HARE, B.Sc.

Introduction

This essay contains a critical summary of existing methods of climatic classification, a subject hitherto almost ignored in British geographical literature. Systems of classification of any natural complex make little appeal to the British temperament, which is rarely inclined to like pigeon-holes or card-indexes. This conservative pragmatism sometimes leads us, however, to ignore real necessities imposed by progress in research and by the elaboration of established knowledge. We are then compelled to adopt with an ill grace the systems of classification produced by foreign workers, especially the Germans and the Americans, who excel in such matters. It has been this way with physiography, where rational systems of nomenclature came from the geologists who were exploring the western interior of the U.S.A., culminating in the genius of William Morris Davis, and from the German and Austrian school, typified by Walther Penck. And it is equally true of climatology, in which the names of Wladimir Köppen and C. Warren Thornthwaite stand out. It is noteworthy that the most widely read British works on climatology, W. G. Kendrew's *Climate* and his *Climates of the Continents*,[1] contain no reference of any length to this important topic.

Climatic classification is an essentially geographic technique. It allows the simplification and generalization of the great weight of statistics built up by the climatologists. These figures do not mean much to the geographer unless a way can be found of reducing them into an assimilable form. The real purpose of classification is hence to define climatic types in statistical terms, in which climate as a geographic factor is to be regarded as having definite and uniform characteristics : only by such a classification can rational climatic regions be defined.

Since climates are faithfully reflected by soil and climax vegetation, all modern systems of classification attempt to give the climatic limits of characteristic soil or vegetation regions. This has the effect of

[1] W. G. Kendrew, *Climates of the Continents*, Oxford, 1922, 1937, et seq.

rendering the classifications suitable for use in a discussion of the economic consequences of climate, particularly as regards the agricultural potential of a region. They are less useful in a study of climate as a factor influencing human distributions directly, through physiological effects. The earliest workers in this field were plant-geographers, and even today, interest in the subject is mainly found among ecologists, soil scientists and hydrologists. During the war years, however, much progress has been made in the study of medical climatology, and it is possible that further progress in classification will have to take account of these developments.

As has already been said, two names dominate all others in this field. Wladimir Köppen, a St. Petersburg-trained biologist, was the pioneer of comprehensive classification, publishing his earliest results in 1900. His scheme has been repeatedly modified, both by himself and his well- or ill-wishers. C. Warren Thornthwaite, of the U.S. Soil Conservation Service, did not publish his first classification until 1931, and his second was deferred until 1948. Although there have been other attempts at classification (such as the recent work of W. Gorczynski) few are statistical, and they cannot in consequence be regarded as adequate for geographical purposes. The three major classifications are discussed in detail below. It is hoped that the discussion will be useful by making the subject more easily accessible and by showing some of the limitations of the existing schemes.

Köppen's Classification of Climate

It is not proposed to make here a detailed review of early attempts at climatic classification ; a good sketch of their history has been given by Thornthwaite [2] in a recent study of Köppen's work. The whole idea of a systematic analysis of world climate is in any case very recent, as the necessary statistical basis has only been available for the past seventy or eighty years. Even today analysis is handicapped by the paucity of observations over much of the earth's surface, especially over the oceans and the continental interiors. Until the nineteenth century, therefore, there could be no effective substitute for the classical concept of the thermal zones. The first world-wide pictures of temperature and rainfall distributions began to appear early in the latter half of the nineteenth century, and it was from these pictures that the classification of climates was first attempted. The work was undertaken by the group of naturalists who were concerned with the world distribution of species, especially of plants. Among these we

[2] C. W. Thornthwaite, "Problems in the Classification of Climates," *Geog. Rev.*, Vol. XXXIII, 1943, pp. 233–55.

can select the work of de Candolle,[3] Linnser [4] and Drude,[5] whose contributions have been admirably summarized in the article by Thornthwaite already mentioned.[2]

Köppen, whose classification of climate has dominated the subject for many years, was one of these biologists, and the earliest motive underlying his work was to reduce the complex facts of climate to a simple, numerically-based classification that could be used in the study of plant geography. This may be regarded as Köppen's major contribution to modern climatology, *that schemes of classification must rest on an objective, numerical basis*, so that different climatologists can refer the same set of climatic figures to the same climatic class. His other important contribution to the ideal basis of the science is the view that the climatic classification should be drawn up with respect to the known facts of phytogeography. Though by no means the originator of the idea, Köppen was the first systematically to investigate the climatic limits of the great regional divisions of the plant cover. The idea that climax vegetation and soil groups are reflections of the climatic régime is now commonplace, and permeates most modern discussions of physical geography. In Köppen's youth, however, this idea had nothing but probability and deduction to support it. By showing that definite climatic values could be assigned to the boundaries between natural vegetation regions, he proved the general validity of the idea.

Köppen's earliest attempt at classification dates from 1900.[6] He has subsequently much changed the scheme, and has greatly added to its complexity and scope. In its latest form, it appears in his masterpiece, the *Handbüch der Klimatologie* (Vol. I, C. 1936), published in 1936, and it is discussed below primarily as it appears in this most recent form. There are several good summaries in the English language, notably those of Leighly,[7] and Haurwitz and Austin.[8] It is advisable to exercise care in using English re-statements of Köppen's

[3] A. de Candolle, "Les Groupes physiologiques dans le régne végétal," *Rev. Scientifique*, Ser. 2, Vol. XVI, 1875, pp. 364–72.

[4] C. Linnser, Review of his scientific work by C. Abbe, *U.S. Weather Bureau Bull.* 36, 1905, pp. 211–33 (original papers are very inaccessible).

[5] O. Drude, Berghaus' *Physikalischer Atlas* (Part 5), Gotha, 1887. Reproduced in many German atlases in early twentieth century.

[6] W. Köppen, "Versuch einer Klassifikation der Klimate," *Geog. Zeit.*, Vol. VI, 1900, pp. 593–611, et seq. The most comprehensive review of his system appears in *Grundriss der Klimakunde*, Berlin, 1931 (2nd edition). This is almost identical with the treatment given in "Das geographische System der Klimate," *Handbüch der Klimatologie*, ed. W. Köppen and R. Geiger, Vol. I, pt. C, 1936.

[7] J. B. Leighly, "Graphic Studies in Climatology," *Univ. Cal. Pub. Geog.*, Vol. II, no. 3, 1926.

[8] B. Haurwitz and J. A. Austin, *Climatology*, McGraw-Hill, 1944, pp. 109–30.

views, as they often refer to different stages in the development of the scheme, and are in many cases made so that the classification can be criticized : a friendly approach has not been too common. The above

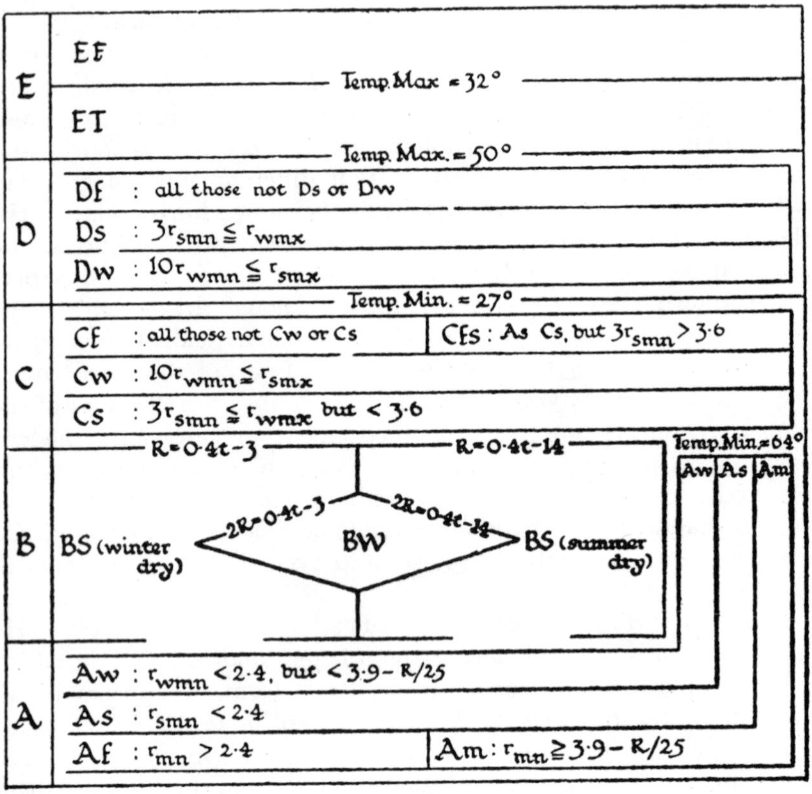

FIG. 30.—The limits of Köppen's climatic types.

Some idea of the complexity of Köppen's scheme is given by this diagram, which goes only as far as the first order sub-division in most cases. The diagram shows the limiting values corresponding to each of the possible boundaries between types. An exception is that the BS climates can also come into contact with the D climates ; the boundary between them is the same as that between the B and C climates. The first-order sub-divisions of the forest climates (A, C and D) are shown as enclosed areas within the space allotted to the major types corresponding. The complexity of the classification is also brought home by the formidable number of climatic elements listed in the key below.

KEY

Temperature Values (° F.)		*Rainfall Values* (inches)
Temp. max. = avg. temperature of warmest month	r_{mn}	= avg. rainfall, driest month
	r_{mx}	= avg. rainfall, wettest month.
Temp. min. = avg. temperature of coldest month	r_{smn}, r_{wmn}	= avg. rainfall of driest summer or winter month
t = avg. annual temperature	r_{smx}, r_{wmx}	= avg. rainfall of wettest summer or winter month
	R	= avg. annual rainfall

writers, however, have been eminently fair. The reader is referred to the Leighly and Haurwitz-Austin versions for nomograms and other aids in deciding the class of a particular climate.

A sketch of Köppen's classification is given in diagrammatic form in Fig. 30, which gives the critical limiting values for each climatic type.

As will be seen from Fig. 30, Köppen envisaged five major climatic types :

Letter	*Type*
A	Tropical rainy climates (Megathermal)
B	Dry climates (Xerophilous)
C	Warm temperate rainy climates (Mesothermal)
D	Cold snow-forest climates (Microthermal)
E	Snow climates (Ekistothermal)

which broadly correspond to the five major vegetation regions of the earth. For a definition of these latter regions, Köppen referred to the work of de Candolle,[3] whose vegetation-groups are indicated in brackets behind Köppen's corresponding climatic type in the above table. De Candolle was a plant physiologist, and his regions were based on the climatic requirements of each group of plants for their internal functions. This approach is essentially distinct from that of the ecologists, who dealt in the field of plant associations. Of these, the greatest was Schimper,[9] whose *Plant Geography* remains a standard work. As Thornthwaite [2] pointed out, it is a pity that Köppen based his scheme on de Candolle's work rather than Schimper's : the latter has dominated ecological thought and method, and Köppen's life work has therefore been constantly out of harmony with that of the ecologists.

Köppen's method can be summed up as the effort to find the climatic correspondences of the boundaries between de Candolle's regions. From the start, therefore, he was dedicated not to synthesis of climatic controls but to the finding of arbitrary climatic limits. The result has been that Köppen has used a wide variety of different elements in defining his boundaries, and has often changed his mind about which element to use in defining a particular boundary.

Köppen's major types are sub-divided several times, and it is a great help in remembering the details to refer these sub-divisions to a hierarchy of orders. The first order sub-divisions usually refer to special features of the rainfall régime, and the second order to thermal characteristics. The third and remaining orders refer to various characteristics of a more variable and local kind, and they are not discussed here. The first and second order sub-divisions are summarized in Table I.

[9] A. F. Schimper, *Plant Geography*, trans. W. R. Fisher, Oxford, 1903.

TABLE I. THE CLIMATIC TYPES IN KÖPPEN'S CLASSIFICATION

1. *Major Type*	2. *First Order Sub-divisions*	3. *Second Order Sub-divisions*
A. (Tropical, rainy)	Af (wet at all seasons) Aw (dry winters) As (dry summers—rare) Am (monsoon type)	Axi (isothermal)
B. (Dry)	BS (steppe) BW (desert)	Bxh (hot) Bxk (cool)
C. (Warm, temperate, rainy)	Cf (wet at all seasons) Cs (dry summers) Cw (dry winters) Csf (rather dry summer, very wet winter)	Cxa (summers long and warm) Cxb (summers long and cool) Cxc (summers short and cool)
D. (Cold snow-forest)	Df (wet at all seasons) Ds (dry summers) Dw (dry winters)	Dxa (summers long and warm) Dxb (summers long and cool) Dxc (summers short and cool) Dxd (ditto, but great winter cold)
E. (Snow climate)	ET (tundra) ES (permanent snowfields)	—
H. (Mountain climates)	—	—

N.B. The cypher "x" in the third column is meant to indicate that the second order sub-divisions can apply to any of the first order sub-divisions—at least in theory.

The major types in the Köppen scheme are always indicated by the capital letters A–E and H. The first and second order sub-divisions are usually indicated by small letters (except in the Dry and Snow climates where the German substantival capital usage requires a departure).

The boundaries between the first order sub-divisions as well as the major types themselves are summarized in Table I. The major types can be sub-divided into the forest (A, C, D), dry (B) and cold (E, H) climates, which are discussed in turn below.

(1) *The Forest Climates.* The forest climates are distinguished from one another (Fig. 30) on the basis of temperature. The A climates are those in which the coldest month has a mean temperature of over 64° F., and the D climates are those whose coldest month has a mean temperature below 27° F. The C climates lie between. These climates then represent the selva, warm temperate rain forest and cold snow-forest régimes (including taiga), though the selected temperatures are justified on the basis of the optimum for human comfort (64° F.) or the possibility of prolonged winter snow-cover, for which 27° F. is regarded as the upper limit.

As shown in Fig. 30, the forest climates are further divided on the basis of rainfall régime. Those whose rainfall is abundant at all seasons are given the first order sub-division f (Af, Cf, Df), though the meaning of the cypher differs slightly from group to group. The w sub-divisions are winter dry, and the s sub-divisions, summer dry.

Here again the quantitative limits indicated in Fig. 30 differ from group to group. Among the C and D climates, summer is considered dry if its driest month gets a third of the rain of winter's wettest month, whereas the ratio is a tenth in the reverse case of winter drought. This is to allow for the greater evaporation and lower rainfall effectiveness in summer.

Special first order sub-divisions are provided in the A and C climates for those climates which have a wet and dry season, but which are humid throughout because of an excessive winter rainfall.

(2) *The Dry Climates.* Köppen's dry climates (B) are defined by an empirical aridity criterion which he derived from a study of the climates of the steppe-forest boundary. The criterion adopted has been changed several times, and has had to meet considerable criticism, especially in the United States, where it fits the steppe-desert boundary rather poorly. Although the B climates are mainly within the warm or hot latitudes, they are not specifically confined to these latitudes by the aridity criterion, which is of the form

$$R \leq 0{\cdot}44\ T - K$$

where R = the mean annual rainfall (inches) and T the mean annual temperature in degrees Fahrenheit. K is an empirical constant depending on the seasonal rainfall distribution. If the rainfall maximum is in summer (and is therefore less effective due to evaporation) K = 3, but if rain falls mainly in winter K = 14. This relationship is based on the observed fact that in warm climates a greater rainfall is necessary to maintain abundant vegetation than farther north in cooler climates.

Within the dry climates a distinction is made between BW (desert) and BS (steppe) climates, the criterion adopted being that

$$2R \leq 0{\cdot}44\ T - K$$

where R, T and K have the same meanings as before.

(3) *The Cold Climates* (*E*). Following Supan, Köppen observed that the taiga of the D climates passed into tundra very close to the 50° F. isotherm for the warmest month, and this boundary has not been seriously challenged. He drew a further distinction between the tundra climates (ET) and the frost climates (EF) : the latter, characterized by permanent snow-fields, lay north of the 32° F. isotherm for the warmest month. This again is a fairly obvious boundary, though Jones [10] has pointed out that permanent snow-fields may occur well south of the 32° F. isotherm in areas of heavy snowfall. A special class of cold climates (H) occurs on high ground in middle and low

[10] S. B. Jones, *Econ. Geog.*, Vol. VIII, 1932, pp. 205–8.

latitudes: Köppen used different computing formulæ for the H climates, which cannot be described in detail here.

The second order sub-divisions. The further sub-division of the climates is a matter chiefly of interest to pure climatologists. Most geographers appear to regard the scheme as adequate as it stands after the first sub-division has been made. Thus Trewartha [11] has prepared a map of world climates on a simplified first order classification. But Köppen himself attached considerable importance to the thermal, second order sub-divisions, which undoubtedly add greatly to the precision of the classification. The third cypher in a Köppen label (A*si*, Cw*b*), etc. refers to the thermal characteristics of the climate within its major division. Table II gives a summary of the principal second order sub-divisions, each of which can in theory (though not in practice) be applied to any first order derivative of the group.

Table II. The Second Order (Thermal) Sub-divisions of the Köppen Climates

1. *Major Type*	2. *Second order sub-division*	3. *Characteristics and Limits*
A x	i	Mean monthly temperatures all within a 9° F. range (isothermal)
B x	h	Hot desert or steppe (mean annual temperature $> 64^\circ$ F.)
	k	Cool desert or steppe (mean annual temperature $< 64^\circ$ F.)
	a	Warmest mo. $> 72^\circ$ F.: > 4 mo. $> 50^\circ$ F.
C x	b	Warmest mo. $> 50^\circ$ F.: > 4 mo. $> 50^\circ$ F.
D x	c	Warmest mo. $> 50^\circ$ F.: < 4 mo. $> 50^\circ$ F.
	d *	As c, but coldest month $< -36^\circ$ F.

* refers only to D climates.
x indicates any first order sub-division.

Further sub-divisions have been suggested by Köppen, but it is extremely doubtful whether the further complication of the scheme can be justified. They will not be discussed here.

The Geographic Validity of Köppen's Classification

For at least two decades the Köppen classification has been widely accepted by geographers, most of whom have used it as the standard system. The great majority of simplified classifications, like Blair's [12] and Trewartha's,[11] are based on Köppen's system. So widespread has it become that some American writers have begun to regard it as an international standard, to depart from which is scientific heresy. There has none the less been very extensive criticism of the classification, some of it constructive, in the U.S.A. America is a land of climatic

[11] G. T. Trewartha, *Introduction to Weather and Climate*, McGraw-Hill, 1943.
[12] T. A. Blair, *Climatology*, Prentice-Hall, 1942.

contrasts, and it has been clear for years that Köppen's boundaries do not correspond to the real natural limits of the major climatic zones as they are expressed by the soil and vegetation. To remedy this (or to try to) many American workers have suggested moving a boundary one way or the other by slightly changing its numerical basis. Others have attacked the scheme on the grounds of its empiricism, and have in some cases attempted to replace its methods by others more logical.

Among the first group, who can be dubbed the line-shifters, we may list Van Royen,[13] Russell [14, 15] and Ackerman.[16] Van Royen constructed a map of the climatic regions of the eastern U.S.A., and showed that there is no sort of fit between the boundaries and those of the natural vegetation. Russell applied the classification to California,[14] and later to the dry western half of the U.S.A.[15] Like Van Royen he found a bad fit, especially along the B/C and BW/BS boundaries. He attempted to improve the fit by using an older expression for the aridity criterion, but it cannot be pretended that this modification has improved matters very much. Russell also introduced the concept of "climatic years." [15] He had been concerned at the substitution on climatic maps of lines for what in the landscape are the broad transition zones between climates. He tried to remedy this by showing how the boundary migrates from year to year, and defining the zone over which it oscillates. He also drew attention to the fact that it is the occasional drought, flood or other climatic extreme which is most significant in limiting development of the plant cover and of land-form : hence Köppen's extensive use of annual and monthly means may not give a true picture of the climatic boundaries. Ackerman [16] prepared a map of the U.S.A. showing the Köppen divisions, but used the 32° F. isotherm for the coldest month as marking the C/D boundary, following Russell. All these writers, however, accepted the classification as the best possible at present, while admitting its obvious defects.

More radical objections have been voiced by Gorczynski,[17, 19] who

[13] W. Van Royen, "The Climatic Regions of North America," *M. Weath. Rev.*, Vol. LV, 1927, pp. 313–19.

[14] R. J. Russell, "Climates of California," *Univ. Cal. Pub. Geog.*, Vol. II, no. 4, 1926, and "Climates of Texas," *Ann. Assoc. Amer. Geog.*, Vol. XXXV, 1945, pp. 37–52.

[15] R. J. Russell, "Dry Climates of the United States," *Univ. Cal. Pub. Geog.*, Vol. V, no. 1, 1931, and no. 5, 1932, and "Climatic Years," *Geog. Rev.*, Vol. XXIV, 1934, pp. 92–103.

[16] E. A. Ackerman, "The Köppen Classification of Climates in N.A.," *Geog. Rev.*, Vol. XXXI, 1941, pp. 105–11.

[17] W. Gorczynski, "Comparison of Climate of U.S. and Europe," *Polish Inst. of Arts & Sciences*, New York, 1945, pp. 239–47.

especially quarrels with the aridity criterion. He himself employs a criterion which depends on temperature range, rainfall variability and latitude. It is of the form

$$A \propto (T_{max.} - T_{min.}) \frac{(R_{max.} - R_{min.})}{(R)} \operatorname{cosec} \phi$$

where $T_{max.}$ = highest monthly mean temperature
$T_{min.}$ = lowest monthly mean temperature
$R_{max.}$ = highest annual rainfall so far recorded
$R_{min.}$ = lowest annual rainfall so far recorded
ϕ = latitude
R = annual rainfall average.

The proportionality constant is chosen so that $A = 100$ under mid-Saharan conditions, and the criterion then becomes a percentage fraction of maximal aridity. Gorczynski introduces this concept into a new classification of climates that he calls the decimal system.[18] His arguments are hard to follow, and the scheme seems even more empirical than Köppen's. His main preoccupation is to allow a differentiation between continental and maritime types ; otherwise his scheme resembles Köppen's closely. The ten types are summarized in Table III.[19]

Table III. Gorczynski's Decimal Classification of Climates

Group	*Decimal Type*	*Description*	*Limiting Factors*
I (Warm group)	1	Wet tropical type	Mean temperature > 68° F. in three coolest months
	2	Savannah type	
II (Arid group)	3	Desert type	Aridity criterion (see above) high throughout year
	4	Steppe type	Ditto, for several months only
III (Maritime group)	5	Summer-rain type	Mean temperature of coldest month between 59° F. and 23° F. ; or less than four months below 32° F.
	6	Winter-rain type	
IV (Continental group)	8	Heavy winter snow type	Mean temperature of coldest month below 23° F., of warmest above 50° F.
	9	Light winter snow type	
V (Sub-polar or mountain group)	10	—	Mean temperature of warmest month below 50° F.

N.B. This table is not intended to define the types precisely, but merely to indicate the general character of the classifications.

[18] W. Gorczynski, " Climatic Types of California according to the decimal scheme of World Climates," *Bull. Amer. Met. Soc.*, Vol. XXIII, 1942, pp. 161–5 and 272–9.

[19] W. Gorczynski, " Uber die Klassification der Klimate," *Gerlands Beit. zur Geophysik*, Vol. XLIV, 1935, no. 2, pp. 199–210.

Far more damaging, far more explicit, is the criticism of Thornthwaite,[2] who bases his attack on the arbitrary nature of Köppen's limits, on their empiricism and on what he regards as their unnecessary complexity. He objects to the use of temperature to define the major types, pointing out that changes in temperature régime do not in fact lead to abrupt boundaries of soil or vegetation. Thornthwaite himself has in the last twenty years developed alternative systems of classification that attempt to remedy the empiricism of Köppen's methods. The whole question of climatic classification has become closely identified with Thornthwaite's name ; his work must be considered the most advanced and most complete as yet available.

The Thornthwaite Classifications

The fundamental idea behind Thornthwaite's methods [20, 21, 30] is that of climatic efficiency, i.e. the capacity of the climate to support the growth of plant communities. The factors which govern this growth, soil and ecological factors being ignored, are the available moisture, the annual march of temperature, and the degree of correlation in time of the optimum periods in both temperature and rainfall régimes. The reduction of these facts to a single, numerical standard is achieved in the earlier classification by the use of precipitation and temperature efficiency indices : a third factor expresses their time-correlation. In the second of his classifications [30] Thornthwaite radically changes his position, while still seeking a measure of efficiency.

(*i*) *The first classification : precipitation efficiency.* The moisture available for plant-growth, in Thornthwaite's view, is conditioned by (*a*) the precipitation and (*b*) the loss due to evaporation. The efficiency of an annual rainfall of 20 in., for example, is obviously low if the corresponding evaporation is 30 in. and high if the latter is only 10 in. The P–E index is essentially of the form, precipitation divided by evaporation, and is unity when the two are equal. In detail, Thornthwaite proposed (*a*) a P–E ratio, which is the ratio of the two elements in a month and (*b*) a P–E index, which is the sum of the twelve monthly ratios. This idea was by no means new ; Transeau [22] had used similar ratios early in the century, and Thornthwaite also acknowledged the

[20] C. W. Thornthwaite, "The Climates of North America according to a new classification," *Geog. Rev.*, Vol. XXI, 1931, pp. 633–5.

[21] C. W. Thornthwaite, "The Climates of the Earth," *Geog. Rev.*, Vol. XXIII, 1933.

[22] E. N. Transeau, "Forest Centres of North America," *Amer. Naturalist*, Vol. XXXIX, 1905, pp. 875–89.

work of Lang,[23] de Martonne,[24] Meyer [25] and Szymkiewicz.[26] All these workers had been handicapped by the lack of accurate and numerous evaporation data. Thornthwaite proposed to overcome this difficulty by expressing evaporation as a function of temperature. He found that the P–E ratios were related in a simple way to the mean monthly temperature. In his final formula for the P–E index, evaporation does not appear : it is replaced by the temperature-function which he considers expresses it adequately. The P–E index is of the form

$$(\text{P–E index}) = \sum_{n=1}^{12} 115\left(\frac{P}{T-10}\right)_n^{\frac{10}{9}}$$

where Σ = summation of twelve monthly ratios
P = monthly mean precipitation (inches)
T = monthly mean temperature (degrees F.)
n = the particular month.

The P–E index was then computed for areas along the boundaries of recognized vegetation regions, and the following humidity provinces determined :

Humidity Province	*Vegetation*	*P–E Index*
A	Rain forest	> 127
B	Forest	64–127
C	Grasslands	32–63
D	Steppe	16–31
E	Desert	< 16

These cyphers A–E are the first members of the three-cypher group used to indicate the climatic type (except in cold climates, see below) : thus the major divisions in Thornthwaite's scheme depend on the humidity rather than on the thermal characteristics of the climate. Although Thornthwaite is well ahead of Köppen in allowing for the evaporation, there is no denying that this P–E index is as empirical as anything in Köppen, especially as regards the functional expression for evaporation. The fit of the provinces with the vegetation zones is also achieved by purely empirical methods. A major arbitrary

[23] Richard Lang, *Verwitterung und Bodenbildung als Einfuhrung in die Bodenkunde*, Stuttgart, 1920.
[24] E. de Martonne, "Regions of Interior Basin Drainage," *Geog. Rev.*, Vol. XVII, 1927, pp. 397–414.
[25] A. Meyer, "Uber einige Zusammenhange zwischen Klima und Boden in Europa," *Chemie der Erde*, Vol. II, 1926, pp. 209–347.
[26] D. Szymkiewicz, "Etudes Climatologiques," *Acta Societatis Botanicae Poloniae*, Vol. II, 1925, no. 4.

element is that the P–E index is regarded as being equivalent to the value for 28·4° F. if the temperature is below that figure, on the grounds that snow is of no value to plants until it melts ; how the value of 28·4° F. was obtained, Thornthwaite does not say.

The summer concentration of precipitation-effectiveness (rightly regarded as a significant element) is indicated by small letters placed third (after temperature-effectiveness) in the climatic labels. They are "r" for all-year-round rainfall, "s" for summer drought "w" for winter drought and "d" for all-year-round drought. Fig. 31 (after Thornthwaite),[21] gives the precise limits of these classes.

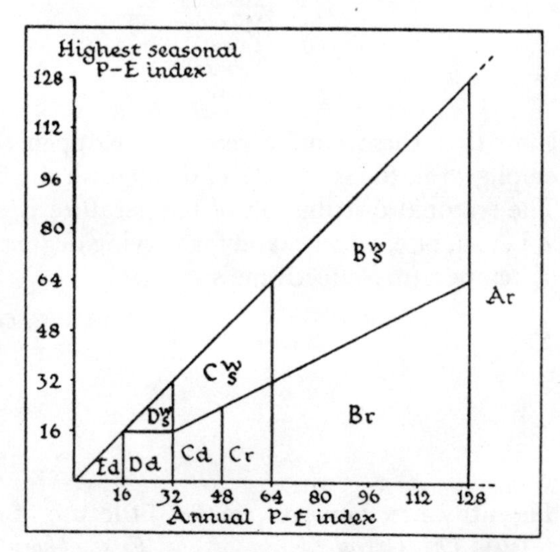

FIG. 31.—The summer concentration of rainfall.

This diagram shows how the various sub-divisions on the basis of summer rainfall-concentration are determined. The scale on the vertical is the sum of the P–E ratios for the three summer months, and that on the horizontal, the P–E index.

(ii) The first classification : thermal efficiency. The thermal efficiency index (also called the temperature-effectiveness) is a totally empirical quantity, computed by analogy with the P–E index. Temperature is highly significant in its influence on plant-growth, operating through many different physiological processes. Thornthwaite claimed that its influence is closely analogous to that of available moisture ; he therefore sought for an arbitrary scale which would give numerical values similar to those of the P–E index. The scale he eventually adopted gives zero effectiveness along the poleward limit of the tundra and a value of 128 along the outer limit of the selva. The index is computed by the summation of twelve T–E ratios, i.e.

$$\text{(T–E index)} = \sum_{n=1}^{12} \left(\frac{T-32}{4}\right)_n$$

where T = monthly mean temperature (degrees F.)
and n = the particular month
Σ as before indicates summation over the twelve months.

The use of this T–E index leads to the definition of thermal provinces as follows :

Temperature (Thermal) Province	*T–E Index*
A′ (Tropical)	> 127
B′ (Mesothermal)	64–127
C′ (Microthermal)	32–63
D′ (Taiga)	16–31
E′ (Tundra)	1–15
F′ (Frost)	0

Note that these closely resemble Köppen's major types. The dash employed in the symbol is to distinguish the index from the P–E index. The seasonal distribution of temperature is expressed by small letters, a, b, c, d, or e, which stand for varying degrees of summer concentration of temperature-effectiveness :

Sub-province	*% Concentration in 3 Summer Months*
a	25–34
b	35–49
c	50–69
d	70–99
e	100

Thornthwaite, however, makes little use of this sub-division.

(*iii*) *The Overall Form of the First Thornthwaite Classification.* The indices described above are combined to give the overall description of the climates. The policy adopted by Thornthwaite is as follows :

(*a*) With T–E indices of 32 or over, the climates are divided on a primary basis of the humidity province, viz. AB′, CC′, etc., the P–E cypher coming first, the T–E cypher (dashed) second. The seasonal distribution of the P–E index is indicated by the third (small) cypher. The seasonal distribution of the T–E index is shown as the final cypher. Thus AB′ra means rain forest, mesothermal, even distribution of precipitation-effectiveness, little summer concentration of temperature-effectiveness.

(*b*) With T–E ratios below 32, Thornthwaite regards P–E as being of little significance ; very little moisture is needed to maintain the limited plant growth possible at these temperatures. Thus the main climatic groups in this classification become :

AA′ BA′ CA′ DA′ EA′
AB′ BB′ CB′ DB′ EB′
AC′ BC′ CC′ DC′ EC′

Adequate temperature-effectiveness

D′ E′ and F′

Inadequate temperature-effectiveness

The equivalence of these to the main vegetation groups is indicated in Fig. 32, from Thornthwaite's original paper.

The Merits of the Earlier Thornthwaite Classification

There can be no question that Thornthwaite's earlier classification was the most far-reaching and theoretically adequate scheme available until recently. In many ways it resembles the Köppen system, but it is a more refined tool for use in geographical analysis. Both classifications rest on the assumption that the world distribution of plant associations is the real clue to the distributions of climates. In both cases the central method is to determine the quantitative climatic limits of the major plant regions. There, however, the parallelism ends. Köppen's limits are essentially arbitrary; he uses a wide variety of climatic elements to define his boundaries, and there can be no denying that the result is an illogical, unnecessarily complex method. Moreover, arbitrary limits cannot be used in extrapolation, and it has often been found that limits determined in one area do not work when applied to another. Thornthwaite, on the other hand, attempted (i) to define the climatic elements which were active in limiting plant growth, (ii) to find logical parameters for the main controls, viz. precipitation- and temperature-effectiveness, and (iii) to determine empirically the numerical scale which completes the fit of these parameters to the natural plant regions. The empiricism of Thornthwaite's methods is due to our imperfect knowledge of soil-moisture relationships, of plant physiology, and of the climatic requirements of specific plant associations. Since we cannot hope to learn these requirements in the near future, it is no criticism of his methods to declare them empirical in this particular. Criticism is better directed at the effectiveness parameters, which rest on premises which appear to be in part fallacious.

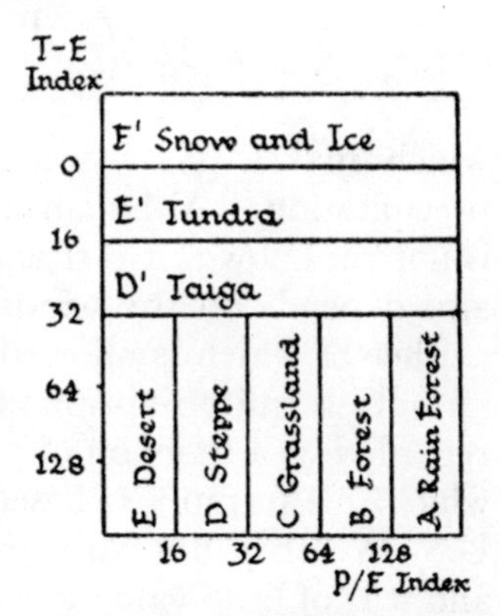

Fig. 32.—The primary types in Thornthwaite's classification.

When the T–E index is below 32, the major divisions disregard the P–E index, and appear on the horizontal F¹, E¹ and D¹ zones. With the T–E index above 32, however, the primary types depend on P–E index first; the thermal divisions become secondary. Hence the primary types appear as the vertical columns A–E.

The most serious criticism refers to the assumptions made about evaporation, chiefly as regards two points: (i) the means of obtaining the evaporation and (ii) the view that evaporation from soil is simply a loss to plant growth. These points will be considered in turn.

(i) The evaporation is expressed by an empirical formula based on the precipitation and temperature. It is of the form

$$\frac{P}{E} = 11{\cdot}5\left(\frac{P}{T-10}\right)^{\frac{10}{9}}; \text{ or } E = \frac{P}{11{\cdot}5}\left(\frac{P}{T-10}\right)^{-\frac{10}{9}}$$

which makes the evaporation a function of the temperature and precipitation. Although the exact functional character of evaporation is not yet known, it is true that it depends on the temperature ; but it also depends on the windspeed and the moisture content of the air, neither of which is allowed for. The presence of precipitation in the formula is quite unjustifiable, for in this sense evaporation cannot be regarded as a function of precipitation. The " evaporation " data on which this formula is based are for losses from open pans run by the U.S. Weather Bureau ; the formula can refer only to such figures, and it is of little value to say that the presence of P can be justified on the grounds that if there is no precipitation there can be no evaporation. This would be true of soil surfaces, but it is not of pans in which water supply is unlimited. There are now many expressions by which evaporation from open water surfaces can be predicted ; all are to some extent empirical, but they at least make some attempt at physical reality. One such expression is due to Thornthwaite himself in collaboration with Holzman.[27] Penman [28] has recently described encouraging results from England.

(ii) It may also be doubted whether evaporation from pans is in any way representative of the evaporation from plant-covered soils. Furthermore, evaporation takes place primarily through the transpiration of the plants themselves, so that it is not really a loss to be deducted from the rainfall, the balance being available for plant growth. In fact the opposite is more nearly true ; the real evaporation from a soil represents almost entirely water that *has* been used in growth. For these reasons it is highly improbable that a simple precipitation-pan evaporation ratio can represent the humidity régime of the soil.

The evaporation from a soil surface is in practice a highly complex process. An upper limit to the loss of moisture to the soil through evaporation is afforded by the capacity of the air to transport the moisture away. This transport is effected almost solely by eddy diffusion. Near the ground there is a very shallow laminar layer in which evaporation is by pure molecular diffusion, the rate being deter-

[27] C. W. Thornthwaite and B. Holzman, "The determination of evaporation from land and water surfaces," *Mon. Wea. Rev.*, Vol. LXVII, 1937, pp. 4–11.

[28] H. L. Penman, " Natural Evaporation from open water, bare soil and grass," *Proc. Roy. Soc. A.*, Vol. CXCIII, 1948, pp. 120–45.

mined by the temperature of the soil surface. Above this laminar layer, however, eddy diffusion vastly outweighs molecular diffusion, the rate of upward transfer depending on the rate of windspeed and the hydrolapse (i.e. the decrease in moisture content of the air with height). The effectiveness of the eddy diffusion hence limits the evaporating power of the air, which in turn limits the capacity of the soil to evaporate its water. The evaporation off an open water surface is sometimes regarded as being a measure of the evaporating power ; even if this proposition is accepted, however, it can be regarded as sound only for unit area of an infinite water surface. The evaporation off small, limited surfaces like that of an evaporation pan is affected by many special boundary conditions, and also by the turbulence set up by the flow of air across the rims of the pan. The evaporation-pan measurements tell us something about evaporating power, but we are not yet sure what it is.

The capacity of a soil surface to take up this limited evaporating power depends on the nature of the soil and on its plant cover. Evaporation from saturated fallow soils is rapid, equalling or exceeding that of an open water surface, but as the soil dries out the rate diminishes ; capillary rise may maintain the supply for a short time, but the rate is very slow indeed after the initial rush. The fallow soil, then, creates for itself an insulating dry cap, retaining the moisture below it almost intact. Penman,[29] for example, showed that loss from the loamy soil of Rothamsted diminished almost to zero after about 0·5 in. had been lost. It is this property of fallow soils that is made use of in dry farming operations.

From plant-covered surfaces, direct evaporation from the soil is much reduced, being largely replaced by transpiration from the plant leaves. The plants are able to withdraw moisture through their roots from a considerable depth, and can hence maintain full rate evaporation for a longer period than can a fallow soil. In the absence of rain, however, they eventually abstract all the moisture not held too firmly by colloidal adsorption, and wilting begins. If rain does not come, the plant cover dies or becomes dormant, and evaporation is once again negligible. Both on fallow and plant-covered soils, then, there is a limit to the amount of moisture that can be removed by evaporation, and the amount of pan evaporation is a poor clue to this limit. The maintenance of full-rate evaporation from either depends on constant replenishment by rainfall.

The idea that evaporation is an inescapable charge that has to be

[29] H. L. Penman. "Evaporation from Fallow Soils," *Quart. Journ. Roy. Met. Soc.*, Vol. LXVI, 1940.

paid before plant growth can begin is also plainly unsound. If the soil stays fallow, evaporation is negligible after a brief initial flush ; and if it becomes plant covered, the plants themselves bring about the evaporation. In most mid-latitude areas there are definite soil seasons. In spring the soil is normally saturated, either because of snow-melt, or because (as in Britain) the winter rains have far exceeded the winter evaporation loss. As temperatures rise, the growing season begins ; rich vegetation springs up on the moisture-loaded soil, and rapidly evaporates off much of the moisture. In the mid-summer months, the evaporating power of the air exceeds the rainfall in all but the wettest areas, and the plant cover is hence able to reduce the soil well below its field capacity. Any rains that fall are greedily absorbed by the drying soil. Vegetation may wilt if the drought is severe. In the autumn the fall of the evaporating power allows the soil moisture content to rise again ; in warm climates (like those of Mediterranean countries) there may even be a second growing season. But finally the cold of winter supervenes and plant growth becomes restricted.

The Second Thornthwaite Classification

It is apparent that Thornthwaite's P–E ratios and indices can give little hint of the complexity of soil moisture relationships, and they cannot be regarded as adequate measures of precipitation efficiency. In an effort to provide a rational basis for classification he has recently proposed a radical change of viewpoint. Recognizing that the task of determining the climatology of individual plant associations is almost hopeless, he has formed his new classification on certain rational assumptions of the greatest interest. As before, he classifies climates on the basis of precipitation and thermal efficiency, but he has entirely altered his way of evaluating these indices. Since the new classification has only recently been published,[30] it is as yet far too early to pass judgment upon it. It has already been extended to Canada by Sanderson,[31] but there has not as yet been much published debate as to its value. The objections raised above to the earlier classification have very largely been met, and there can be no doubt that the second scheme is by far the most refined analytical tool yet devised. The treatment given below is necessarily brief, and for a more extended review the student is referred to Thornthwaite's original paper.[30]

[30] C. W. Thornthwaite, "An Approach toward a Rational Classification of Climate," *Geog. Rev.*, Vol. XXXVIII, 1948, pp. 55–94.

[31] M. Sanderson, "Drought in the Canadian Northwest," *Geog. Rev.*, Vol. XXXVIII, 1948, pp. 289–99.

Evaporating Power

A cardinal feature of the earlier Thornthwaite classification was, as we have seen, the use of computed " evaporation " in the calculation of precipitation efficiency. In the foregoing paragraphs exception has been taken to this step on the grounds that (i) the computation is based on insecure physical grounds and (ii) the actual evaporation off a plant-covered soil is rarely the same as the evaporating power of the air. In his second scheme, Thornthwaite proposes as a fundamental quantity the evaporating power of the vegetation over a moist soil surface, i.e. one in which the supply of moisture is unlimited, and evaporation is limited only by the capacity of the vegetation for vigorous transpiration. For this quantity he suggests the term "potential evapo-transpiration," which is similar in objectives to Penman's [29] " evaporating power " except that it includes a term allowing for the effect of daylight on transpiration.

It is necessary to be very clear as to the meaning of this new concept : Thornthwaite in effect rejects the attempt to estimate real evaporation, and bases his classification on its upper limiting value. This limit is almost wholly dependent on plant agencies, though it depends on the capacity of the atmosphere to carry the moisture away, and on the length of day. It is hence an unfamiliar measure climatically. The main difficulty it raises is that the measurement or computation of evaporating power is an obscure problem. Thornthwaite makes no attempt at a theoretical treatment of this difficulty, and computes the potential evapo-transpiration as a function of temperature alone :

$$e = 1{\cdot}6(10t/I)^a$$

where e = 30-day evapo-transpiration (for 12-hour days) in centimetres,
t = mean monthly temperature in degrees C.,
a = an arbitrary constant varying from place to place,
$I = \Sigma_0^{12}(t/5)^{1{\cdot}514}$, the sigmoid indicating summation over the twelve months.

This equation is empirical and rather unwieldy. The arbitrary constant a is determined by a manipulation based upon the so-called heat index I, mentioned above. Its place to place variation is very natural, for evaporating power is a function of windspeed, the hydrolapse and the eddy diffusivity of the atmosphere, none of which appears in the equation. Insofar as the latter works, it does so as a regression equation expressing the degree of correlation between e and t : it cannot be regarded as expressing a complete functional relationship. It must be noted also that e has still to be corrected for length of day and month.

The Moisture Index

Having thus defined the *capacity* of the atmosphere to absorb evaporation, Thornthwaite goes on to define a moisture index relating the *water need* of the climate (i.e. that amount of rain necessary to meet the demands of potential evapo-transpiration) to the available water (viz. rainfall together with approximately 10 cm. of soil-water). The steps he uses to get his index are briefly these:

(i) At certain times of year the rainfall r exceeds the water need n; there is hence a surplus of available water

$$s = r - n$$

and the climate is humid. A humidity index can be defined as follows:

$$I_h = \frac{100s}{n}$$

which expresses the surplus as a percentage fraction of the water need.

(ii) In other months the rainfall may very well be less than the water need, and there is a deficit

$$d = n - r$$

and the climate is arid. An aridity index comparable with the humidity index defined above is

$$I_a = \frac{100d}{n}.$$

Both I_h and I_a can be computed for the year by adding together the twelve months values. In normal North American conditions the summer months have moisture deficits, and the winter months moisture surpluses. Both I_h and I_a can therefore be computed for a typical station. At some stations in the humid south, however, and at many in Highland Britain, rainfall exceeds water need throughout the year, and I_a is zero.

(iii) The moisture index I_m is the difference between I_h and I_a. Since, however, deep-rooted perennial plants may be able at times of drought to draw upon subsoil moisture, a surplus at one season may be able to compensate for a somewhat larger deficit in others. Thornthwaite claims that a surplus of 6 cm. in one season may compensate for a deficit of 10 cm. in another, but no evidence is adduced in support of the claim. The index of aridity is hence weighted by a factor 0·6, and the moisture index is defined as

$$I_m = I_h - 0{\cdot}6\, I_a = \frac{100s - 60d}{n}.$$

This index is the primary basis of classification in the new scheme, replacing the P–E index in the earlier classification. The two are related by the simple formula

$$(\text{P–E index}) = 0{\cdot}8\, I_m - 48$$

provided that E is computed in the same manner in both schemes. A similar set of climatic types can be defined on the basis of the new classification :

Climatic Type	*Moisture Index*
A Perhumid	> 100
B_4 Humid	80–100
B_3 Humid	60–80
B_2 Humid	40–60
B_1 Humid	20–40
C_2 Moist sub-humid	0–20
C_1 Dry sub-humid	− 20 to 0
D Semi-arid	− 40 to − 20
E Arid	− 60 to − 40

The line dividing dry from moist climates ($I_m = 0$) plainly occurs where $I_h = 0{\cdot}6\, I_a$, not as one might expect where $I_h = I_a$.

A sub-division of these primary types can be devised on the basis of the summer or winter concentration of water surplus or deficit. These limits are rendered quantitative by reference to the I_h or I_a values :

Moist Climates

Symbol		*Value of I_a*
r	Little or no deficit	0–16·7
s	Moderate summer deficit	16·7–33·3
w	Moderate winter deficit	16·7–33·3
s_2	Large summer deficit	> 33·3
w_2	Large winter deficit	> 33·3

Dry Climates

Symbol		*Value of I_m*
d	Little or no surplus	0–10
s	Moderate summer surplus	10–20
w	Moderate winter surplus	10–20
s_2	Large summer surplus	> 20
w_2	Large winter surplus	> 20

Thermal Efficiency

One of the most surprising features of Thornthwaite's second scheme is his treatment of thermal efficiency. Since potential evapo-transpiration is expressed as a function of temperature and duration of daylight, Thornthwaite claims that it will also serve as a measure of thermal efficiency ; the thermal divisions are hence based on the values of *e* computed from the relationship discussed above. The T–E index of the second classification is expressed in inches or centimetres of potential evapo-transpiration. For a boundary between the megathermal and mesothermal climates Thornthwaite chooses the 23° C. (73·4° F.) annual isotherm, which is equivalent on the equator to a potential evapo-transpiration of 114 cm. (45 in. approximately). The remaining divisions (listed below) are made in a descending arithmetic progression :

Annual Potential Evapo-transpiration (T–E Index)		*Climatic Type*	
cm.	*in.*	*Symbol*	*Title*
		E′	Frost
14·2	5·6		
		D′	Tundra
28·5	11·2		
		C'_1	Microthermal
42·7	16·8		
		C'_2	”
57·0	22·4		
		B'_1	Mesothermal
71·2	28·1		
		B'_2	”
85·5	33·7		
		B'_3	”
99·7	39·3		
		B'_4	”
114·0	44·9		
		A′	Megathermal

Summer concentration of thermal efficiency is also regarded as a prime factor. The fourth digit in the label of a Thornthwaite class is a small letter indicating the degree of summer concentration of the T–E index :

Type	*Summer Concentration of T–E Index* %
a′	< 48·0
b'_4	48·0–51·9
b'_3	51·9–56·3
b'_2	56·3–61·6
b'_1	61·6–68·0
c'_2	68·0–76·3
c'_1	76·3–88·0
d′	> 88·0

The label describing a climate therefore has four digits, as in the first classification. The label

$$B_3\ B'_2\ s\ b'_4$$

means that the climate is third-order humid, second-order mesothermal, with a moderate summer moisture deficit and a low summer concentration of thermal efficiency.

Conclusion

We have briefly traced the progress of overall climatic classification from the pioneer work of Köppen through to the very recent work of

Thornthwaite in the U.S.A. Little or no reference has been made to the work of the many biologists who have sought on a more limited scale to find climatic equivalents for the major structural or formational boundaries in the plant cover. The work described in this essay stands alone in its attempt to define a comprehensive classification sufficient to divide the world into unique climatic regions. Only Köppen and his successor Thornthwaite have maintained this completely geographic outlook.

The geographer may well wonder whether the basis used by these climatologists for classification is the most adequate for geographic purposes. Köppen, following de Candolle, began with the assumption that the living plant—the living individual plant species—is, as Thornthwaite puts it, " a meteorological instrument which integrates the various factors of climate " ; Thornthwaite himself, working in a day when plant ecologists stress the structure of the great vegetation formations rather than the relations of the individual plant, first of all sought empirically to find climatic equivalents for observed formational boundaries, and later to find a rational classification independently of observed biotic phenomena. In his own words, " vegetation is regarded as a physical mechanism by means of which water is transported from the soil to the atmosphere ; it is the machinery of evaporation as the cloud is the machinery of precipitation."

This stressing of vegetation as the best measure of climate is plainly of geographic value : the " natural vegetation " of a region is at least in some measure a measure of its economic potential for settlement, especially for farming economies. But we may wonder whether or not a comprehensive classification should take account of that other great branch of climatology allied to the field of environmental physiology. The direct effect of climate on man's health and comfort has never been reduced to a set of simple parameters analogous to those of the classifications described in this essay. Most of us are familiar with Griffith Taylor's simple but effective " climographs " and " hythergrams " ; today the environmental physiologists are seeking, under the stress of military necessity, measures of the climatic stresses of the human frame of the same type, though based on experimental evidence. Windchill, for example, has been reduced to a numerical measure [32] of outstanding value in assessing the direct effect of low temperatures on the human frame.

In any ultimate synthesis it seems inevitable that this little explored field must be somehow incorporated. The geographer cannot ignore

[32] P. A. Siple and C. F. Passel, " Measurements of Dry Atmospheric Cooling in Sub-Freezing Temperatures," *Proc. Amer. Phil. Soc.*, Vol. LXXXIX, no. 1, 1945.

a specification of the direct environmental influence of climate: the latter does not work wholly on his activities, but also on man himself.

Acknowledgement

The writer would like to thank his friend, Captain H. Anda, for help in the translation of some of Köppen's work. He must also thank Dr. Kenworthy Schofield and Dr. H. L. Penman of the Rothamsted Experimental Station for an education in soil moisture relationships.

Reference has also been made to those two remarkable books, *Soils and Men* and *Climate and Man*, Yearbooks of the U.S. Department of Agriculture for 1938 and 1941 respectively. These books are mines of information that no geographer can afford to neglect.

VIII

TRANS-SAHARAN RAILWAY PROJECTS

A Study of their History and of their Geographical Setting

By R. J. HARRISON-CHURCH, B.Sc. (Econ.), Ph.D.

In the history of French expansion in Africa and in discussions of national security, the idea of a trans-Saharan railway has had an outstanding place. At most times in the last ninety years it has been supported by strategists and by those who favoured closer contacts between North and West Africa, and between these and France. But it has been opposed by most economists and by left-wing politicians. Whatever the viewpoint, the idea of such a line has received wide attention in France, but little notice in England. This essay is an attempt to outline the major proposals that have been made and to evaluate from a geographical viewpoint the prospects of such a line.

Trade between North Africa and the margins of the Sahara was considerable in Carthaginian and Roman times. Similarly, commerce between ancient West African Empires and the southern margins of the Sahara goes back to quite 2,000 years ago, but its details are less known. It was the introduction of the camel into Africa from Asia and its general use from the middle of the fourth century that led to systematic trans-Saharan traffic, as distinct from trade between the coasts and the margins of the desert. It came to be of great economic and political importance. Those who were engaged in the trade—and they were mainly Barbary Arabs—kept it as a highly organized and closed monopoly. Islam and Arabic culture spread south by this means and the Tuareg people of the Saharan massifs and oases derived their political and economic power from their control over this trade. There were many major routes, with cross-links between them which later developed into west–east routes to the eastern Sudan, followed by the pilgrims to Mecca.

It is the north–south routes with which we are here concerned. Among the most famous were four routes from Fez and Marrakech in Morocco. One went via In Salah to Agadès (the Tuareg capital) and on to the great Mohammedan cities of what is now northern Nigeria. Two others via In Salah or Tindouf served Timbuktu and other cities of the Niger Bend such as Ségou, Sansanding and Gao. The fourth passed along the inland margins of what is now Spanish

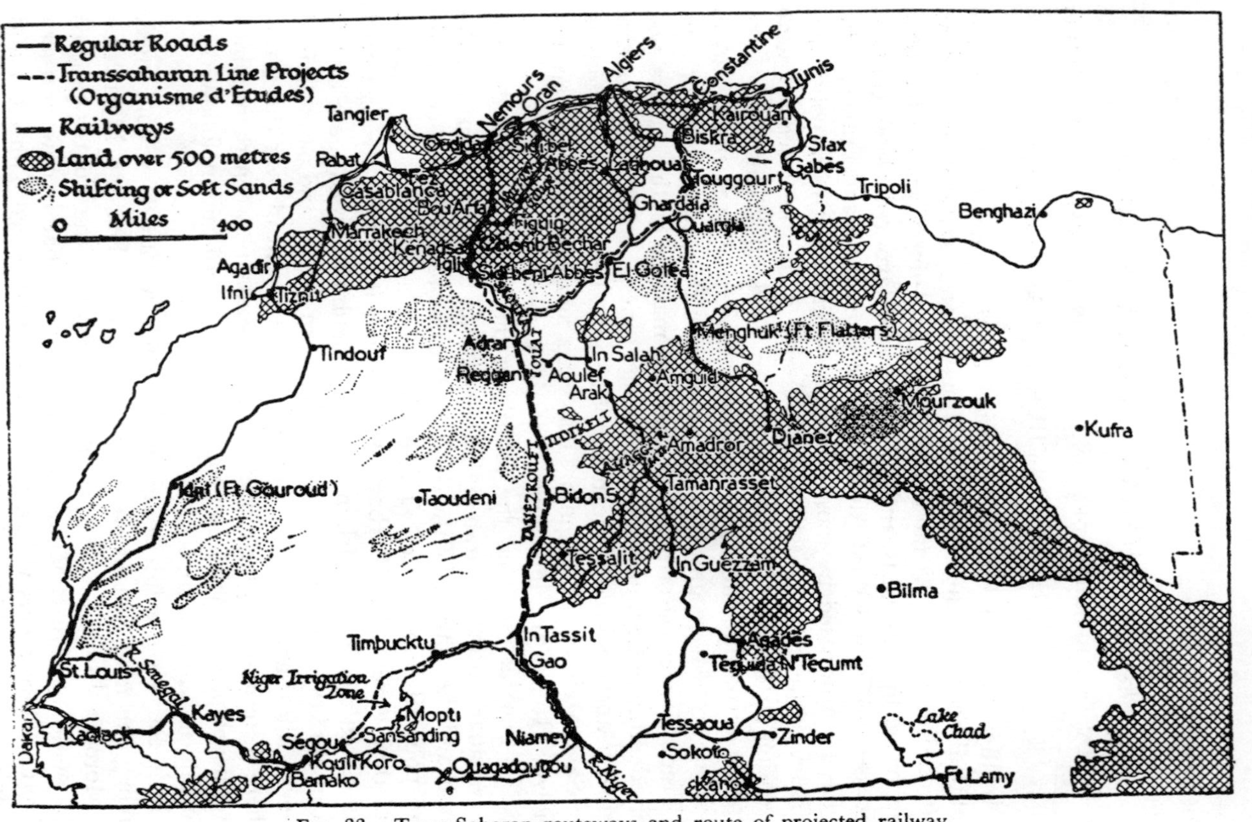

FIG. 33.—Trans-Saharan routeways and route of projected railway.

Rio de Oro and Mauritania to cities of the interior of modern Senegal and French Sudan. All but the latter districts were also served from Constantine, Tunis, Kairouan, Sfax and Gabès. Finally, there was also exchange between the cities of northern Nigeria and Tripoli via Bilma and Murzouk and with Benghazi via Kufra. Terminal cities on both sides became most important markets and frequently also cultural centres such as Kairouan, or manufacturing centres such as Kano. Venetian glass and copper, French and Spanish wines and cloths, silk and woollen cloths from Morocco and Tunis, sugar, olive oil, coral, sponges, brass and pottery ware and even horses were all taken southward. In exchange there travelled northward gold, spices and slaves as the most lucrative traffic and also ivory, ostrich feathers, ebony, skins and leather. Indeed, the latter taken from Kano to Morocco became known as "Moroccan leather." Journeys in each direction normally took 90 to 150 days. Routes were often devious, since water and settlements had to be found en route. Leaders were changed at various stages and goods were bought and sold on the way.

In addition to the true trans-Saharan trade, much quarrying and traffic in salt took place from the Sahara to its southern confines. There were five well-known salt rock quarries—Idgil (Fort Gouraud) in northern Mauritania, Taoudeni in French Sudan, Amadror in southern Algeria, Téguida N'Técumt, south-west of Agadès in French Niger, and Bilma, also in French Niger. Of these Taoudeni was the most famous and still produces small amounts.

But decay has come to both these types of traffic, due to the advent of Europeans from the Guinea coasts. Factories on the coast date from the sixteenth century and they started a movement of trade away from the trans-Saharan routes towards the West African seaboard. By the end of the nineteenth century effective penetration of the Niger Valley, the suppression of most of the slave trade and the construction of railways from the ports all combined to hasten the decline of these ancient traffic routes. Some also consider that the physical conditions of the Sahara have seriously worsened in historical times, and the lack of slaves to cultivate the oases certainly led to the decline of many of them.

Although the penetration of the interior by the French led to the virtual elimination of trans-Saharan trade, it led the French themselves to consider the strategic and possible economic advantages that might accrue from a railway link across the desert. There have been several distinct phases in their thought and plans, which have changed with altered techniques, new lines of economic development and changing political and economic fortunes in France and in the world.

Discussion of a possible railway link goes back to before the occupation and pacification of the Sahara. Probably the first printed suggestion for a line came in 1860 in the preface of a Tomacheq grammar by Hanoteau where he says " who knows if one day a line linking Algiers with Timbuktu will not make it possible to reach the tropics from Paris in six days ? "[1]

After 1871 there were many enthusiasts among whom Duponchel was the leader. After studying everything that had been written already on the subject, and himself making surveys in Africa, he produced a report in 1876 *On the Trans-Saharan Railway Plan—Algiers to the Sudan.* He advocated a route from Algiers via Laghouat, El Goléa and the Touat to Timbuktu and thence to Lake Chad. In 1879 he raised the matter in the Chamber of Deputies, outlined his proposals and urged them as necessary to confirm French occupation and to open up commerce from the Niger. He secured a considerable measure of support and the valuable interest of Freycinet, then Minister of Public Works. The French public, still chafing under the defeat of 1870–1, was interested in colonial enterprise and the Germans encouraged this interest. By a decree of July 13, 1879, three survey missions were appointed. Pouyanne was to examine routes south of Oran, Choizy from Laghouat to El Goléa and back to Biskra, and Flatters from Ouargla to the Sudan. Meanwhile, there was to be an examination at home of economic, technical, geographical and political aspects of the plan. Soleillet was also sent to survey a Senegal–Niger line which was generally regarded as a necessary complement to the trans-Saharan scheme. The various survey teams carried out their missions, save that of Flatters, which was halted by the Tuaregs at Menghuk and forced to return. In 1890 this same mission returned to its uncompleted work. The Mission was annihilated by Tuaregs in the Ahaggar Mountains on February 16, 1881. This tragic event is a turning-point in the history of schemes, for it led to general disappointment and a decline of interest in the Sahara. Effort was thenceforth concentrated on the survey and construction of railways from Dakar to St. Louis and between the Senegal and Niger rivers. Penetration from the west and south was substituted for that from the north, although military operations for the complete annexation of the Sahara continued. But in spite of the prosecution of these military operations with considerable vigour, French Governments showed little interest in a possible railway.

From 1881 to 1912 there was a period of mild public interest which

[1] *Afrique Occidentale et Septentrionale*, A. Bernard, *Géographie Universelle*, Tome XI, 2e partie, 1939, p. 351.

was momentarily quickened by the diplomatic defeat of France after the Fashoda Incident in 1898 and by the defeat in 1903 of the Tuaregs, the last peoples to hold out against the French in the Sahara. The Fashoda Incident of 1898 led to a new and rather jingoistic public interest in the Sahara and the Sudan. Paul Leroy-Beaulieu was the leader of this and set forth his views in many articles in *L'Economiste Français* and *La Revue des Deux Mondes.* He insisted that after this diplomatic reverse of France at Fashoda, she ought to be secure in her colonies and to develop them economically. To secure French West Africa, the individual colonies ought to be linked by a trans-Saharan Railway which would be a re-insurance against sea attack and which would at the same time help the economic development of the middle Niger valley. He thought a considerable traffic was assured—meat, wool, rice and other products northwards and phosphates, sugar and salt southwards. He favoured construction by private enterprise and hoped to see a line with branches west to the Niger and east to Lake Chad.

Meanwhile, a number of exploratory missions were working in the Sahara sponsored either by scientific bodies or, as in one case, by a newspaper *Le Matin.* The defeat of the Tuaregs in 1903, the publication of a book by Leroy-Beaulieu in 1904, the reports of the various missions which came out at about the same time and the publication of A. Bernard's book on the Sahara in 1907 all helped to stimulate interest. In general, a route to Biskra and Algiers was advocated, but sometimes one to Gabès or Oran was suggested as being shorter. Meanwhile practical experience was being obtained in the construction of lines through semi-desert in southern Algeria.

The arrival in Dakar in 1902 of Governor-General Roume, " the economic founder of French West Africa," brought to the fore a man who believed that efficient transport was the essential prerequisite for economic development. He found a receptive public, since he put forth his views in the years immediately after Fashoda. His objects were to build lines from the sea through each of the West African colonies to reach the Niger and then later to link them by a transversal. This plan, with many lines crossing varied belts of country, had much better economic prospects and was also more relevant at a time of political rivalry between the French and British in West Africa. Thus attention was diverted to the construction of new railways in French Guinea, the Ivory Coast and Dahomey on the one hand and to the improvement of the Senegal–Niger line and its extension towards Dakar on the other. French West Africa received three loans between 1903 and 1910 totalling 179 million francs, of which 125 million were spent on railways. Another loan of 167 million francs in 1913, of which 150 millions were for railways, was not used owing to the outbreak of war.

Just before the war the trans-Saharan scheme was revived in a new form. It was put forward by André Berthelot as part of a scheme for a standard gauge trans-African Railway, with few gradients, wide curves and diesel traction. The idea was undoubtedly copied from the earlier one of Rhodes for a Cape to Cairo Railway. It was intended to publicize the merits of Algiers, as a terminal much nearer to Europe than Cairo, and of a line which would also develop French West and North Africa as well as linking them to the Belgian Congo and South Africa, thus enhancing French influence and prestige in Africa. In 1911 Berthelot formed "La Société du Transafricain" and organized two surveys. As a result, various routes were suggested, but eventually the cost of such a grandiose scheme, the difficulty of estimating traffic prospects and the great political complications were made evident to most of the supporters and the war stopped all further scheming.

On the conclusion of the 1914–18 war the idea was again raised in the form of a military conception to bring thousands of African troops safely to France should she ever again be attacked. U-boats had shown the dangers of the sea route and the Moroccan revolts had lent emphasis to the desirability of the line on local strategic grounds. The "Comité National du Rail Africain," founded in 1918, put forward these arguments with great insistence and also urged it as a means of developing the "inland delta" of the Niger Bend below Ségou, and as a shorter route to South America via Dakar. In this connection, it is interesting to note that Marshal Lyautey advocated another route via Mauritania from Morocco to Senegal, with branches to Dakar and Kayes. Such a route would be shorter, cross better areas, but would be near sea competition and could hardly help the new areas of the Niger Bend.

The French proceeded after the war to the effective re-occupation of the Sahara and there was most certainly a renewed public interest in colonial development. In 1920 Carde, the Governor-General of French West Africa, stated that:

> it had still only developed its coasts. Senegal groundnuts, forest and cocoa exports from the Ivory Coast and palm kernels from Dahomey account for four-fifths of the Federation's exports and those areas took the same proportion of imports, although they are only one-fortieth of the area and have only a quarter of the people.

In 1922 a start was made with the active development of the Niger Bend by means of irrigation under the newly established "Office du

Niger." This scheme has now made considerable progress and an analysis of it and its prospects is made on a later page. Optimistic and not too extravagant hopes have been entertained as to traffic which might be provided for a trans-Saharan line from these newly irrigated and settled lands. Such hopes have, since 1922, been coupled with the strategic, administrative and time-saving advantages it might confer.

It is evident from all this that the war of 1914–18 may be taken as a most important turning point in the whole history of trans-Saharan transport schemes. The development of diesel engines made the line easier of realization, since the prime difficulties of a long stretch without water and the great heat to be endured with steam engines could now be overcome by other techniques. There were also the added incentives, particularly strategic and economic, now so much more insistent than heretofore. But at the very same moment two other technical advances acted as a deterrent to the construction of a railway, namely the possibility of trans-Saharan motor-coach and air services.

In 1920, General Laperrine, then Commander-in-Chief of the Saharan Territories of Algeria, led an air survey flight across the Sahara. One plane got across, but the other crashed with Laperrine and he was killed. Ever since 1900 he had played an outstanding part in the exploration and military control of the Sahara. Nevertheless, further air survey work was undertaken and ultimately a regular service was established by "Air France" and "Air Afrique" from Algiers or Oran via Colomb Bechar, Beni Abbès, Reggan, Bidon 5 to Gao from where branch services served Timbuktu, Ségou, and Bamako, while the main service went on from Gao via Niamey, Zinder, Fort Lamy and thence to the Belgian Congo and Madagascar. From Niamey there was also a link to Cotonou on the Dahomey coast. The planes followed the Tanezrouft track, which is now also used by buses and is approved for a railway, should it ever be built. The Belgian "Sabena" Company also had services following this route or the Ahaggar one via Laghouat, Ghardaia, El Goléa, In Salah, Arak, Tamanrasset, In Guezzam, Agadès and Fort Lamy. A third route via Biskra, Touggourt, Ouargla, Fort Flatters, Djanet, Bilma and Fort Lamy was also used, although less frequently. These services carried considerable numbers of passengers, mail and some freight.

Meanwhile cars had also entered the Sahara. During the war the Italians had reached Murzouk, 375 miles from the coast. In July, 1916, one of two cars reached In Salah from Ouargla, 470 miles, and by 1918 a weekly service of lorries was established. In 1919 the Saoura-Tidikelt mission covered 1,560 miles with seven caterpillar tractors. In December of that year, several cars reached the Ahaggar

Mountains and two went 190 miles farther south. It was during the winter of 1922–3 that the famous Citroën Expedition, led by Haardt and Audouin-Dubreuil, made the first car crossing of the Sahara with five tractor cars from Touggourt to Timbuktu and back. In November, 1923, the Estienne brothers crossed in an ordinary sports touring car, via Colomb Bechar, Adrar, Bidon 5 and Tessalit to the Niger. It was realized that caterpillar cars were too slow, too expensive on fuel and in any case unnecessary. Six-wheeled cars were found to be ideal. In 1925 the Tanezrouft track via Bidon 5 was marked out and in 1926 Lieut. Estienne was the first person to cross and re-cross alone by car. His route was Oran, Colomb Bechar, Bidon 5, Gao, Fort Lamy and thence to Cotonou on the Dahomey coast and the return via Niamey, Gao and so across again to Algiers.

The results of all these crossings made it possible for the " Compagnie Générale Transsaharienne " (a subsidiary of the " Compagnie Générale Transatlantique ") to establish a regular trans-Saharan Coach Service, which has achieved very marked success. A weekly service was maintained from October to May from Colomb Bechar via Beni-Abbès, Adrar, Reggan, Bidon 5, Aguelock and Gao to Niamey. At Adrar there was a connection for In Salah, at Gao one for Mopti and Bamako and at Niamey for Zinder, Tchaourou and Cotonou. The Company established hotels at Reggan, Gao, Niamey and Beni-Abbès to lodge travellers overnight and at Bidon 5 and Aguelock in mid-Sahara were depots with a watchman and a camping site. The journey took a week including a day's rest at Reggan. Another service was maintained by the " Société Algérienne des Transports Tropicaux " from Algiers via Laghouat, Ghardaia, El Goléa, In Salah, Arak, Tamanrasset, In Guezzam, Agadès, Tessaoua, Zinder and thence by connections to either Niamey, Kano or Fort Lamy. This is a more picturesque route, cooler, better watered and more direct for the French Niger and French Equatorial Africa, but the route is rougher. This service was a fortnightly one and took thirteen days, including five days' rest at various places en route. Hotels were built at all the above-mentioned places. The first company has also experimented without much success in this case, with a service from Marrakech to Dakar, via Tindouf, although the French only formally occupied the latter place in March, 1934. It also maintained a cross-link from Algiers via El Goléa and Aoulef to Reggan.

But there are heavy expenses in maintaining these services and especially the tracks. The passenger fares, including board and lodging en route amounted to about £35 single from Colomb Bechar to Niamey. A considerable number of passengers, mainly officials and tourists out for a novel experience, were carried, but such traffic could only be taken during two-thirds of each year. Nevertheless, the air and road services (which are in most ways essentially complementary in that the air services follow the ground tracks, use the same stopping places and can continue the service all the year round) may be deemed sufficient for passenger and mail traffic for many years.

Although air and motor services have proved their capacity to carry these traffics, it is possible that with any considerable increase, costly regular roads would be necessary, as the present tracks soon deteriorate with heavy usage. Permanent roads would be as expensive as a railway, yet could not serve for the cheap transport of large quantities of bulky raw materials over such great distances. Such observations obviously apply also to air services. Thus it is, that although the value of a railway for the transport of passengers has diminished with the advent of the car and the aeroplane, it may be even more necessary for the full development of the irrigated Niger Bend, if the latter proves even "a minor Egypt." And a railway would almost certainly recapture any present passenger traffic by road. Thus although great interest was shown in the development of road and air connections across the Sahara, railway schemes were continually being put forward.

A very important move was made in 1923 when, encouraged by Senator Mahieu, the "Conseil Supérieur de la Défense Nationale" considered various projects and declared in favour of one from Oran via the Tanezrouft to the Niger and beyond with a standard gauge capable of bulk traffic. It was to be financed by the P.L.M. Railway and constructed with German Reparations. But the scheme collapsed when these failed to come adequately and interest reverted to private propaganda, especially that from the "Comité du Rail Africain," the "Société du Transafricain et Transsaharien" and the writings of Doumergue, Messimy, Estienne, Tardieu, Steeg and Maître Devallon.

In 1927 agitation on the subject again became strong. Many Chambers of Commerce and the "Comité de L'Afrique Française" re-urged it. M. de Warren, an ardent enthusiast of the "Comité du Rail Africain," persuaded the Chamber of Deputies to grant 12 million francs for an "Organisme d'Etudes d'un Chemin de Fer Transsaharienne." This body has its headquarters at Algiers and was created on August 1, 1928. Maître Devallon became its Director and Fontaneilles and Steeg were also on its Directorate. During the ensuing winter, four missions covered nearly 20,000 miles and three Commissions worked at home on the financial and economic aspects. Very detailed reports were made in the winter of 1929–30, suggesting an easy western route from Algiers or Oran via Bou Arfa, Colomb

Bechar, Beni Abbès, Adrar to In Tassit, whence one branch would go to Gao and down the Niger to Niamey and the other to the Niger at Timbuktu and then up the river to Ségou and Koulikoro. Both branches would later link with local lines. In the north there was to be a later link from Reggan via El Goléa, Ouargla, Touggourt and Biskra to benefit Constantine and Tunisia. The main line would be 2,160 miles in length, of which 1,200 miles would be desert. With diesel traction and normal gauge it would take two and a half days to cross and the construction costs were estimated at £30 million to £50 million. It was proposed to establish a new company with money from the French Government, North and West African Governments and the French Railways, although a later suggestion was for an appeal to private investment with a state guarantee. Traffic was very optimistically estimated at one million tons export and 700,000 tons import, with 70,000 passengers per annum. Such traffic would cover the capital cost in sixty years. The zone of attraction was said to be almost all West Africa for passengers, and for goods as far west as a north–south line through Bamako and as far south as Dédougou, Ouagadougou, Gaya and Sokoto. With two to three thousand men at each end, the line would take not more than eight years to build. There were no constructional difficulties and Diesel locomotives might be developed to run on African vegetable oils, with steam engines as far as Colomb Bechar using Kenadsa coal. The cost of transport would be £2 to £5 per ton. The line already existed to Bou Arfa on normal gauge and it would serve the palm belts of Saoura and Touat, 375 miles farther south.

The various Governments and French Ministries spent most of 1930 and 1931 in considering the plans. Their comments are unknown except for the Duchêne Report from the Ministry of Foreign Affairs, which declared that from the economic point of view the line was not vital, but that politically it would unify West and North Africa, provide a new outlet and help the Niger Irrigated Zone. By the time a final verdict could be given by the French Cabinet it had to be negative, owing to the world depression of 1931.

There was little further discussion until the Chamber of Deputies again discussed it in early 1938 mainly on strategic grounds, as a weapon of economic and military value. Its great protagonist was again M. de Warren. M. Mandel, Minister of Colonies, and M. Monzie, Minister of Public Works, agreed to new discussions. It was believed that it could be built in two years and at much less expense than previously estimated. In March, 1939, a Conference was held of representatives of all interested ministries and preliminaries were put in hand on the basis of the earlier plan. Had it been built in time it would have materially helped France and Britain between 1939 and 1940 and again after 1942. On the other hand, it would have strengthened the hand of Vichy between 1940 and 1942 and might have led the Germans to occupy North Africa in 1940 to secure it. Such action might have taken them into West Africa with serious consequences to the Allies in Egypt and the Middle East.

By a decree of March 22, 1941, the Vichy Government decided to start construction on a " Chemin de Fer Méditerranée-Niger " to join the railways of Morocco and Algeria to West Africa. Operations began with native civilians and soldiers, political and other prisoners from France and the African colonies, four hundred men from the former Spanish Republican International Brigade and some former Foreign Legionaries. Conditions in the camps were grim and a number of men died on the works. In Kenadsa there were four camps of approximately 300 men each, mostly Central European Refugees and Spaniards, and there were other camps elsewhere.

The existing normal gauge line from Oran and Nemours via Oudjda to Bou Arfa in Morocco was extended to Colomb Bechar and Kenadsa and brought into operation. From Colomb Bechar to Kenadsa it duplicated the narrow gauge Algerian line from Oran. Work from Kenadsa onward was undertaken mainly in 1942 and the works were ultimately halted at Igli, some 82 miles south of Colomb Bechar. The whole future of the line is once more uncertain.

The effect of these latest works is to define finally the northern end of any future trans-Saharan line and to make certain that it would at least be based on a large and modern port, namely Oran, and be throughout of normal gauge—a real merit. Before turning to a review of its future economic prospects, a description of the existing lines and of the country traversed and to be traversed is appropriate here.

Oran may be considered practically ideal as a terminus. Algiers would have involved a longer railway journey and it has no standard gauge line so far into the interior. As far as Sainte Barbe du Tlélat, 19 miles from Oran, the line is double and crosses a plain thickly colonized by Europeans, especially around La Sénia, although the soils are often saline or alkaline. The railway passes the Sebkha d'Oran, a large salt marsh, and enters the Tlélat plain. It then follows the Tlélat valley into the mountainous Tell Atlas, using a former course of the Mékerra, which had cut a gorge in the mountains but now flows through a plain to the south. The plain and the river are met at Les Trembles. Sidi bel Abbès, 48 miles from Oran, is a large and prosperous town and the centre of one of the most fertile and best colonized plains of Algeria. Cereals are very important, especially barley and wheat, and the vine is largely grown. There are many industries, mainly connected with food processing.

The line then runs westward to Tabia, where there is a branch line to Ras-el Ma. Thereafter, the line runs along the northern edge of the Tlemcen Mountains, a section which involved much heavy engineering. At Tlemcen, 103 miles from Oran, it serves another rich plain whose products are similar to those from around Sidi bel Abbès, but also include maize, vegetables and the temperate fruits. The town is a great market, an outstanding place for the study of Moorish art and civilization and has a silver-lead-zinc mine nearby, with oak and pine forests away to the south. The line continues along the southern edge of this plain into the similar one of Marnia. Before the frontier, the Nemours branch, opened in 1936, joins the main line. Nemours might be developed as the trans-Saharan outlet instead of Oran, thereby shortening the rail journey by 112 miles. It is already important as an outlet for Morocco, especially for the minerals and sheep brought by the section of the line from between Oudjda and Kenadsa.

Fig. 34.—Existing railways.

Oudjda, 157 miles from Oran and 45 from Nemours, lies within the Angad plain of Morocco and is a town of considerable importance, being nearly a thousand years old. There is much agricultural traffic from the local market and there are silver-lead mines nearby. Ten miles beyond Oudjda, the line turns south away from that to Fez and leaves the relatively well-populated agricultural plains to cross a gap in the western end of the Tlemcen Mountains. South of

Oudjda, the line was built to develop anthracite deposits at Djerada and manganese at Bou Arfa and has only recently been opened to passengers or envisaged as a part of a trans-Saharan line. For the remainder of its route the population is very sparse. Anthracite is mined at Djerada, west of the railway and evacuated by a cable-way to Guenfouda station. The anthracite is used on the railway and in industry, especially when mixed with Kenadsa coal, and a large export may develop.

Beyond Guenfouda, the railway climbs by many curves and valleys to the Tiouli Pass, 3,575 ft., and then drops to Berguent (229 miles from Oran and 117 from Nemours) in a valley near the Ouedi el Hai, a tributary of the Moulouya. Lead, zinc, and silver ores are railed here from mines 35 miles to the west. The railway climbs again out of this broad valley on to the rather featureless prairie-like High Plateaux to Tendrara at 4,495 ft. in the Chott Gharbi region. At Bou Arfa, 337 miles from Oran and 225 miles from Nemours, the railway crosses a highly mineralized zone, to serve which the line was originally constructed. Manganese, lead and silver are sent away. The section beyond Bou Arfa was opened in 1941 and passes over two ridges of the Saharan Atlas—very poor country—and then once more into Algeria to Colomb Bechar, at the foot of the Saharan Atlas, on the edge of the desert, 417 miles from Oran and 305 from Nemours. The town has some 5,000 inhabitants, is a route centre and market and is also served by a narrow gauge railway from Oran via Figuig. Colomb Bechar has great palm groves and important craft industries in leather and precious metals.

The line then turns south-west to Kenadsa, doubling the narrow-gauge railway from Oran. Kenadsa, 14 miles from Colomb Bechar, has six coal-mines which produced about 240,000 tons in 1946, and the coal may now be taken by standard gauge line to the several mineral fields aforementioned. Beyond Kenadsa the railway turns sharply south-east back to the Oued Saoura, the richest point of the north-western Sahara and ends at Igli, a collection of villages, 499 miles from Oran and 387 from Nemours. The line might soon be extended to Beni Abbès, another 22 miles, as there are large date palm districts in the Saoura valley at this point. Thus far the line would serve a useful purpose throughout its length, but beyond Beni Abbès, or Reggan at the most, practically no further traffic could be encountered until the Niger were reached 860 miles away.

Beyond Beni Abbès the line would skirt the Great Western Erg and some engineering would be necessary before the next considerable settlement is reached at Adrar, 257 miles from Beni Abbès. After

Adrar there are many palm groves in the Touat country as far as Reggan, 348 miles from Beni Abbès. Reggan is the last oasis and a small route centre. Beyond lies the waterless Tanezrouft some 500 miles broad, with the added difficulties of occasional loose sand or dunes. Once the Niger were reached new traffic would be found.

The reports of 1929–31 made the serious error of assuming that the

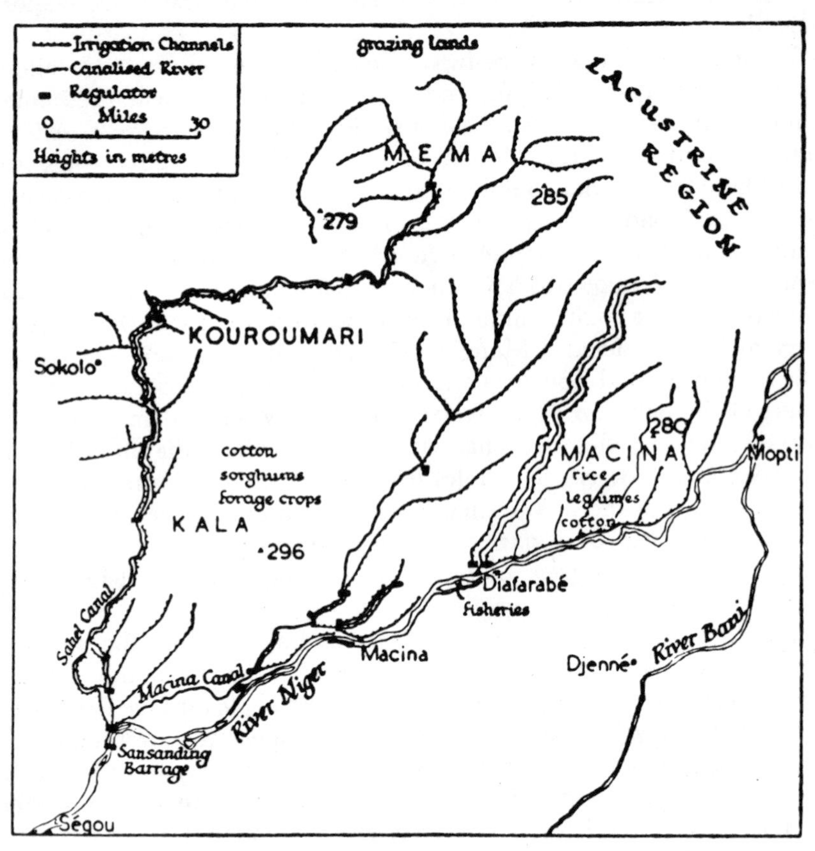

FIG. 35.—The inland Niger delta.

line would create traffic "as railways always do." They only do this in temperate and rich tropical lands or in mineralized zones. Nevertheless, it might seriously help the Niger Bend Irrigated Zone for which the navigable Niger and the Dakar–Koulikoro railway might in future prove insufficient, in view of the break of bulk between river and rail, their low capacity and devious route. It was over-optimistic to compare the Niger Bend with Egypt, which has such a long history of irrigation, nor does the Niger carry such quantities of rich silt as does the Nile. The Niger Irrigations only started in 1929, on the completion of the Barrage des Aigrettes at Sotuba, 2½ miles below Bamako, which irrigated 27 square miles. The Sansanding Barrage below Ségou has now brought about 400 square miles under irrigation and ultimately 5,860 square miles are envisaged. This is a small area, even when in full production, to be the mainstay of such a line. All crops are to be produced by peasant proprietors under Government contracts. Optimistic estimates of yearly crops which have been suggested are 60,000 tons of rice and other cereals, 150,000 tons of cotton, 84,000 tons of animal products, 200,000 tons of groundnuts, 35,000 tons of vegetable oils, 30,000 tons of kapok and 100,000 tons or more of other products. These could not be produced for a long time and not all would be for export.

It may well be argued that the Niger Bend cannot greatly develop without better communications, but it is doubtful if a trans-Saharan railway is the best way of securing them, at least in peacetime. It might well prove to be more economic in the strict sense to realign and double the Dakar-Koulikoro line and extend it to the Niger Bend. It would still be 500 miles shorter and, if necessary, Kaolack could be further developed as an alternative outlet to Dakar. This cheaper rail haul would compensate for the longer sea journey and there would be far less capital outlay. Such new capital outlay would develop good areas en route, unlike the trans-Saharan. What expense was essential would be for a certain benefit, and to a well-developed and natural outlet.

Nevertheless, a direct rail link between North and West Africa might well develop trade between those areas rather than between each of them and France. In 1936, Algerian products valued at about £200,000 went south and rather less went north from West Africa, almost entirely by sea. A railway would doubtless increase this traffic, but would deprive the sea routes of some traffic. North Africa would send phosphates, agricultural equipment, manufactures such as bags and cork materials, cottons, wines, tobacco, dried fruits, flour, dates, salt and sugar, which would find a ready market among the Niger Bend colonists and West Africa would reciprocate with products from the Niger Bend and perhaps some bananas, coffee, cocoa and other valuable tropical products appreciated in North Africa. Wheat, which West Africa now imports by sea from North and South America and South Africa, might come by rail from North Africa. External trade would be kept more in French hands. North African peoples might go south and so spread better technique. It is along such lines that the most hopeful economic future of the line lies. In addition there would be a small victualling trade in the Saharan oases—cereals, wool, butter and manufactures for dates, millet and salt.

With a settled outlet to the north, the depopulation of the Niger lands towards the Gold Coast might be arrested. In any case, the irrigation scheme vitally depends upon sufficient colonists. It should be recalled that the Egyptian population has increased from two to fifteen million in a hundred years, so that the same might theoretically be possible here. Further afield, by means of feeder roads, the line might indirectly serve the Niger Colony and free it from virtual dependence upon Nigeria.

Thus it is clear that the railway, now more of a practical possibility than ever before, might find its justification in two fields. First the strategic one of providing a safer overland route for the transport of African troops to Europe. In view of the position of France in Europe since the war of 1870–1 it is easy to appreciate why so much thought has been given to the line from this aspect. The administrative value of such a link is more apparent on the map than effective in reality, since air and road services already provide most of these requirements. The second group of considerations is economic—whether the line is vital to the development of the Niger Bend. As the new irrigated lands come into full productivity within the next few years the final answer should be available.

IX

THE INADEQUACY OF THE REGIONAL CONCEPT

By GEORGE H. T. KIMBLE, M.A., Ph.D.

Introduction

We geographers are men of many creeds and tongues. We have plenty to say, but we seldom say it in unison or in harmony. Our labours of a generation notwithstanding, we are still without an agreed testament of faith. No doubt if we were asked for a single-sentence definition of our subject, most of us would talk of "regions" and contend that the highest form of geographical enquiry was a kind of "hunt-the-region" game, in which (provided we were offered enough clues) we were bound to discover that life in a given land or continent resolves itself into a neat pattern of cultural entities we call regions. But having said this much, we should need to add hastily that we were still debating the exact connotation of the term: that there were no less than one hundred definitions of it in our geographical literature, and that those who did agree about their terms and techniques were not altogether certain what to do with the region when they had eventually identified it and described it.

Unable (or unwilling?) to present a united front to the academic world, it is not altogether surprising that our external impact has so far been slight. Nor is it entirely surprising that some of our near neighbours in the other social sciences are beginning to ask us whether a phenomenon which has thus far escaped definition can have any real existence. And not our neighbours only. In the light of present social and economic trends, even some geographers are beginning to wonder whether the regional concept is in fact as satisfactory a mould for the fashioning of geographical studies as they have been wont to regard it.

The Meaning of the Concept

Reasons for this growing dissatisfaction are not hard to find. First, there is the historical fact that the region is an eighteenth-century concept (if, indeed, it is not much older): the world that fathered it now lies "mouldering in the grave," and it has yet to be shown that the ponderous ideological fashions of the eighteenth century become the much shrunken and infinitely more fleet-footed world of the twentieth.

It will be well for us, perhaps, to call to mind the original signification of the term. From the time of Strabo onwards, geographers repeatedly expressed concern over the treatment of earth knowledge on a basis of political divisions. Some writers, notably Varenius, had proposed alternative organizations of the subject, but before the eighteenth century none of them attracted any great following. However, in the second half of that century, Gatterer in Germany, following a lead given him by Buache in France, proposed a physical division of the world into lands and regions (*Gebiete*). This scheme was later taken up by Hommeyer, who developed the idea of a real unity based on a single phenomenon, namely, land-forms. The concept of a composite unity, the integration of all phenomena (natural and human) of an area into an individual unit distinct from those of neighbouring areas, did not appear before the nineteenth century. It finds emphatic expression about 1810 in the writings of Zeune and Butte. For Butte, the individual units were organisms which, like any organism, included a physical side—inanimate nature—and a psychical side—animate nature. "The unit areas (*Raüme*)," he averred, "assimilate their inhabitants, and the inhabitants strive no less constantly to assimilate their areas."

Needless to say, so radical a notion did not fail to arouse opposition. Wilhelmi and Selten, while ready to recognize the existence of the "natural" unit, both argued that its boundaries could not be determined on the basis of any one kind of phenomenon, and that, in those many instances where the boundaries were not clearly defined by nature, the selection of one would necessarily be arbitrary. The criticisms of Bücher were more fundamental. He claimed, and he supported the claim with an impressive array of material culled from some forty current textbooks, that it was futile to look for "natural divisions," since nature had for the most part not thought it necessary to administer her affairs as neatly as civil servants. Systematic geography was, in his view, the important thing: areal studies on a unitary basis were "only needed for special purposes, for which the areas concerned could be arbitrarily bounded in any convenient way."

But, by and large, the critics of the new concept did not attract as much attention as the sponsors of it. Most of the "classical" and "post-classical" writers accepted the idea of the region without challenge, and even without curiosity. Indeed, Hartshorne (whose work on the origins of modern geography cannot be too highly regarded) goes so far as to say that in the century and a quarter since Butte first spoke of areas as organisms, nobody has seriously attempted to establish the organic character of a single region and in the century since Bücher's scientific integrity led him to announce his failure to establish a regional division of lands, no one has seriously attempted to show the error of his method or argument, or to produce what he found impossible. Yet many modern writers of geography textbooks continue to predicate the existence of such organic units, and to assume that the whole world is tidily parcelled up into " unitary entities." Not to put too fine a point on it, this looks uncommonly like intellectual laziness—if not dishonesty !

Broadly speaking, present-day supporters of the regional concept are divided into two camps. Those who, following Herbertson, make the natural region serve as the plastercast of a specific kind of human economy : and those who focus attention on cultural distributions, and regard the region—to cite only two of the hundred-odd current definitions—as an area in which " a functionally coherent way of life dominates," [1] or an area which is " dynamically homogeneous in respect of certain inter-related characteristics in the make-up of its society, whether past or present." [2] The first school of thought is not as influential as the second, possibly because of the difficulty it has in demonstrating that a climatico-vegetal region (or climax area) is capable of enslaving its inhabitants, whether in respect of their movements, occupations or *mores*. Too often in modern society, and in earlier societies as well, it is the State rather than the environment that does the enslaving ! As Dickinson shows in his *German Lebensraum*, there is absolutely no evidence for supposing that a natural region, even when possessed of the most clearly defined frontiers, gives rise to a corresponding unit in the human modes developed within it.

The trouble with the followers of the second school of thought is that they are united more by their general disagreements than by their beliefs. They are still hunting for a formula of words which will satisfy everybody's requirements. So far, the most common usages are the least meaningful. After all, who can tell us when an area is, or is not, " functionally coherent " ? Is " coherence " a measurable quantity ? Is it something that is achieved when the industrial output per man-hour reaches a certain level ? Or when there is full employment ? Or when local supplies of men, goods and services are exactly balanced by local demands ? And if it is not measurable by unequivocal standards, how can its presence, or absence, be confirmed ?

Similarly, how may we define " dynamic homogeneity " (or even " static homogeneity," for that matter), and what is the precise connotation of " inter-related characteristics " and " entities of circulation of

[1] C. O. Sauer, " Foreword to Historical Geography," *Ann. Assoc. Amer. Geog.*, Mar. 1941, pp. 11–12. [2] R. E. Dickinson, *The German Lebensraum*, p. 19.

goods and persons," to use another term favoured by Dickinson?[3] And precisely how stable is such an entity, or state of homogeneity? If an area is, at a given point of time, a homogeneous product of the fusion of various elements or element-complexes, then it must follow that as soon as one of those elements changes in time or place, the whole regional complex changes. And how often some of these elements—industrial output, export and import trade, supplies of raw materials, labour-management relations, etc.—do change! (Is the manufacturing belt of north-eastern U.S.A. "functionally coherent" when it lies palsied under the hand of John L. Lewis, or the Palestinian coast plain, when it is in the grip of a Terrorist campaign?)[3a] Unity and coherence can scarcely be more than an instantaneous condition of cultural relationships.

Again, how must we regard those areas which lack even this fleeting coherence? For we do not need to be told that there are plenty of regions in the world where there is no fundamentally coherent way of life, and no "sufficient concordance of common traits."[4] At the same time there are areas which have so many interwoven strands that they produce confusion rather than cohesion: colour in plenty, but little form: unlimited activity, but little pattern. (It has always seemed to us that the capital-heavy countries of the Southern Hemisphere are troubled in this way. With 75 per cent of its population cooped up in five cities, Australia is hardly a model of regional coherence: and much the same applies to South Africa and South America.)

At the same time, there is no reason to over-stress the significance of "homogeneity" and "common traits," etc. Not all homogeneity spells unity or coherence: the dominant historical quality of the Asiatic steppes has been uniformity rather than unity. *Per contra*, unity is frequently the product, not so much of *uniformity* of terrain, climate and resources, as of *contrasted* physical and economic conditions, integrated by commercial movements and/or by political ideals. This would certainly seem to be the case with the Paris Basin, and (in so far as it is possible to speak of nation-wide functional unity) with Nazi Germany and Soviet Russia as well.

[3] Dickinson, op. cit., p. 172.

[3a] This essay was written in 1946, before the creation of the state of Israel.

[4] C. O. Sauer, op. cit., p. 12. It is worthy of note in this connection that Rodwell Jones never once succumbed in his lectures or writings to the temptation of "regionalizing" the whole of a continent or country. In the preface to *North America* he and Dr. Bryan put it on record that "though . . . regions exist, it by no means follows that a whole continent can be divided up into equally well-marked areas . . ." p. 41 (7th Edition).

With the academicians unable to agree upon the content of the word, it is little wonder that the general laity continue to think of " region " as little more than an alternative unit of area which can be administered with slightly less fuss than tiresome political units.

Our doubts about the adequacy of the regional concept increase when we come to look into the nature and constitution of the region, as represented in current geographical literature.

The Constitution of the Region

Consider in the first place the problem of delimiting regional boundaries.

Self-evidently, if there is such a thing as a real region, it must be capable of mensuration. Granted that there may be alternative ways of carrying out the measurement and of determining the exact lineaments of the region at any one instant, there can be no question of indeterminism. No amount of sophistry will ever persuade our critics that an " organism," a " coherent whole," a " unitary entity," etc., etc., can have substance without form, or quality without quantity. And it is no defence to say that we are dealing with dynamic boundaries, or zones which are not amenable to measurement, because it must surely be impossible to tell how dynamic the boundaries are unless we have a measured datum line. Nor is it of any avail to argue that the homogeneous centre of the region matters more than the diversified circumference. As geographers we are just as much concerned with diversity on the earth's surface as we are with homogeneity : and who shall say that the diversities of the boundary zones are less rich in meaning for us than the uniformity, or unity, within the region ? [5]

But what tools can we use for the measuring operation ? Contours ? Isotherms ? Population densities ? Linguistic distributions, etc. ? The truth of the matter is that the geographer's world is so full of a number of unlike things—some of them mobile, others incalculable, and nearly all of them sporadic in occurrence—that it is impossible to use a *universal* yardstick for the operation. A unit area as determined by one criterion does not necessarily coincide with a unit area as determined by another. In one area, as, for instance, Southern California, it is the climate that seems to define the limits of the " region " : in another, the Laurentian Shield, for example, the circumscribing feature would seem to be the rock structure : in yet a third, e.g. the Pacific North-west, we might argue that it is provided

[5] Sauer (op. cit.) has recently reminded us that the centres of culture areas are often near the *boundary* zones of physical regions.

by the topography. The problem would not be so complex if the boundary lines of these natural phenomena coincided. But this is far from being always the case. A mountain crest line frequently lies astride a geological outcrop or a vegetation belt: in the Alps and Pyrenees, flocks and herds freely wander over the mountain-top pastures flanking *both* sides of the international boundary—to the perennial embarrassment of their owners.

Which brings to mind a further consideration, namely that, owing to their greater versatility, humans are wont to show even less regard for the significance of boundaries than nature. Were their actions ruled by respect for logic, for statistics and the niceties of ecological adjustment, no doubt they would have realized long before now that they were mere eddies in the stream of time, doomed to extinction the moment they wandered from the main channel of social evolution. But they have shown an amazing ability for "wandering," and for surviving the impact of new, and suddenly changed, conditions. If their native environment ever had such a thing as organic unity, of which they formed an integral part, they seldom seem to have noticed it—or its absence.

We have only to recall how hard it is to get any of the usual climatic and vegetation regions to "jibe" with human landscape types, with economic, ethnic, linguistic and commercial distributions in order to realize how little respect man has shown for his alleged organic unity with his environment. The truth would seem to be that none of these basic natural distributions is sufficiently determining to tether men in perpetuity to specific parts of the world and to specific modes of living. To superimpose two or more of the basic distributions may help the student to see broad correlations between place, people and their occupations, but the more elements he attempts to correlate, the greater, generally speaking, the fragmentation of the land into element patterns, few of which can be found to bear any close resemblance to the distribution of distinctive ways of human life. Indeed, we should be lucky to emerge from such an exercise with anything more than a "doodle" of intertwining boundaries. Attempts to reduce these to significant order would lead us into the field of arbitrary compromise, and be likely to give a manifestly false picture for a greater number of places than those for which the picture proved to be even approximately true. Regional boundaries cannot be fixed by ballot: no region is a region merely because a quorum of geographers have voted it to be so.

Nor do we fare much better by attempting to define regional boundaries with the aid of only those element complexes for which man is responsible, e.g. land-use patterns and settlement types. Thus the limits of the Cotton Belt are not coterminous with the " South " : the *milieu* of the Kentucky cotton farmer is very different from that of the Texas cotton farmer. Similarly, the Corn Belt is not coterminous with the Midwest or with the distribution of any special brand of politics, religion or newspaper. If it is impossible to equate economy with culture in localities which have a dominant (almost homogeneous) economy, we need not expect to be able to do it in those far more numerous areas where men have elected *not* to use their land in any such unitary fashion : for example, in the Great Plains, where ranching (stock-raising) and grain farming intermingle, yet produce two distinctive and independent economies? Or where, in cases equally numerous, we find intrusive social features which fit oddly into the general picture, and which suggest a dichotomous, rather than a unitary organization of living patterns? Often, as Fleure has recently reminded us, the geographical features we observe are not responses to the present environment of the group studied, but instead are " imports " : many aspects of Maori life and tradition indicate this clearly, as also does Islamic culture in Nigeria and the transplanting of European tradition to the new world and Australasia.[6] In such cases we can often define our boundaries between the intrusive and native groups with great precision, but that does not mean we have " corralled " a region in the sense of having isolated a unit whole, or a coherent economy.

We must face the fact that it is far more easy to devise seemingly logical systems of classification than it is to produce, from these systems, patterns of culture that have reality on the ground. And, furthermore, we must face the equally disturbing fact that natural boundaries can sometimes be less meaningful than man-made ones. For what proof is there that an isotherm or a contour is any more restrictive of human ways than a political frontier? Does the ten-inch isohyet act as more of a determinant in the economy of its contiguous areas than does the European frontier of the U.S.S.R.? To the best of our knowledge, traffic in ideas and commodities continues to move freely back and forth across the Great Plains. Nature's " curtains " are fashioned of more malleable material than iron!

To persist in making these classifications (whether political or natural), and endowing them with the authority of law, is to pay a lip service to the " logic of geography " for which the makers of history have so far shown much less regard than the writers of it, and to

[6] H. J. Fleure, " Geographical Thought in the Changing World," *Amer. Geog. Rev.*, Oct. 1944.

construct a strait-jacket which man has shown little or no inclination to wear. As Hettner observed years ago, the difficult question as to precisely which criterion should be selected for the determination of regions finds no answer in nature. The choice must be made by the individual geographer,

according to his subjective judgement of their importance. There is no universally valid regional division which does justice to all phenomena : we can only endeavour to secure a division with the greatest possible number of advantages and the least possible number of disadvantages.

It is small wonder, then, that Hartshorne, viewing the labyrinth of so-called regional boundaries, should be constrained to admit that " our regions are merely fragments of land whose determination involves a considerable degree of arbitrary judgement " ; that it would be unwarranted to expect geographers to come to even an approximate agreement on the specific limits of these fragments, or even on their central nuclei ; and that, at best, we may only hope to contrive a fragmentation of land that has a little more significance than is expressed by the bare phrase " arbitrarily selected . . . ' Because the different areas of the world have not been separated (either as individuals or types) in their development, but rather the development of each particular characteristic of an area has been a part of the development of that element elsewhere, we know that this process has produced no simple system of classification of areas, whose general outline can be recognized on the basis of our present knowledge of the field."

Just how " unwarranted " it is to expect geographers to agree on such matters would quickly become apparent were they to do more of their travelling about by air. As a class, we geographers are still far from being air-minded. (Some of us do not yet regard aerial surveys and photographs as part of our stock-in-trade.) The more's the pity ! For as Mary de Bunsen observes in a recent article, from the air it is possible to catch a glimpse of the larger pattern of man's life, and to gain an insight (not otherwise to be had) into the nature of man's ecological achievement. " From this viewpoint," she writes, " man is judged, not at his own valuation, but . . . as part of a universe which is greater than himself." [7] From the air, surface valuations of spatial phenomena are frequently invalidated. Having, in the course of the past two or three years, flown over many of the " textbook " examples of regions, I readily confess to experiencing real difficulty in locating from the air their precise position and lineaments. The Wheat Belt of the American prairies merges almost

[7] *Geographical Mag.*, Aug., 1946, p. 158.

imperceptibly into the Hay and Dairy Belt to the east, and the Corn Belt to the south. The cotton economy of the Deccan is subtly woven into the rice economy of its north-eastern and western perimeters. In the Kenya Highlands, there is no telling where the zones of the white man's culture ends and the African's begins. Similarly, in Australia, there are no landmarks separating the sheep-ranching country from the desert.

From the air it is the *links* in the landscapes, the rivers, roads, railways, canals, pipe-lines, electric cables, rather than the *breaks* that impress the aviator: and if he flies low he cannot fail to be impressed by the animation of the " arteries " which couple area with area—by their solidarity, in fact. In India, during the past war, two of the commonest sights seen from the air were the endless files of people padding along the highways in search of employment, and the " hobo " encampments fringing airfields and factories under construction. Here, in the aggregate, were millions of displaced persons whose wanderings were fast disrupting such regional unity as India's economy had ever known.

Our suspicions that regional geographers may perhaps be trying to put boundaries that do not exist around areas that do not matter are further reinforced when we turn our attention from the circumference to the centre. For then it becomes apparent that, as far as present-day distributions are concerned, we are certainly not dealing with a world of neatly articulated entities, unitary in nature and integrated in activity. Whatever degree of functional autonomy existed between one area and another before the advent of the Air Age, practically none survives today. The airplane has made the whole world a neighbourhood, and made it just about as private. The effects of the new mobility are legion, and none more significant than the accelerated tempo of cultural diffusion. The purveyors of culture—whether in the form of merchandise or ideas—are already storming the last strongholds of autarchy. Even Tibet is now open to the blandishments of foreigners selling everything from corrugated iron to ladies' slacks; while Solomon islanders chew gum and ride jeeps along tarmac roads where, until 1942, there was nothing but jungle paths; and Kabyles replace the immemorial burnous by Parisian modes, and make paper-knives for tourists instead of sabres for assassins.

Of no single part of the earth can it now be said that the pattern of living is a straightforward product of evolution brought about by local and internal forces. Rather is the pattern—if pattern it be—the interim product of partially related, and partially independent, changes set in motion by such forces as the will and energy of the people living both within and without the area, and by alterations in the individual natural elements composing the area. On balance these act more often in conflict with one another than in unity. In other words, a geographical area should perhaps be viewed more as a register of internal and external conditions, than as the derivative of a predestined evolution. We must, however, be wary of employing figures of speech in this connection. Thus the familiar analogy of a " terrestrial canvas " produced by the combination of different colour designs each applied by different artists working more or less independently and each changing his plan as he proceeded, does little to convey the dynamism of geographic reality. If anything, we are dealing with a " cinema," not a " canvas " : the true complexity of our picture can never be apprehended by merely card-indexing the various colour designs and their combinations, or by determining the exact contribution and period of activity of the several artists. Geographic reality is not trapped by such naïve devices, any more than the reality of man is apprehended by an anatomical dissection of his body.

Naïvété should have no place in geographical research : yet, if we are frank with ourselves, we must confess that many of us are reluctant to credit geographical man—that is, man living in specific areal groups—with anything like as much wit and foresight, sophistication, artifice and irrationality as we know he manifests as an individual. Few of us would make so bold as to say that we *know* ourselves physically —let alone say that we have explored the " dark caverns of the mind " —yet we do not hesitate to set about the infinitely more complex task of knowing geographical areas, armed only with an eye for scenery, a head for figures, and unlimited enthusiasm. Despite many illustrations of the geographic complexity of even the smallest area, some of us still cling to the notion that one person, even a newly graduated student, is competent to study, understand and interpret all phases of a given locality. Not a little of our dogmatism concerning, and unquestioning acceptance of the reality of the " region " would seem to stem from the fact, as Ackerman points out in his indictment of amateurish tinkering with geography, that few of us have been properly apprenticed on the systematic side of our calling.

With the complexity that is now evident in the study of settlement forms, agricultural occupance, industrial location and governmental structure to mention only a few—we can hope for very little more than we have received from the single " regional " worker. He has not, and cannot report adequately on all phases of anything more than a very simple element complex. Where many elements are involved, he usually has an appreciation only of the more obvious features of each. The holistic investigator has even less foundation for the complete correlation which he seeks, since accurate correlation requires a deeper understanding than observation alone.[8]

But the reality we are investigating is even more subtle and many-sided than this, since the whole life of a given area is greater than the sum of all the measurable parts, whether dynamic or static. Ideas cannot be measured or card-indexed or dissected, yet without doubt they can be a most potent force in shaping the geographical ensemble. An engineer conceives a plan, the upshot of which is the T.V.A., that not only alters the whole areal pattern from drainage schemes to settlement, but fires the imagination of people living in other areas, even in other continents. Or to take another instance. When the Italians moved out of Cyrenaica, the Arabs moved in : today they live in the same houses, farm the same land, raise the same crops, travel the same roads, buy and sell the same goods, as did their Fascist predecessors. But Arab Cyrenaica is a very different geographical phenomenon from Italian Cyrenaica. The Italians were intruders—aliens—and they lived in cultural isolation, and in contempt for native ways. The Arabs " belong," and today in spite of its Italian framework, Cyrenaica is, in culture, ideologies and loyalties, just another part of Islam.

Living, as we do, within the sound of the atomic bomb, it should not be difficult for us to realize that ideas and associations, memories and prospects, can be just as influential in energizing culture groups as they often are in fashioning individual lives : and that these ideas do not necessarily stem from environmental constraints or from biological inheritance. Nor are they subject, of necessity, to the ordinary geographical laws of movement : frequently they transgress political boundaries as freely as those of land and sea. So strong is the power of an idea that even a veiled threat of a shift in the political frontier or a change in the form of government can lead hundreds of thousands to live, in a sense, on both sides of a boundary—their homes on one side and their bank accounts on the other—a precaution that does nothing to promote functional harmony in the areas in question. Can we doubt, further, that the new post-war styles in European and Asiatic economy, indeed in the whole pattern of living relationships, are being tailored to fit ideological designs quite as much as geographical ones? So far the great powers have paid no particular heed to the words, or blueprints, of geographers : where security, diplomacy or strategy demand it, regional considerations are discarded.

[8] " Geographic Training, Wartime Research and Immediate Professional Objectives," *Ann. Assoc. Amer. Geog.*, Dec. 1945.

The partition of Germany, the amputation of Hungary, the Danubian reshuffle and the internationalizing of Trieste can more readily be rationalized by reference to power politics than geographical discernment. Trieste, in particular, demonstrates how intricately the web of ideas and associations has been woven into the woof of European geography, and how dismal is the prospect of achieving an abiding synthesis of habitat, economy and society in that troubled part of Central Europe. For how can an area possibly attain unity of work and outlook, all the time it is denied its national affiliations, and deprived of the right of self-determination, even of the use of its cultural symbols, but which at the same time is daily bombarded by the propagandist weapons of fascist and democrat, communist and catholic; and an area, furthermore, which is almost certain to be sidestepped commercially by its neighbours? The only unity it is likely to know is that of caged prisoners scheming their escape. And yet politicians (and even geographers?) continue to speak of geographical societies as though they were geological fossils. Unfortunately for their peace of mind, these "fossils" have a knack of hitting back!

This last consideration raises yet another question. Is it possible for an area possessing an appreciable degree of unity to co-exist alongside areas lacking even the very vestiges of it? That is, can we circumscribe areal unity to the extent of being able to say, for instance, that a lasting solution to the Trieste problem could be arrived at, without at the same time concerting the discordant geographical economies and ethnic distributions in its hinterland? All the evidence goes to show that geographical unity is one and indivisible. To deny a locality its national affiliations is to give it no greater expectation of life than that of a cut-flower. It may, so to speak, "live off its own fat" for a while, but sooner or later it will show symptoms of internal disorder, economic stringency, and sub-normal, ersatz living standards. Rather does geographical unity stem from an appreciation of the complementary nature of human environments—no single area has enough of all the ingredients of the good life for all its people all the time—and the consequent necessity of cultivating close "arterial" relations between differently endowed neighbours, each serving the other, and in turn being served by it. (In a sense, geographical unity represents the "climax association" of human activities, in that, like a plant climax, it is the end-product of a process of selection by elimination.) If neighbour areas are not organized on an optimum basis, if, for example, region A is not producing the grain crops for which it is better endowed than region B, and B is not producing the meat and wool for which it is better suited, the economy of both areas will suffer.

"If one member of the body suffers, all the members suffer." In other words, it is impossible for optimum geographical relationships in one area—the prerequisite of regional unity—to co-exist with sub-optimum relationships (*alias* regional disunity) in another. Judging by the number of functionally unintegrated, characterless areas in the world today, it would seem either that we are still a very long way from organizing society on a regional basis, or that we gave up the attempt in the eighteenth century when the Industrial Revolution got out of hand.

Were it simply the tangible, measurable phenomena that were unintegrated, there would perhaps be good ground for believing in an ultimate regionalization of our world: but the present disunity and confusion arises quite as much from imponderable ideologies and feelings as from measurable economies. In many parts of Europe and the Near East the feelings of two or more nationalities about the same piece of land are seemingly irreconcilable: the French, we may take it, are never likely to "feel" as the Germans do about the Saar Basin: nor are the Arabs likely to "feel" the same way as the Jews do over the Jordan valley: or the Italians, as the Yugoslavs, in regard to Venezia Giulia. The overlapping sovereignties and remembered wrongs of the past effectively prevent the regional integration of activity in these areas. Which brings us to our next topic, namely, precisely how widespread is the occurrence of the region?

The Incidence of the Region

In their writings upon the subject, geographers frequently make two cognate assumptions: first, that the "region" is the ultimate, inevitable synthesis of all the folk-work-place relationships in a given unit area—much in the same way as Darwin regarded man as the consummation of a slow evolutionary process: second, that the whole world is organized (or in process of becoming so) on a "synthetic" regional basis, which it is the primary function of the geographer to identify, describe and explain.

Is such optimism justified? Is it possible to separate the whole world into neat areas exhibiting even the barest rudiments of "unity," "coherence" or "organic entity"? We must confess to a difficulty in naming and delimiting, unequivocally, a dozen such areas outside Europe and the British Isles. And even in Europe it is by no means easy to accommodate certain areas into a generally acceptable and meaningful scheme of regions, for some of those which are generally regarded as classic examples of regions, give the impression of being the product of a past, rather than a contemporary, synthesis. Their survival, in as far as they do survive, would appear to be due partly to the persistence of the earlier cultural *forms*, e.g. house and road patterns, after the decay of the cultural *modes* ; and partly due to the momentum of the " going concern." To substantiate this thesis, it will be necessary to remind ourselves of some of the features of the European world during the epoch of its cultural differentiation.

As we noted earlier, the regional concept is of European provenance, and was first propounded to a world as yet untouched by the Industrial Revolution. As far as the general pattern of life was concerned, eighteenth-century France and Germany *looked* very little different from their sixteenth- or even fourteenth-century prototypes. The turnpike and the stagecoach were beginning to affect the tempo of life in the towns, but beneath this ripple of change the great currents of eighteenth-century society continued to flow on quietly in their mediæval channels. What struck the writers of the time was the stability, even the permanence, of the cultural landscape : the unchanging farm practice—they could point to thousand-year-old triennial rotations : the Romano-Gallic origins of the farmsteads of Picardy and Flanders—farmsteads which had often been in the possession of the same family since Norman times. They were constantly impressed by the harmonies between land and economy : the localizations of dialect and of distinctive domestic and ecclesiastical architectural styles. Most communities in Germany, France, England, etc., were, from the continuing exigencies of communication, still practically self-supporting. They had to live " off the land " and the local land at that.[9] Roads were mostly paths and often impassable because of mud. (Even to this day the countless little hamlets in the southern half of the Massif Central, the farmsteads of the Basin of Rennes and of Brie are distributed quite independently of roads : they communicate with one another by means of a network of byways.) While this meant that the basic structure of society was much the same throughout Europe in as far as most areas needed to raise their own requirements

[9] In eastern France and in the agricultural plains of Germany it was possible until recently to see vestiges of this autarchical economy. " From a nucleus of farm houses, the fields, under a system of communal rotation, extended in long, parallel bands so that sowing, tilling, harvesting and consignment to pasture succeed one another in regular sequence, and are completed simultaneously. Originally there were village streets between them in the form of narrow strips, either grass-covered or lying fallow for the convenience of the peasant. This local system of communications is characteristic of a village community. It is sufficient unto itself. Roads may have been added later in order to communicate with the external world, but the social unity, the cell so organised has to be self-supporting, unless we already provide for its own circulation." Vidal de la Blache, *Principles of Human Geography*, pp. 280–1.

of grain, fuel, and raw materials, there was plenty of local diversity. Beauce soon acquired a reputation for its grain ; Berry, for its wool and meat ; and so on. A further landscape differential was provided by the fact that these early communities, being substantially self-supporting in constitution, had to learn to fashion all their cultural artifacts out of local materials : farmsteads, churches, manors, as well as fences and bridges were built in the media most readily available. In the clay vales of the Severn and Thames it was brick : in the Cotswolds, stone : in the Weald, timber : on the Downlands, thatch and flint. Likewise only those industries could flourish which could draw upon local raw materials and local sources of power. Even in eighteenth-century Europe, mineral workings and factories serving more than their immediate localities were very few and far between. Neither the demand for coal and iron nor the technique of their production on a large scale yet existed. Similarly the trade of Europe was still almost entirely parochial in character, and small in amount. No efforts had then been made to provide suitably constructed roads for internal transport. As for the Roman roads, these for the most part had been plundered of their stone and allowed to fall into disuse. The remarkable sub-division of middle Europe into no less than three hundred distinct states involved the continual crossing of frontiers and payment of customs. Under these conditions, the movement of goods was bound to be costly : it is estimated, for instance, that wood for fuel could be economically carried by road for a distance of only ten to twelve miles.[10] It is perhaps impossible for us, living in an age of all pervading mobility, to appreciate how very small a world these little communities administered, and how ignorant of the outer world they were. To find a modern-day parallel we should need to go far afield ; for example, to Annam, concerning which country one twentieth-century writer has spoken as follows : " Because of the great variety of communal institutions, we Annamese imagine ourselves in China or America as soon as we leave our village."

Restricted severely in their external contacts by intervening forests and marshes, by the absence of roads and multiplication of political boundaries, European communities had early become distinguished from one another, in physical appearance, loyalties and outlooks, if not in the organization of their social life. This rudimentary regionalization of life owed much to the mediæval Church. As the paramount influence in the community, the Church was able to direct the secular as well as the spiritual activities of her followers. Not only did she enforce regular attendance at mass and take tithes

[10] *Vide*, W. G. East, *Historical Geography of Europe*, p. 406.

of the farmers' crops (thereby putting a premium upon sedentary habits and good husbandry, and a discount on landless nomadism), she also actively promoted land settlement. The widespread clearing of forests and founding of new villages in France in the twelfth century was to no small degree the effort of the Cistercian Order. With the consequent increase in rural population and the Cistercian-inspired improvements in methods of cultivation went a growth of the towns and a development of those industries needed to supply the elementary wants of the rural population. While the appearance of the artisan class did something to break down the self-sufficiency of the manor, it substantially strengthened the solidarity of the parish and diocese, in that it gave the inhabitants of town and country a sense of common cause—of being "workers together." This sense was assiduously fostered by the Church, which had no wish to promote the prosperity of the town at the expense of the countryside. On the contrary, she had early discovered that good Christians were more easily made from landed farming folk than from landless urban folk. Nomadism and paganism have long gone together; and even though it might be unkind to draw a parallel between the godlessness of the horse-riding Magyars and the car-driving, Sabbath-breaking New Yorkers, it remains a fact that itinerant souls have always presented a serious ecclesiastical problem.

So town and countryside came to be looked upon as a unit whole, with the town subserving the country. If need be, the country could do without the town, but not the town without the country. This mutuality of interest made for close liaison between the two, and did much to promote the feeling of local self-consciousness and local patriotism.

Over large parts of Europe this areal differentiation of the culture pattern can be traced far back into the Middle Ages.[11] By the time the eighteenth-century thinkers were beginning to enquire into its nature and origin, the mould had been set for centuries, and the dominant features of life had already acquired a strong traditional flavour. Many signs of this persistence of acquired characteristics survive until this day: indeed it would almost seem that the regionalization of Europe is most conspicuous in those areas where the hand of history lies most heavily. For if we call to mind almost any of the textbook examples of regions, such as the Weald of Kent and Sussex, the Fenlands, the Scottish Highlands, the Paris Basin and its constituent *pays*, the Bohemian Diamond or the Tyrol,

[11] In France, most of the *pays* were already well differentiated by the twelfth and 13th centuries: similarly with the German *gaus*.

what are the features that strike us most forcibly? The distinctive topography? Mountains have been laid low and valleys exalted, but not within recorded times. The characteristic weather pattern? There have been few significant changes (whether cyclical or non-periodic) in the European climate since the fourteenth century.[12] Is it the appearance of the cultural landscape? The general disposition of forests and farms, villages and roads has changed little in the last half millennium: many English fields and lanes are of more ancient lineage than England's nobility, while the clustered villages of Normandy, Artois and the middle Neckar, the dispersed hamlets of the Weald, the Cotentin Peninsula and the Allgäu, have retained the same basic morphology since the days of their founding. Is it the style of the secular and ecclesiastical architecture? Apart from those in the newer industrial and dormitory towns, most of Europe's churches were built before A.D. 1600, and her most typical farmsteads, inns and noble houses are of similar antiquity. Is it the social and human economy? Here again the distinctive features of local folk-lore, dress, dialect, music and art, and even of agricultural practice, have changed little since the late fourteenth or early fifteenth centuries. What interested Arthur Young most notably when he made his famous survey of European agriculture in the eighteenth century was its traditional and primitive character. The open field system and simple rotation of crops and fallow, was still general in the arable regions, and serfdom, in all but a few areas, was still the order of the day.

It is true, of course, that life in the Middle Ages had not entirely stagnated: that there had been periods of economic and social unrest, especially during the fourteenth and early fifteenth centuries, when there were popular uprisings in the Netherlands, in England and elsewhere: that there had been an increase in urban activity and in specialization of labour with a consequent increase in the role of capital and of the middle class; but as agencies of attrition upon a custom-controlled world, their influence was neither profound nor enduring. Feudal institutions, chivalric and autarchical ideals continued to be all-powerful in the life of the continent.

In other words, many of the most unequivocal, highly individualized geographical areas in Europe and Great Britain today are in the nature of historical survivals from the pre-industrial age—relics, as it were, from a world that was practically self-contained and self-supporting.

[12] The last major change seems to have taken place towards the end of that century, when "the ancient dry phase of British climate [gave place to] the present cool phase" (E. Huntington, *Mainsprings of Civilisation*, p. 595). That an agriculturally significant change did occur at that epoch is predicated, *inter alia*, by the disappearance of the grape, as an outdoor crop, from about A.D. 1350 to 1400 onwards.

Is there any evidence that the Machine Age has produced a cultural pattern comparable in significance and coherence with those developed in feudal times? Some writers claim to see a new pattern of geographical areas crystallizing out from the crucible of industrialization. For instance, in place of old static entities, Dickinson saw emerging in pre-war Germany a new, more dynamic regional pattern, based on economic activity. These regions he called " entities of circulation and persons, with respect both to actual movements and to regional organization." But even if there were such a thing as " an entity of circulation " in Germany (which we take leave to doubt, for entity connotes functional independence and/or separate existence, and quite clearly the aim of Nazi Germany was to organize the *whole* Fatherland, rather than its component parts, on a unitary basis—thus the *Autobahn* system served national strategy rather than local economy), we cannot agree that the areas Dickinson chooses to call " circulation regions " had in fact anything more than a statistical existence. For what unitary quality did Silesia possess, with its three well-marked physical sub-divisions, set as it was in the midst of one of the most hotly-contested march-lands of Europe, with its borders on three sides arbitrarily fixed by treaty, not to speak of the ardent and opposing nationalisms of its peoples? And how can we regard the Rhine-Ruhr region as an " entity of circulation "? As every statesman in Europe now knows, this region is the very nerve ganglion of Germany, and has no more power of circulation, when severed from the rest of the body politic, than its human counterpart. If the only kind of regionalized unity possessed by modern (pre-war) Germany was its circulation pattern, then the day of the region is nearly at an end, since mobility is the catalyst of regional diversity, and must sooner or later dissolve the whole compartmental structure of our civilization.

It is not without interest to note in this connection that wherever men are sensible of the permeation of society by a monotype culture, they are busy organizing associations for the encouragement of the old regional ways. In France, where the differentiation of the country into *pays* (*alias* " regions ") is probably more strongly accented than in any other part of the world, staunch endeavours are being made to resuscitate dying folk-habits, local dialects and traditional arts and crafts. The French know (who better?) that it is the " silent backward tracings ", the feelings of cultural identity and group experiences surrounding place and home that compose the stuff by which regional consciousness is sustained and fostered.

We see, therefore, that the standard-model region is essentially a phenomenon of the European continent: that it was sired by feudalism and raised in the cultural seclusion of a self-sufficing environment, that it owed almost as much to history as to geography, and that it does not appear to thrive in the more turbulent atmosphere of modern times. This being so, are we justified in assuming that it will transplant freely in the radically different *milieux* of the Americas and Australasia? [13]

From a purely environmental point of view, Europe had a flying start over all the other continents in the matter of localization of culture patterns. Hume remarked upon this in the eighteenth century: "Of all parts of the earth, Europe is the most broken by seas, rivers and mountains, and . . . most naturally divided into several distinct governments." [14] Modern geographers might perhaps choose to express this implied relationship rather more academically, but they are unlikely to dissent from the general thesis that Europe's national and cultural fragmentation could not have persisted so long and been so radical, had it not been for the physiographic fission—to borrow a rather over-worked word—of that continent. The distribution of land and water, the dovetailing of plains and mountains, and the contiguity of steppes and forests all have had their share in producing a compartition of peoples and societies which no other region on earth possesses to the same degree.

Then, from the historical standpoint, it needs to be remembered that the ideologies which nourished Europe in the era of its cultural differentiation were already passing away, if not extinct, when the New World was being opened up. Not even the oldest settled parts of North America, with the possible exception of the Lower St. Lawrence Valley, succeeded in transplanting the regional idea in its pristine European likeness. From the time of the Pilgrims onwards (the very name Pilgrim is suggestive of a different attitude—one of detachment from regions and places), almost all the immigrants were animated by a sense of involvement in a new, un-European kind of life. [15] "A spirit of brotherhood," writes Commager, "transcending class, race

[13] By general consent, Asia has always been more conspicuous for the widespread uniformity of its cultural life and *mores*, than contrariwise. One reason for this may be, as Vidal de la Blache opines, that "in China as in Japan and in India, there has been no methodical attack on new agricultural areas . . . Human establishments continue to be confined to particular zones . . . Europe is the most humanized part of the world. No other presents so rich a field with such a hierarchy of types" (*Principles of Human Geography*).

[14] *Of the Rise and Progress of Arts and Science.*

[15] William Bradford put it—rather too pietistically, perhaps, for our modern ears—in these words: "So they left that goodly and pleasant city, which had been their resting place near twelve years: but they knew they were pilgrims, and looked not much at those things, but lifted up their eyes to the heavens, their dearest country, and quieted their spirits" (*Of Plimoth Plantation*).

and religion, a feeling that all dwellers within these states are partners in a common enterprise, is the peculiar quality that brought the American Republic into being." [16] At the Stamp Act Conference of 1765, a delegate declared : " There ought to be no New Englander, no New Yorker, known on this continent, but all of us Americans. . . ." Seventeen years later St. John de Crèvecoeur could write :

> Individuals of all nations are melted into a new race of men, whose labours and posterity will one day cause great changes in the world . . . The Americans were once scattered all over Europe : here they are incorporated into one of the finest systems of population which has ever appeared, and which will hereafter become distinct. . . .[17]

To realize this new mode of life, it was soon found necessary to eschew thoughts of areal self-sufficiency ; to do business with pagan Indian, Anglican Virginian and even catholic Cavalier ; to welcome men, irrespective of their cultural and national prejudices, and, above all, to keep the frontier moving. Where the frontier rolled back apace, and the immigrant flood ran strongly, as in the Midwest, Central California, and the Pacific Northwest, European regionalism never had a chance to take root.

> The triumphant achievements of the great railways gave the continent physical unity and overcame the natural effects of geographical sectionalism . . . A single type of life spreading from a single centre created the economic unity of the country . . . Free trade over the whole area assimilated the mechanism of daily existence, the habits of the people, their food and dress. Wherever the immigrant went, he entered into the same kind of life and was caught up into its activities.[18]

Is it not significant that almost the only parts of North America where distinctive, areally circumscribed modes of life still persist are those which were sidestepped by the highway and the railroad during the epoch of the retreating frontier ? It is only in the comparatively quiet backwaters of the continent, like the Lower St. Lawrence Valley, the higher Appalachians, the Ozarks and the Indian reserves that we can still see relics of a dissident cultural tradition, and of a distinctive regionalism. But every year sees a further shrinkage in the size and influence even of these " islands."

In the light of the foregoing, we need not wonder that it is the uniformity of American civilization, rather than the diversities of its peoples and countrysides, that impresses the modern traveller most deeply. When one hundred and fifty millions of people, enjoying, in the mass, a higher standard of well-being than any other people

[16] *The Growth of the American Republic*, Vol. II, pp. 591–2.
[17] *Letters from an American Farmer.* [18] Benians, *Race and Nation in the U.S.*

on earth, are coaxed into buying the same soap, eating the same canned goods, reading the same magazines, listening to the same radio programmes and joining the same service clubs, it is difficult to see how distinctive folk-habits, dialects and arts and crafts, which have in the past done so much to sustain regional self-consciousness, can possibly survive. Areal variations in the pattern of living, in all categories except perhaps land-use, modes of employment and population densities, would seem destined to attenuation, if not gradual extinction. Messrs. F. W. Woolworth, Henry Ford and Sears Roebuck have already gone quite a long way towards ensuring this end.

We have only to reflect upon the phenomenal achievements of the Tennessee Valley Authority to realize something of the efficacy of modern technology and propaganda in breaking down the barriers of cultural non-conformity. Some writers tend to regard the transformation newly wrought in the hills and valleys of Tennessee as merely the substitution of an antique regional pattern by a more modern pattern ; but the T.V.A. has done far more than integrate the economy of the upper Tennessee Valley. It has altered the whole conception of group living in neighbouring states, and threatens to disrupt the balance of the agricultural economy even as far afield as Wisconsin. Thus we read in a recent article [19] that, thanks largely to cheap electricity, cheap fertilizers and almost year-round outdoor grazing, dairy farming is expanding in Tennessee to such an extent that there is talk of moving a part of the vast dairy industry of Wisconsin to the Valley. So far from being a rather inconspicuous brick in the geographic fabric of the nation, it now bids fair to become the cornerstone of a new social and economic structure which may well alter the entire look of the continent. We should, therefore, be singularly out of step with the times to over-emphasize the significance of distinctive distributions of human modes and economies, whether localized in time or place.

As Dr. Bowman reminded us in a recent paper,[20] scientific invention is a perpetually unsettling factor. Every time a new implement or machine is invented or a new technique is devised, a new appraisal must be made of every scrap of territory and the possibility of a new orientation of human activities be predicated. Areal distributions are essentially impermanent. Even while we are, so to speak, photographing it, the picture changes. The mechanism of civilization is so delicate that it can be disrupted almost over-night. Ancient desert civilizations like that of Palmyra (and what is a civilization but an expression of cultural unity over a given geographical area ?) have

[19] *New York Times*, Sunday, July 28, 1946. [20] *Foreign Affairs*, Jan., 1946.

vanished like the smoke of a nomad's fire when some internal tumult or foreign unrest has dislocated the organization of the caravan routes. Much the same is true of Tyre and Sidon, Petra and Ur, and a hundred other cultural centres of the past.

Modern civilizations manifest rather more resilience to disruptive forces than those of antiquity : it must be confessed, however, that this does not seem to have extended their expectation of life appreciably ! None of the Axis empires lasted even a generation : Fascist Africa was destroyed almost as easily—and speedily—as Palmyra. But, thanks largely to the greater versatility of twentieth-century man, there has been no abandonment of the desolated lands to the wild beasts. Even the conservative, tradition-ridden Arab has learned to make himself almost as much at home in Benghazi as in a Bedouin encampment, and to be almost equally happy irrigating crops as watering flocks. Indeed, Miss Freya Stark contends, in her *East is West*, that "the desert . . . no longer gives the essential picture of Arabian life."[21]

What is happening in Islam is, of course, also happening (and has already happened) in many other parts of the world. The ferment of change is universal. What with the desolations of war, the imperfections of the peace treaties and the ever-accelerating peace of scientific progress, the structure of every human society is now in process of being rebuilt. The russification of Eastern Europe means, among other things, the introduction of collective farming and planned industrial economies in Poland, Eastern Germany and elsewhere. The transference of twelve million Germans and the liquidation of eight million Jews from Poland will do much more than alter the ethnic composition of the country : similarly the exchange of populations between Sudetenland and Saxony can hardly be achieved without making a major impress on the life of both regions. The peace treaties provide a new order in Italy's old African colonies ; while the allied government of Germany is fashioning a very different kind of economic unity from that divined by writers a few years ago. And what is true of the Old World is also true of the New, for the process of renovating the environment to suit the cultural fashions of our day extends to the remotest isles of the Pacific.

Conclusion

We must, then, face the fact that the old order is changing, and that we should only be deceiving ourselves to say as the French have taught us to do : " Plus ça change, plus c'est la même chose." That

[21] *East is West*, p. xiii.

epigram was coined before the invention of the internal combustion engine, sponsored radio programmes, totalitarian propaganda and jet-propelled bombs. Whatever the pattern of the new age may be, we can be sure that there will be no independent, discrete units in it —no " worlds within worlds." There will be no neatly demarcated " regions " where geographers (or economists or sociologists, for that matter) can study a " fossil " community. Man's " region " is now the world. This does not render superfluous the continued organization of geographical studies on a systematic areal basis. On the contrary : the very increase in the interpenetration of human modes would call for more, rather than fewer, such studies. But it does mean that we should be well advised, when making these studies, to refrain from searching for " unitary patterns of living," " entities of distribution," and from assuming, in the manner of determinists, that " regional unity " is the goal towards which civilized society is moving, and that if we but had the wit, we could not fail to see signs of it emerging. Let us rather admit that the ways of man, like those of his Maker, are frequently inscrutable, that while the regional idea may look very promising on paper, it does not enjoy the wholehearted support of the facts (any more than does the Darwinian idea of biological evolution with which the regional concept has obvious parental affinities). At best, a regional study can be only a personal work of art, not an impersonal work of science—a portrait rather than a blueprint. As such, it can have substantial value, but its value will lie in the realm of illumination and suggestion rather than of definitive analysis and synthesis. It will do well if it catches the dominant traits of the area : it cannot hope to carry out a distillation of all the compound elements (physical, economic, social and political) which are present. To presume otherwise would be grossly to under-estimate the complexity of our world. For the understanding of areal differentiations, we must, as a *sine qua non*, be able to draw upon the analyses provided by men trained in the several contributory disciplines. At present such analyses are woefully few in number, and flimsy in character. But even if this were not the case, it is still uncertain whether we could contrive to establish regional geography on a rigorously scientific foundation. With Finch we find ourselves inclining to the view that " the complexities of man and nature are too much for his abilities at rationalization." [22]

Instead of continuing to add to the already impressive pile of chorographic studies—many of which amount to little more than a

[22] V. C. Finch, " Geographical Science and Social Philosophy," *Ann. Assoc. Amer. Geog.*, March, 1939.

laborious transferring of dead bones from one coffin to another—should we not rather devote our energies to the pursuit of more worthwhile, if more restricted, objectives? Ackerman has recently indicated where we shall find such objectives. "Human geography," he says, "will never be accepted as a mature scholarly discipline until a more thorough systematic literature begins to take shape in it." [23] And he reminds us that geomorphology, climatology, plant and soil geography are not the whole of systematic geography: that land-use, conservation, the geographies of manufacturing, transportation, settlement and resources are equally important, though in professional esteem (judging by the amount of attention they have received over the past twenty-five years) they are a bad second to the more physical disciplines. Carl Sauer has also recently indicated a number of worthwhile objectives: as, for example, the study of man as an agent of physiographic change, of dwelling patterns, house types and field systems, of "cultural climaxes," successions and receptivity. These, of course, put the emphasis on the more historical side of systematic geography. [24] Nor is this list in any way exhaustive: a reading of Van Burkalow's work on the distribution of fluorine in U.S. water supplies, [25] suggests almost unending possibilities for research in the geography of disease: while our current refugee problems should surely cause many of us to bestir ourselves in the neglected field of pioneer settlement and the geography of marginal areas.

Here are subjects worthy of our most earnest and sustained consideration: they bear closely on the social and political problems which bedevil our times: and what is more, they are, for the larger part, amenable to definitive treatment. To ignore them, and spend our days "regionalizing," is to chase a phantom, and to be kept continually out of breath for our pains.

[23] Ackerman, op. cit., p. 141.
[24] Sauer, op. cit., pp. 17 et seq.
[25] *Vide Geog. Rev.*, Apr., 1946.

X

THE BEGINNINGS OF THE AGRARIAN AND INDUSTRIAL REVOLUTIONS IN AYRSHIRE

By J. H. G. LEBON, B.Sc.(Econ.), Ph.D.

(i) *Coal-Working.* To satisfy a demand for coal in Ireland, especially Dublin, the first real coal-pits in Ayrshire were sunk near Saltcoats just before 1700. Others followed at Kilwinning, Irvine, Ayr and near Girvan, enabling shipments to grow steadily throughout the century. Yet it must be emphasized that coal was not the basis of eighteenth-century industrial development. Largely won for external consumption, it was little utilized for non-domestic purposes in the regions from which it was hewn, except for lime-burning. This process, adopted after 1760 to provide the new farming with a means of sweetening the soil, required a fuel, obtained from numerous small outcrop workings operating inland.[1] In the comprehensive description of economic life provided by the *Statistical Account of Scotland*, only the contributor for Kilmarnock mentions that an abundant coal supply was aiding the progress of manufactures. But this was coupled with cheap provisions as an asset to a rising industrial town, and no industrial process was cited requiring coal as a fuel or to furnish motive power.[2] Its function was evidently indirect : a community of nearly 6,000 persons, situated in a region devoid of forests and remote from peat mosses, could be assured of warmth in homes and workshops by the neighbouring collieries. We may also infer that coal fires were employed to heat water needed for the washing, dyeing, etc., of wool, cloth, etc. ; for a limited iron-working ; and also for soap- and candle-making ; but these were only a part of contemporary industry. The Muirkirk Iron Works, founded late in the century (1787), was the sole Ayrshire establishment depending directly upon the energy of coal, and this antedated the main coal-using phase of the Industrial Revolution by more than a generation.[3] Accordingly, at the end of

[1] J. H. G. Lebon, " The Development of the Ayrshire Coalfield," *Scot. Geog. Mag.*, 1933, pp. 138–54.

[2] *Statistical Account of Scotland* (hereafter abbreviated to *S.A.*), edited by Sir J. Sinclair, 1791–9, Vol. II, pp. 84 ff.

[3] " Fire engines " (i.e. steam engines) were used for pumping at collieries near Newtown, Saltcoats and Newmilns. See *S.A.* Vol. II, pp. 262 ff., VII, pp. 1 ff., III, pp. 103 ff.

the eighteenth century, there was no tendency for industrial distribution to coincide with coal-working, a correspondence increasingly apparent during the next century. The principal coal-mines were confined to the coast until 1780 ; but modern forms of manufacturing became widespread, albeit most advanced generally in the north.

Apart from exports and lime-burning,[4] the last thirty years of the eighteenth century witnessed the substitution of coal for peat as a domestic fuel, at least in the Ayrshire Plain. Thus, in Fenwick parish, about 5 miles north of the new Kilmarnock mines, coals in 1793 were being used by " tradesmen " in preference to peat, although farmers (in proximity to large peat mosses) still burnt the traditional fuel.[5] In Girvan, coal was " plentifully supplied from Dailly."[6] In Craigie, the (southbound) carting of coal and lime was notable.[7] Negatively, we read of three parishes where the new fuel was expensive or difficult to obtain. For in Largs both coal and lime were scarce ;[8] in Beith high prices supported arguments favouring the connection of Saltcoats with Glasgow by canal,[9] and in Ardrossan the unimproved parish roads hindered the distribution of the new fuel and the new means of improving the soil.[10] Indirectly, we also learn that Dailly coal was used throughout the neighbouring parish of Kirkoswald, for it is stated that although tenants had been relieved of many feudal servitudes, they were still obliged to cart coals at their laird's behest for a stated number of days each year.[11]

A colliery started about 1780 and situated about a mile south-west of Kilmarnock had within ten years attained an annual output of 8,000 tons and gave employment to 120 miners. 3,200 tons were exported in carts from Irvine ; the remainder was consumed in Kilmarnock and the surrounding countryside.

The swelling economic momentum, of which developing coal-mining was but a part, was due ultimately to a double political impulse. The Act of Union, in 1707, brought Scotland into the English commercial system and opened trade with the colonies. From this, the west of Scotland, especially Glasgow, was quick to profit. Both commerce and industry rapidly advanced. Peace also succeeded to strife ; for the '45 was too brief to undermine the new stability which was enabling

[4] This use is stated explicitly in the accounts for Loudoun (*S.A.* Vol. III, pp. 103 ff.) and Dailly (Vol. X, pp. 34 ff.).

[5] *S.A.*, Vol. XIV, pp. 53 ff. In the uplands of Kilwinning (XI, pp. 142 ff.), and Dalry (XII, pp. 90 ff.) peat was still largely cut.

[6] *S.A.*, Vol. XII, pp. 335 ff.

[7] *S.A.*, Vol. V, pp. 369 ff.

[8] *S.A.*, Vol. XVII, pp. 503 ff.

[9] *S.A.*, Vol. VIII, pp. 314 ff.

[10] *S.A.*, Vol. VII, pp. 42 ff.

[11] *S.A.*, Vol. X, pp. 474 ff.

enterprise to reap a reward. Thus was the pace of economic development forced ; and the changes wrought between 1700 and 1800 were more far-reaching and profound in the western Scottish Lowlands than generally in England.

(ii) *The Revolution in Transport.* Improved means of communication greatly contributed to the rising coal production by facilitating its wider distribution. Greater mobility of goods and persons constituted the foundation of agrarian and industrial progress. For between 1760 and 1790, the horse, formerly employed as a mount or as a pack animal, was given a new role, and became harnessed to the stage-coach or cart. In short, the use of the wheel became general and revolutionized economic life.

Till this era, road-building in Scotland was much neglected. Muddy tracks connected the scattered farms and few towns ; the pedestrian, horseman or pack animal crept slowly forward, fording lesser streams, ferrying rivers, except at rare bridges, e.g. Ayr, (thirteenth century), Alloway and Kilmarnock. Thus the circulation of persons and goods was severely limited. Many simple tools, utensils and textile goods were made on the spot from local or home-produced raw materials ; or were obtained from craftsmen in the burghs (which exercised the waning privilege of a territorial monopoly for their products in the surrounding district). Usually the movements of raw material, finished goods, producer and purchaser, were confined to a few miles, and consignments were small since the unit for goods traffic was the pack-animal. Thus, till at least the beginning of the eighteenth century, when a collier entered Irvine harbour, a horn was blown as a signal to the country-folk that coal was to be brought to the quays. Panniers containing coal were thereupon loaded on horseback and taken to the vessel.[1]

Of the poor state of the roads in the middle of the century there is evidence both in the *Statistical Account* and contemporary maps. In Ballantrae there were no made roads before 1774 ;[12] in Kirkmichael parish no improved roads existed before 1769 ;[13] in Kilmaurs " 20 or 30 years ago " (before 1791) " there was not a single made road."[14] The MS. maps of the *Military Survey of Scotland* (1755–67)[15] at a first inspection (Fig. 36) might seem to belie these assertions, because a fairly comprehensive network is represented. But (to vary a statement used on modern Ordnance Survey maps), " the representation on this map of a road, track or footpath is no evidence "—of a made surface ; and closer scrutiny reveals that the system served the needs of the unwheeled

[12] *S.A.*, Vol. I, pp. 103 ff.
[13] *S.A.*, Vol. VI, pp. 102 ff.
[14] *S.A.*, Vol. IX, pp. 350 ff.
[15] British Museum, Maps 115 h.

age in transport. For example, south of Girvan, the traveller might select one of numerous interlacing tracks on his way southwards to Ballantrae and Portpatrick, offering no better than a Hobson's choice ; but in their multiplicity perhaps in slightly better state than if all traffic were concentrated upon a single route. A similar indeterminacy existed in the bundle of routes leading south-east from Ayr, as well as in the country between Maybole, Barr and Dalmellington. Ways over high ground were commoner than today, although many are shown on the original by single instead of by doubled lines, and were doubtless less frequented. These included the " Drove Road " from Kirkconnel to Crawfordjohn ; the track from the Nith bridge by New Cumnock kirk to the same Lanarkshire terminal point ; the routes from Dalmellington to Old Cumnock, from Mauchline to Darvel, from Barr to Bargrennan, and from Darvel to Eaglesham. Further limitations were imposed by the paucity of bridges, and the corresponding frequency of fords and ferries. For the whole of Ayrshire, but twenty-four bridges were indicated on the *Military Survey*. These were well-sited to ensure continuity (across the principal rivers), of the two main routes traversing the country from north to south. Bridges at Irvine, Ayr and Alloway carried the coast road across the Irvine, Ayr and Doon respectively ; bridges at Stewarton, Kilmaurs, Kilmarnock, Riccarton, Mauchline, Ochiltree and New Cumnock spanned the chief rivers along the inland route to Nithsdale. Ways leading to the coast, and thus parallel to the drainage lines, could in general dispense with bridges ; but apparently alignment to avoid lesser river crossings was often adopted. The contrast east and west of the Garnock River is illustrative. In the former direction, right-bank tributaries to the lower Irvine Water, e.g. Dusk Water, Lugton Water and Annick Water, flow from the watershed separating Ayrshire from Renfrewshire, and roads from the next county could keep between these streams and avoid bridging most of the way to Irvine or Saltcoats. But, from the west, the Garnock River receives numerous fast and substantial affluents flowing generally south-east from the uplands between Dalry and Largs, and athwart the line of potential traffic. Accordingly, the road following the valley from Paisley and Lochwinnoch (see Fig. 36), turned at Carse, just short of the ferry across the Maich Water, and passing through Beith, avoided the sequence of river crossings which a direct continuation to Saltcoats would have demanded. After passing through Beith, the road remained west of the River Garnock till Dalry, leaving only one more tributary—the Caaf Water—to be bridged.

Clearly, at many main river crossings, and at all lesser streams, the traveller employed a ferry-man or made his way across a ford. In many places, the *Military Survey* was positive in the matter: the

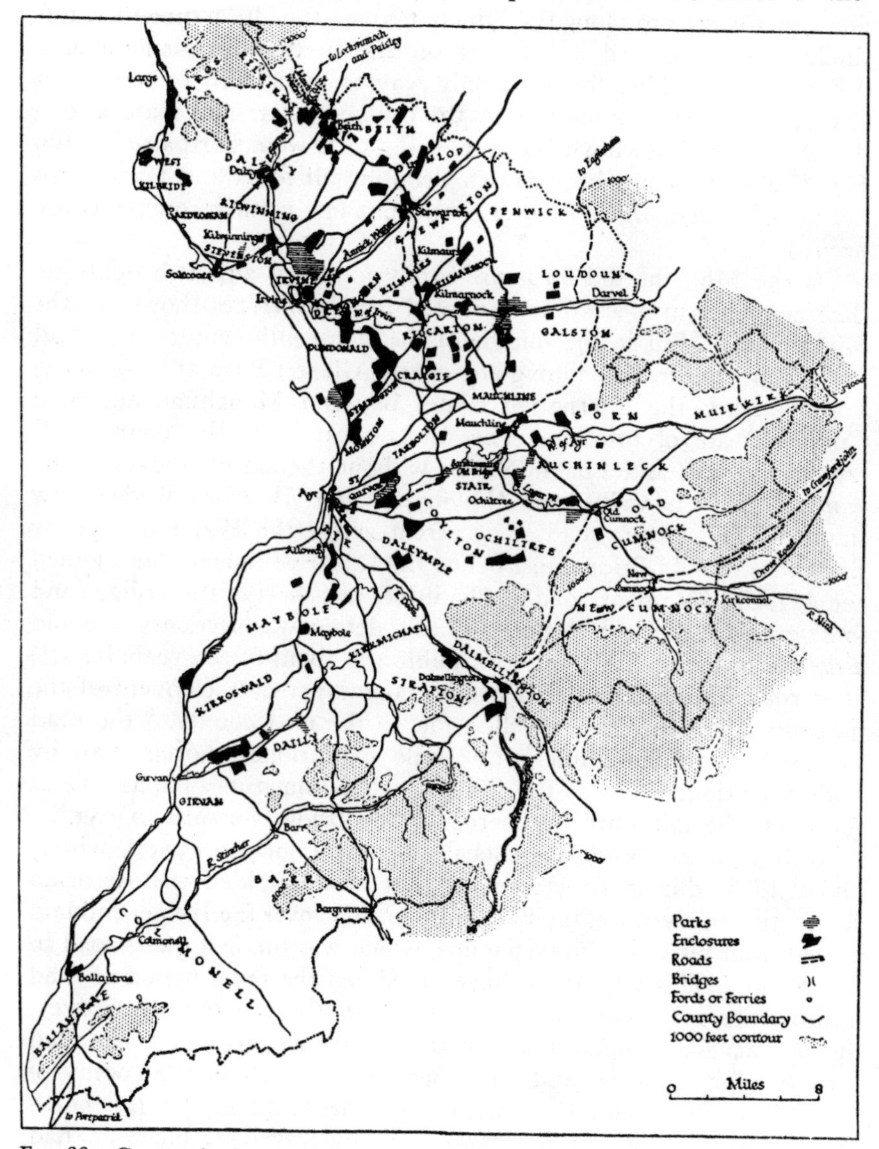

FIG. 36.—Communications and Enclosures in Ayrshire in the middle of the eighteenth century (based upon the *Military Survey of Scotland*, 1755–67).

doubled lines denoting the road were drawn as far as one bank, then broken short and resumed on the opposite side. Thus, at Ballantrae, to specify the most striking instance, a broad estuary is shown between the broken ends of two stretches of the coast road. At Colmonell, four miles upstream along the Stinchar River, the alternative road was similarly treated, and a building on the south bank denominated "Fordhouse." Therefore, the only way to avoid ford or ferry over this river in 1760 was to cross by the bridge at Barr and make a long detour through Newtown Stewart, if the goal were Portpatrick. On Fig. 36 a symbol has been drawn denoting all fords and ferries thus indicated on the *Military Survey*; and but few further comments are needed.

On the MS., the old fords are rarely named; although doubtless their nomenclature was later adopted for the bridges shown on the 6-inch plans. But in the middle of the eighteenth century, the road from Stewarton to Kilwinning crossed the Glazert Burn at "Galloway Ford." Also, the alternative routes between Mauchline and Old Cumnock, are of interest. The direct road (now the main road) crossed the Ayr and Lugar Waters without the aid of bridges; (the word "ford" is written on the map where Howford Bridge now stands). The longer road was carried across the Water of Ayr by Barskimming Old Bridge, just below the Lugar confluence, and joined the Ayr-Old Cumnock road at Ochiltree, just short of the bridge (and ford) over the Burnock Water. Thus a detour was necessary to avoid a double bridging. A road book published about fifteen years later,[16] after road improvement had begun, contains a note eloquent of the difficulties besetting the traveller, for, in the next county, "the road by the boat at Hamilton is one mile six furlongs shorter than by Bothwell Bridge; but as the River Clyde is sometimes impassable at the Boat, the miles are numbered by the Bridge forward to Ayr."

Reconstruction began very locally on the Loudoun estates, where, about 1733, during developments that also heralded the Agrarian Revolution in Ayrshire, the Earl built a bridge over the Irvine, and laid a road from his Castle to Newmilns, which was the first made road in Ayrshire.[17] This appears to have antedated the main period of road improvements by nearly thirty years. Just after the Military Survey, the first act for turnpike roads in the county was passed.[18] Entitled "an Act for repairing and widening several roads in the county of Ayr," it recited a total of twenty-two roads, mainly in the north (Fig. 37) which "were much frequented by travellers, but . . . had become impassable in winter for wheel carriages and horses, and

[16] Taylor and Skinner, Survey and Maps of the Roads of North Britain or Scotland, London, 1776.

[17] *S.A.*, Vol. III, pp. 103 ff.

[18] 7 Geo. III (1766–7), c. 106.

several bridges" were "in ruinous condition." On the petition of numerous landowners, whose names were included in the preamble, it was enacted that the specified roads should be turnpiked, tolls levied,

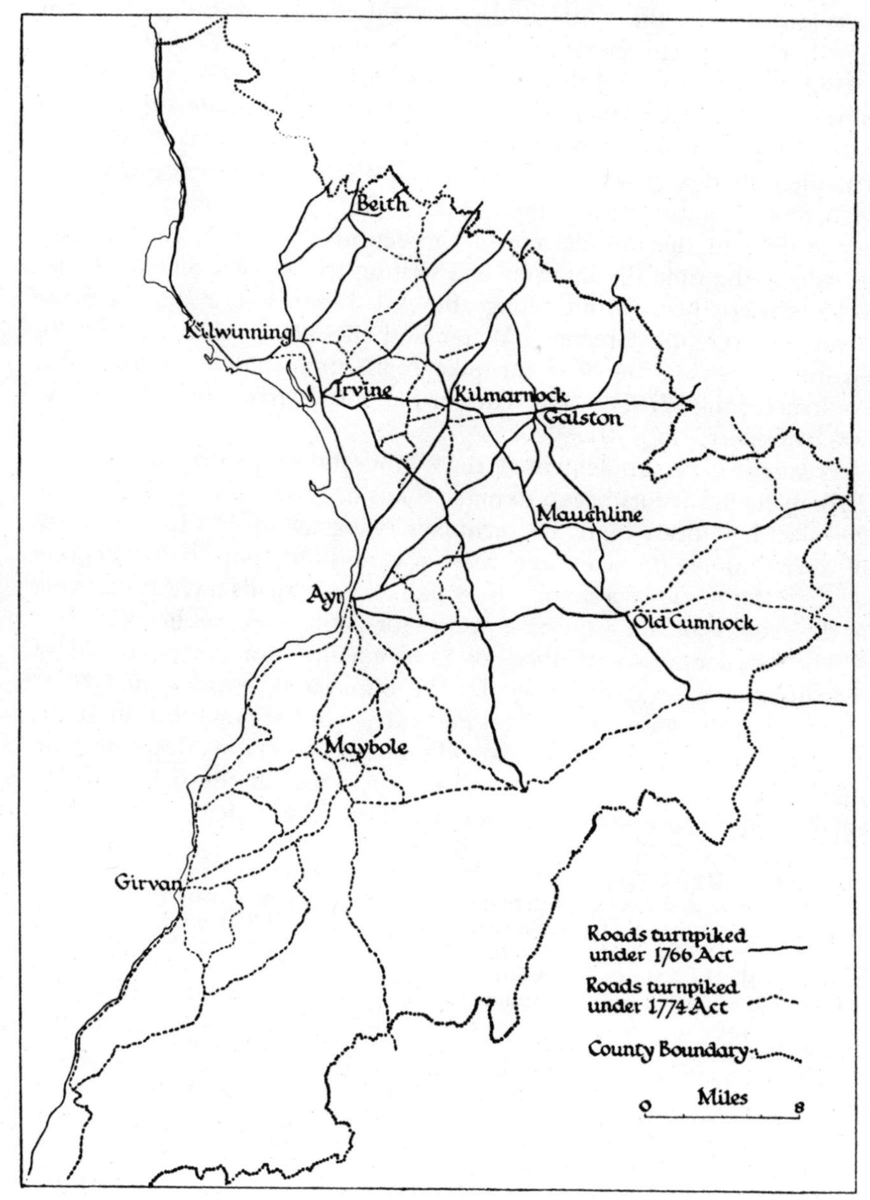

Fig. 37.—Roads which were improved in Ayrshire during the second half of the eighteenth century.

and the monies so obtained applied to improvement under the direction of a Board of Trustees. Statute labour was also to be commuted. A second act followed in 1774,[19] which extended the Board's authority to main roads in the south, and to several connecting by-roads in the north, chiefly between Irvine and Kilmarnock (see Fig. 37). This second act also enabled the Trustees to borrow capital on the security of the tolls, for experience since 1766 had shown that unless this power was granted, works proceeded but slowly. These acts practically complete the list of works undertaken by the Ayr Trustees during the eighteenth century,[20] for under both acts, improvements were still proceeding in the last decade of the century.[21] Thus, in Muirkirk parish, at the time the minister was writing to Sir John Sinclair, three bridges were being built along the Ayr–Edinburgh road, over the Water of Ayr, the Greenock Water and the Water of Garpel.[22] In Ardrossan parish, the first turnpike road, under the 1766 Act, was constructed in 1779.[10] The Girvan-Ballantrae road, via Colmonell, was completed in 1791.[23]

Private works supplemented those undertaken by the Board. In Kirkmichael, heritors began to make roads in 1769 ; and by 1791 some 20 miles had been built, no turnpikes being set up.[13] In the parish of Sorn "about 25 years ago there was nothing that could properly deserve the name of a road. Now half a dozen roads have been made at the expense of the respective proprietors, and three public roads."[24]

The contemporary method of road-making—an early version of macadamizing—was described by the minister of Symington. After stating that the public road from Portpatrick to Girvan and Edinburgh ran through the parish, he explained that "it was made of very durable materials. Land or whin stones were collected off the fields, beaten small, laid on to a great thickness and kept in excellent repair."[25]

[19] 14 Geo. III (1774), c. 109.

[20] Power was granted to a different board (including representatives of Glasgow Burgh, Renfrewshire and Lanarkshire), under 29 Geo. III (1789), c. 79, and 31 Geo. III (1791), c. 95, to turnpike the road from Glasgow to Sanquhar via Muirkirk. From just short of Muirkirk onwards for much of the way to Sanquhar, this road ran over the uplands within Ayrshire and near the county boundary. How far works had proceeded by 1800 is not known. (Later this stretch of the road was abandoned, though its course is shown on Ordnance Survey maps.)

[21] It was not until 1805 that the Ayr Trustees applied for further powers (45 Geo. III, c. 28), when 35 more roads were turnpiked. Later acts were 49 Geo. III (1809), c. 32, 58 Geo. III (1818), c. 3, 7–8 Geo. IV (1827), c. 109 (a consolidating and culminating act, which virtually completed the road network). 10–11 Vic. (1847), c. 213, is as much concerned with widening and similar improvements as with a few last turnpikes.

[22] *S.A.*, Vol. VII, pp. 598 ff.

[23] *S.A.*, Vol. II, pp. 57 ff.

[24] *S.A.*, Vol. XX, pp. 143 ff.

[25] *S.A.*, Vol. V, pp. 394 ff.

Ministerial approval of these measures, and lengthy descriptions of the improved state of communications, amount to a chorus in the *Statistical Account.* In Loudoun, there were formerly no carts in the parish ; by 1791 there were 250 ; [17] in Colmonell there were only two carts 30 years previously, but at the time of writing " every farmer " had " two or three and some even more." [23] Kirkoswald, in 1750, had no communication with Ayr. Some families gave business to a carrier plying fortnightly from Maybole to Edinburgh. But by 1791, the riding post from Ayr to Girvan passed daily, and a carrier weekly from Girvan to Glasgow. Post horses and chaises were available at any time.[11] There was a daily stage-coach from Kilmarnock to Glasgow.[2] From Irvine, there was much traffic along the four great turnpike roads to Kilmarnock, Ayr, Greenock and Glasgow. A fly ran to Glasgow three times weekly, and a stage-coach twice-weekly to Greenock.[26] Thus, although parish roads often remained unimproved,[27] and were the subject of adverse comment, by the end of the century there was general satisfaction at the progress achieved during the previous generation.

Apart from the roads, no means of internal communication existed. The rivers, even the Irvine, Ayr, Doon and Garnock, are too small, shallow and rapid, to permit navigation. Harnessed for several centuries to a succession of water mills, the weirs alone would have impeded the path of improvers. After 1760, proposals for canal construction were several times advertised, to connect Kilmarnock with Irvine and Saltcoats with Paisley ; [28] but works were never begun. Only in Stevenston, to carry coal from inland pits to the shore, was a short canal dug in 1772, 2¼ miles long, 12 ft. wide and 4 ft. deep, on which barges carrying 12–15 tons were towed.[1]

(iii) *Agriculture.* In Ayrshire, the Agrarian Revolution took nearly a century to run its course. Beginning about 1755, it had made very rapid progress by the end of the century.

To summarize this rural transformation in the briefest terms, it may be said that the principal changes were : (1) The enclosure of farmland that was previously wholly open. (2) The abolition of surviving co-tenancies and feudal servitudes, to be replaced by modern individual leases. (3) Consolidation and enlargement of many holdings. (4) The reduction or elimination of cottagers. (5) The general adoption of

[26] *S.A.*, Vol. VII, pp. 169 ff.

[27] E.g. in Beith (*S.A.*, Vol. VIII, pp. 314 ff.), Ardrossan (Vol. VII, pp. 42 ff.), Kilwinning (Vol. XI, pp. 142 ff.) and Ochiltree (Vol. V, pp. 446 ff.)

[28] See W. Aiton, " General View of the Agriculture of the County of Ayr " (*Board of Agriculture Report*), London, 1811, pp. 85–6. Cp. S. O. Addy, *Evolution of the English House*, London, 1898, c. 2.

liming. (6) Soil drainage. (7) Moorland reclamation. (8) Stock improvement, with the general spread of the Ayrshire and Galloway breeds. (9) Elimination of sheep from lowland farms except for wintering. (10) The extension of dairying at the expense of rearing. (11) Adoption of three-course farming in place of infield-outfield; with the introduction of improved implements and machinery. (12) Appearance of arable-mixed farming along the coast. (13) Rebuilding of steadings, farmhouses and cottages. (14) Increase and enlarging of country houses with expansion of policies and widespread ornamental planting.

Some of these processes were virtually completed within a few decades, either early or late in the great century of change; the remainder proceeded more steadily and continuously. Thus the first five aspects enumerated above were completed in much of the county by 1800. Soil drainage and moor reclamation (6 and 7) were chiefly accomplished after 1825, and the coastal belt of arable farming (12) did not become differentiated till nearly 1850. The remaining operations (8–11, 13–14) were in progress throughout the Revolution.

A just appraisal of the course of agrarian change down to 1800 is made easier by a clear conception of the state of the countryside in the middle of the eighteenth century. This the *Military Survey* permits; for in spite of its early date, its large scale (approximately two miles to the inch) and close detail delineate the country in a way not surpassed until the Ordnance Survey traversed the county a century later. Analysed in its entirety, it would demand many pages to describe its wealth of local detail; here only general statements applicable to much of the county will be formulated.

A copy (with certain modifications) of the *Military Survey* plan of Kilbirnie and Beith parishes, is reproduced here as Fig. 38. The latter lies below a culminating height of 690 ft., which is a little lower than the usual upper limit of settlement in Ayrshire, and actually only Lochlands Hill was devoid of the close-dotted farmhouses, otherwise dispersed over the whole area at an average distance apart of about one-third of a mile. The north-west of Kilbirnie, however, extends above the 750 ft. contour on to the "Renfrew" Uplands, and the inhabited area terminated abruptly along this critical contour, in a fashion clearly depicted on Fig. 38. The extent and limits of settlement were therefore coincidental in 1755 with those of 1860 or even today.

But this is less remarkable than the lay-out of farmland. Little of Kilbirnie was enclosed; and not much more of Beith. In the latter parish, small paddocks and gardens surrounded the town, enmeshing as in a web; but the countryside generally had only witnessed the sporadic beginnings of enclosure, by single farms or parts of farms. Otherwise the unenclosed lands were mainly covered by the criss-cross symbol for " arable " land graphically depicting the furrows and " rigs " into which the soil was thrown by the current methods of cultivation. It is improbable that the area thus ornamented was actually under cultivation in a single year. The land nearest the farmhouses (the " infield " or " croft "), upon which the meagre supply of dung was cast in order to maintain continuous cropping, is certainly

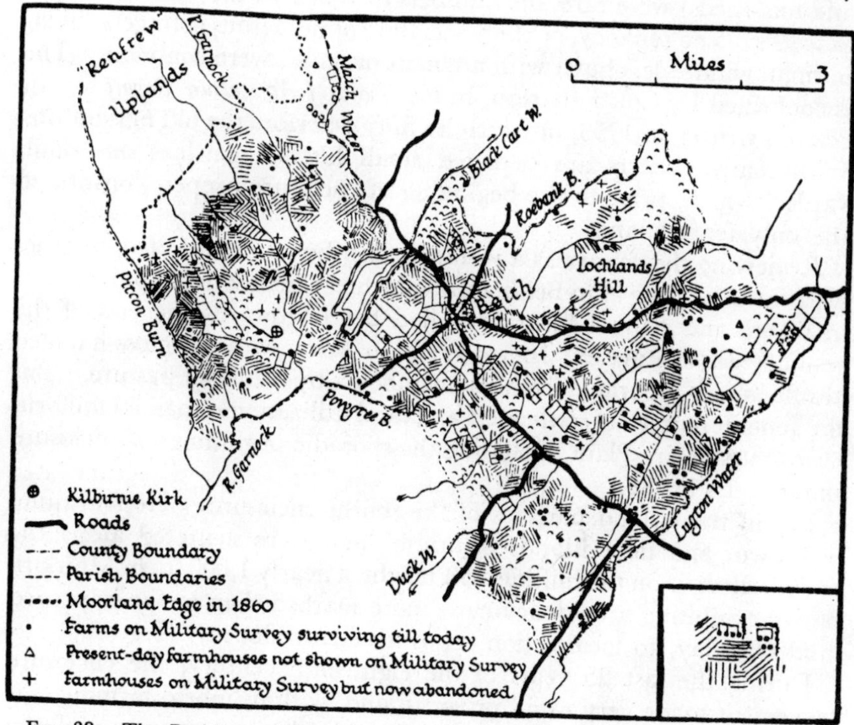

FIG. 38.—The Parishes of Kilbirnie and Beith, after the *Military Survey of Scotland.* The cross-hatching employed here, as on the original, represents " arable " land. In the inset, on a large scale, a representative cluster of steadings is shown without suppression of detail.

included ; and probably also much " outfield " land, which, subject to temporary tillage without manuring, might be deemed arable whether bearing a crop or merely the marks of a recent cultivation. Only small areas, farthest from steadings on farms more widely spaced than normal, were in " permanent " pasture—to borrow a modern term.

In the redrawing and reduction it was impossible to preserve all the separate buildings of the clusters which are shown at most steadings on the original; only an enlarged single example has been given in the inset. These associated houses often number as many as six or eight, although three to five is more usual. Since both farmers and cottagers lived (with their cattle in winter), in these one-storeyed hovels corresponding to the "black houses" still surviving in the Hebrides, with walls of stones gathered from fields, turf, or mud plastered on stakes,[28] with roofs of thatch supported on couples or crutches, and outhouses in the modern sense (including byres, barns and tool-sheds) were rare, the numbers included in the clusters suggest that in 1755 co-tenancy, or the association of numerous cottagers (living in small windowless huts) with a tenant or feuar, were common. This is confirmed by Col. Fullarton, in his *Board of Agriculture Report* for the county, written in 1793, in which he fully describes the old agriculture. With many clusters are depicted small kitchen gardens or "kail-yards" which, prior to the beginning of enclosure proper, constituted the only fenced land.

Reviewing the county as a whole, it may be said that in few parishes were farms so close as in Beith and Kilbirnie, where some improvements in tillage and animal husbandry had begun towards the end of the seventeenth century.[29] In many, each farm with its cross-hatched arable land is separated from its neighbours by a zone of pasture. But the general pattern of settlement and land utilization remained uniform over the Ayrshire Plain, including the sporadic beginnings of enclosure shown on Fig. 36.

But in the upland parishes of the south, enclosures were still quite unknown, and little islands of arable land were scattered along the main valleys or on the hillsides to a height of nearly 1,000 ft., in Muirkirk parish exhibiting a tendency even more marked than in the improved land of today, to localization on south-facing slopes. (Fig. 39).

During the last 25 years of the eighteenth century, the enclosure movement made very rapid progress, and, with it, infield farming was practically abolished. A Mr. Fairlie, who became manager of the Eglinton estates about 1770, and who had seen improved farming in the Lothians,[29] introduced a new type of lease which stipulated that not more than a third, or in some instances a quarter, of the land was to be ploughed in any one year, and none was to be cultivated more than three years in succession. Combined with enclosure, this measure effectively destroyed the infield-outfield system, and ensured that all land was cultivated in rotation. The author has been shown MS. "crop books" in which estate factors kept a record of each tenant's annual use of his fields, so that the terms of lease could be enforced.

[29] J. H. G. Lebon, *The Land of Britain*, Part I, "Ayrshire," London, 1937, c. 2.

Artificial grasslands (mainly rye-grass with clover) had been introduced about 1735 on the Loudoun estates [17] and in the parish of Stair,[30] but did not make much headway for a generation. Then, stimulated by the heavy liming which was integral to the new farming, it greatly enhanced the stock-carrying capacity of the land, this in turn permitting the slow improvement of breeds of cattle.

The parish *Statistical Accounts* written between 1791 and 1795 reveal that much of the Ayrshire Plain was enclosed and had adopted three-course farming (another term for the Fairlie rotation) between 1775

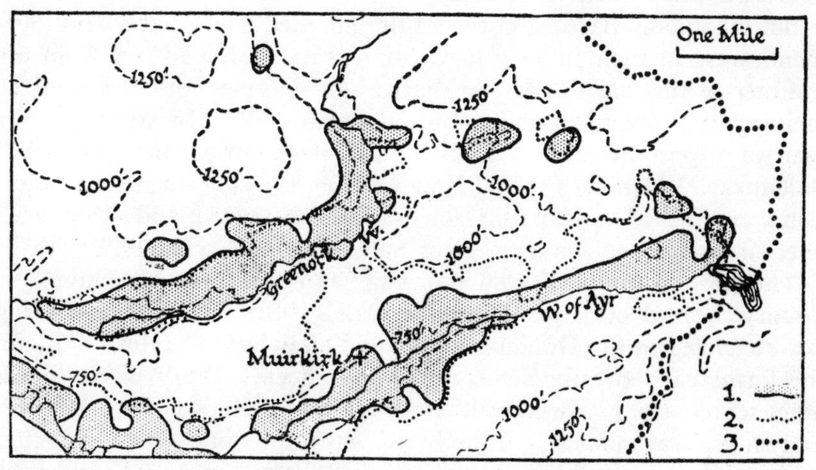

Fig. 39.—The extent of arable land in part of the Parish of Muirkirk, in the middle of the eighteenth century and about 1930.

1—Arable Land shown on the *Military Survey of Scotland*; 2—Moorland Edge or Head Dike, from the Popular Edition, Ordnance Survey of Scotland, One-inch map; 3—County Boundary.

and 1790. Thus, in Loudoun, the land was " nearly all enclosed " ; [17] in Dreghorn enclosure had been continuous since about 1760 ; [31] Dalrymple,[32] Riccarton,[33] Stevenston,[34] St. Quivox,[35] Dundonald,[36] Kilmaurs,[14] Dunlop,[37] Kilwinning,[38] and Symington [39] had all been largely enclosed since about 1770 ; in Kirkmichael only two or three farms remained to be enclosed.[13] In Ardrossan enclosures were making progress ; [10] in Fenwick the upper part of the parish was still open ; [5] in Dailly the arable land (near the Water of Girvan) was fully enclosed, but pastoral farms only partly so ; [40] in West Kilbride there was still

[30] *S.A.*, Vol. VI, pp. 112 ff.
[31] *S.A.*, Vol. IV, pp. 280 ff.
[32] *S.A.*, Vol. IV, pp. 305 ff.
[33] *S.A.*, Vol. VI, pp. 113 ff.
[34] *S.A.*, Vol. VII, pp. 1 ff.
[35] *S.A.*, Vol. VII, pp. 353 ff.
[36] *S.A.*, Vol. VII, pp. 615 ff.
[37] *S.A.*, Vol. IX, pp. 537 ff.
[38] *S.A.*, Vol. XI, pp. 142 ff.
[39] *S.A.*, Vol. V, pp. 394 ff.
[40] *S.A.*, Vol. X, pp. 34 ff.

a rural conservatism to be overcome before the benefits of the new agriculture could be enjoyed.[41]

In a few areas, development during the previous generation had taken another course ; thus in Dalry, grazing farms had been enlarged in the previous 30 years ; [42] and the same was true of the next parish of Ardrossan.[10] In Old Cumnock farms on the Dumfries estates were being let annually in grass,[43] and it seems that this was a recent development, because all these parishes reveal the normal amount of arable land (for 1755) on the *Military Survey*. This unorthodoxy was to be banished soon after 1800.

But the upland parishes of the south remained backward. Improvements were in their infancy in Colmonell in 1791, and it is from the minister of this parish that we derive our lengthiest discussion of the social and economic implications of enclosure.[23] In Muirkirk the new ways were the subject of talk but had not been put into practice.[22] Ballantrae,[12] Ochiltree [44] and New Cumnock [45] continued their traditional rearing of black cattle (for the English towns), and sheep, with only limited infield cultivation of oats, bere and potatoes.

The new farming allowed dairying, and the " Cunningham " or " dairy " breed of cattle, to extend widely. In 1760, cheese-making was restricted to the Dunlop district, and elsewhere, even in the Plain, the farms carried only black cattle and sheep. Dairy produce was often scarce, e.g. in Kirkoswald.[11] By 1790, in Craigie " much butter and cheese was made," and the breed was improving ; [7] in Symington " butter and cheese " were largely made " for sale in Ayr, Kilmarnock and Glasgow." [39] In Stevenston,[34] Dreghorn,[31] Dundonald,[36] Kilmaurs,[14] Kirkoswald [11] and Dalry [42] dairying was advancing, and reducing the numbers of black cattle and sheep.

In the Ayrshire Plain, then, by 1800, about a fifth or a quarter of the land was in oats ; perhaps a tenth in peas, beans, and potatoes. The rest was artificial grassland, part for mowing, the rest grazed. Turnips and wheat were being tested near the coast and along the Irvine River. A little flax was commonly grown, for home spinning and weaving of this fibre had been encouraged by the *Board* established soon after the Act of Union.

The rebuilding of farmhouses was but local before 1800. In Kilmaurs, the advance was exceptional ; [14] but in the next parish, Kilmar-

[41] *S.A.*, Vol. XII, pp. 404 ff.
[42] *S.A.*, Vol. XII, pp. 90 ff.
[43] *S.A.*, Vol. VI, pp. 407 ff.
[44] *S.A.*, Vol. V, pp. 446 ff.
[45] *S.A.*, Vol. VI, pp. 398 ff. In Ochiltree parish, in addition to homespun weaving, snuff-boxes, toddy-ladles and reaping-hooks were made in the farmhouses, and were widely distributed in Scotland. See A. Murdoch, *Ochiltree*, Paisley, 1921, p. 172.

nock, the minister deplored the poor quarters of the country folk.[2] In Sorn "in the last 10 or 12 years most farmhouses have been rebuilt with considerable improvements"; [24] and in Dailly [40] "low huts had been replaced by neat and commodious houses for farmers, and cottages were more cleanly and comfortable." But many authors of the parish accounts are silent about housing, although voluble on the subject of improvements in agriculture.

(iv) *Industry*. Only domestic and craft manufactures existed in Ayrshire in 1750. These were diffused equally in the farmhouses, the early small "kirktouns" (villages clustered around the parish church) and the few towns. Kilmarnock was the only town primarily engaged in manufacturing, owing its eminence to the patronage of the Boyd family (who obtained the charter erecting it as a Burgh of Barony in 1591), and to its importance as the bridging point on the Marnock Water where the Nithsdale road met the Ayr–Glasgow road. At Stewarton, a corporation of bonnet-makers, selling their products in Glasgow, had flourished since the mid-seventeenth century. At Beith, where but a few houses had stood in the shelter of the kirk at the Revolution, linen manufacture had brought the village precociously forward after 1740, this branch of production attaining its maximum about 1760.[9] The ports—Ayr, Saltcoats, Irvine and Girvan—were chiefly engaged in commerce, above all the coal export, in fishing, a little ship-building, and in Girvan at least, smuggling.

Much more widespread was domestic rural craftsmanship, flourishing in many districts till after 1800. The minister of Maybole could report in 1793 that there were 80 looms for wool in town and country employing 300 persons; [46] in the neighbouring parishes of Dailly [40] and Kirkoswald [11] coarse woollen cloth, blanketing and plaiding were manufactured in farmhouses and cottages from raw material brought from Argyllshire and Galloway. Lingering in Dreghorn in 1791, in a district already deeply involved in the new economic currents, "were a few weavers . . . employed in weaving such kinds of cloth as are used by country people." [31] Linen cloth was still made in the farmhouses of Monkton parish; [47] and was just introduced and making headway in New Cumnock.[45]

Accordingly, with cloth-making largely distributed in the farmhouses, and but a few craftsmen (smiths, tailors, wrights, candlemakers, shoemakers, carpenters and the like) in the villages or even scattered in the country, we may readily visualize that "industry" was essentially small-scale and widely diffused. This scattering was fostered by the state of communications. With the impediments to circulation which

[46] *S.A.*, Vol. III, pp. 219 ff.   [47] *S.A.*, Vol. XII, pp. 394 ff.

existed in 1750, specialist craftsmen could not become concentrated, or they would lose touch with their customers. Tailors travelled from one farm to the next, making clothes for each family from a bundle of homespun.[45]

From about 1775, Glasgow and Paisley rapidly developed as the centres of new textile manufactures ; especially of cotton and silk, employing new spinning machinery. Around these central stars in a new industrial system were clustered satellites large and small, including towns such as Beith or Girvan, also villages scattered afar in the Carrick dales. The new industry was almost from the first dichotomous in its organization ; the carding and spinning branches, under the influence of Arkwright's and other inventions, were already being developed in factories ; but weaving remained a craft, the unit of production being the few looms in a master-weaver's home. Some smaller workshops containing spinning machinery were operated in this pioneer period by methods that were soon abandoned, e.g. in Dundonald, thirty persons were employed in 1791 by a carding machine turned by a horse ; but in general power was obtained from running water, and the new " jeanies " were set up on the banks of fast-flowing streams.

But both mills and weavers operated only where roads allowed carriers to bring raw fibre or spun thread and take away the finished product. The localization of the new textile manufactures were thus greatly influenced by the new main roads (Figs. 37 and 40). Kirktouns grew rapidly as roads were improved ; and at cross-roads or bifurcations, villages often quickly appeared in previously open country. The new village of Whitletts, near Ayr, is a good instance. Here, two main roads—Ayr–Muirkirk and Ayr–Galston—bifurcate. No houses existed at this point on the *Military Survey* ; but by 1800 300 persons lived here, supported by weaving or coal-mining. Although contributors to the *Statistical Account* were usually content to describe the rise of weaving, with perhaps spinning, and append occupational statistics, at least one perspicaciously observed that the new public road from Edinburgh to Ayr through his parish (Mauchline)[48] had encouraged the establishment thereon of two new factories—the iron works at Muirkirk and the cotton works at Catrine.

Accordingly, from about 1785, silk and cotton weaving, with tambouring, spread rapidly. In Straiton,[49] 12 weavers with their journeymen and apprentices, engaged till that time in making woollen webs for the Ayr and Maybole markets, had by 1791 changed to muslin for the Paisley merchants. The recent appearance of silk-weaving in

[48] *S.A.*, Vol. II, pp. 109 ff.

[49] *S.A.*, Vol. III, pp. 586 ff.

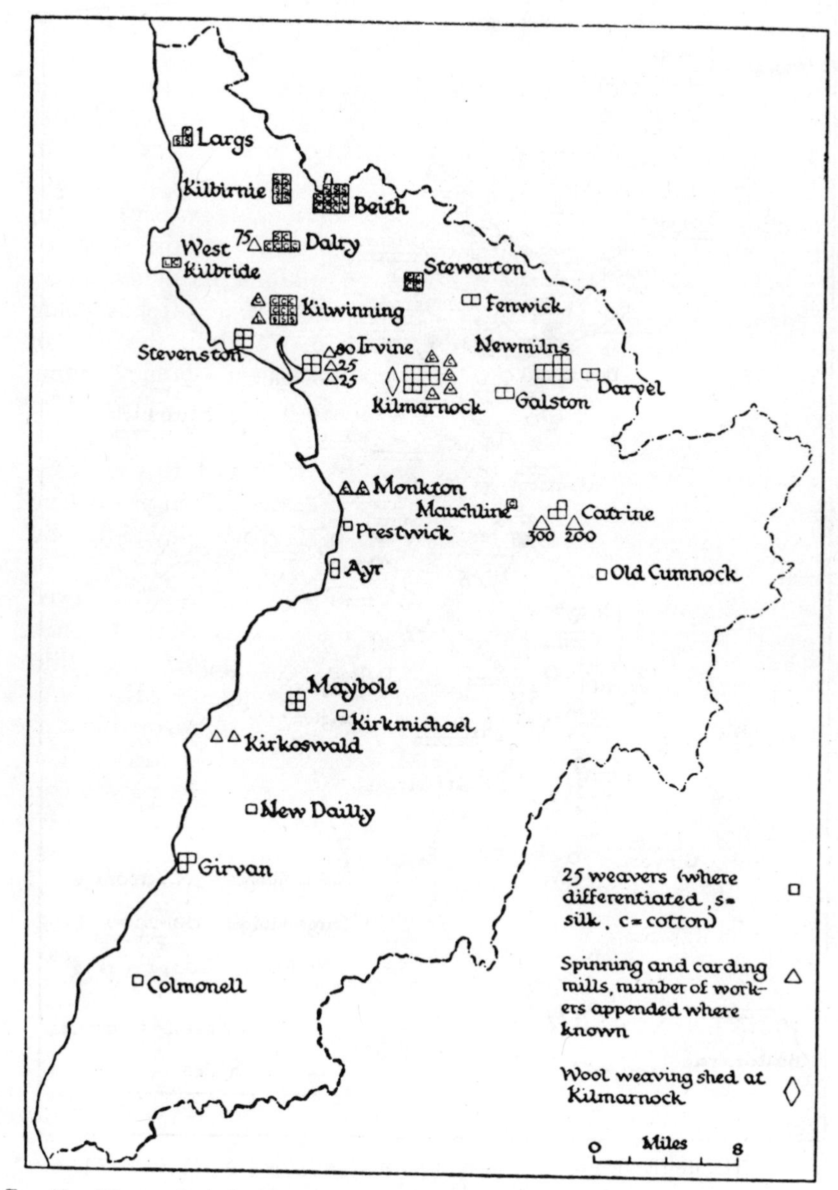

FIG. 40.—The textile industries of Ayrshire in 1790–5 (based upon the returns given in the *Statistical Account of Scotland*, ed. Sir John Sinclair).

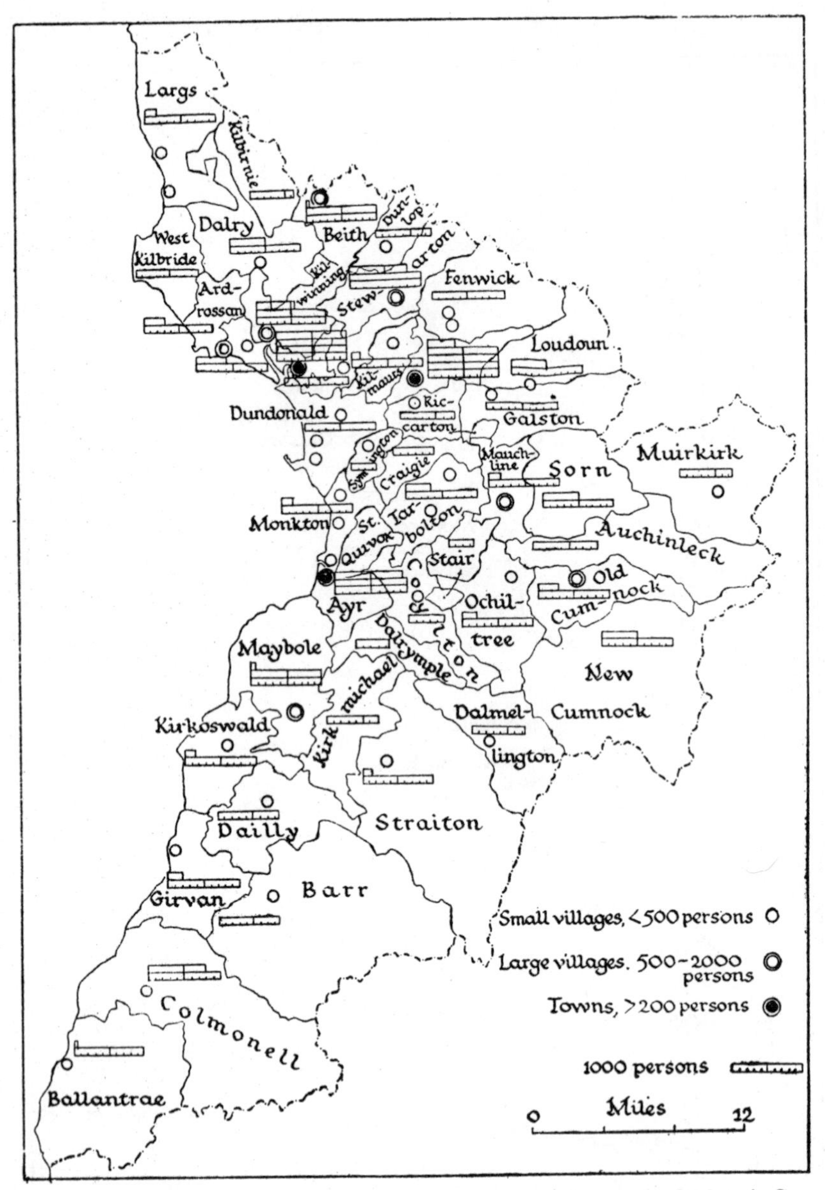

Fig. 41.—The distribution of population in Ayrshire, according to Dr. Webster's Census of 1755.

The classification of villages and towns was derived from the *Military Survey of Scotland*; it should be borne in mind that some of the "villages" were legally burghs, and the caption to the second ornament in the legend might accordingly read "Large Villages and Small Burghs." The figure in the caption to the third ornament should be 2,000.

Fig. 42.—Changes in the Distribution of Population in Ayrshire between 1755 and 1801.

The caption to the second ornament should read " Large Villages or Small Burghs " ; to the third, " Burghs and Towns."

Stevenston,[34] Irvine,[26] Beith,[9] Dalry [42] and Monkton [47] (1787–90) was fresh in the minds of those describing parish life for the *Statistical Account* (Fig. 40). In Galston, as a satellite of Kilmarnock, the older shoe-making industry declined as the weaving of lawn and gauze was established in 1789–91.[50] The full extent of the development of new manufactures, the appearance of new villages, and the expansion of old, is depicted in Figs. 40, 41 and 42.

Two instances are of especial interest. It has been previously shown that Kilbirnie parish was abnormally inaccessible, owing to the sequence of unbridged rapid streams draining from the "Renfrew" Heights to the River Garnock. The memory of its remoteness, for "it was seldom visited by strangers, and was rather noted for the primitive roughness in speech and action of its inhabitants," lingered till the nineteenth century.[51] But (Fig. 37) a main road was laid under the 1766 Act, and the old water mill by the Garnock bridge, shown on the *Military Survey*, became the nucleus of an industrial community rapidly growing by 1800 (Fig. 43). In 1791, silk weaving was active; shortly afterwards, a cotton mill was established, followed by flax mills.

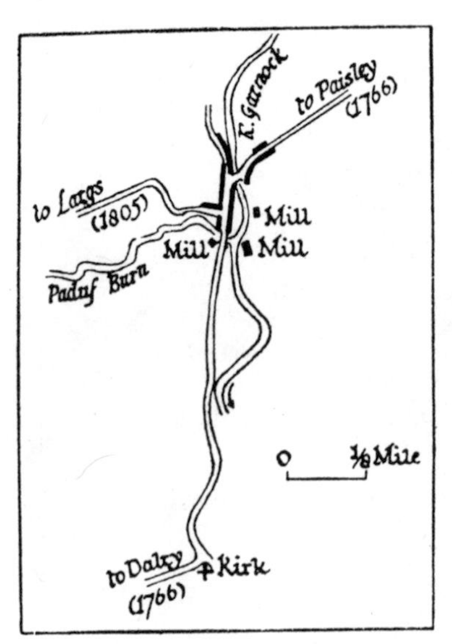

Fig. 43.—The village of Kilbirnie, early in the nineteenth century.

Under the 1805 Act, a branch turnpike road was constructed from Kilbirnie to the Largs–Dalry road, completing the range of geographical advantages encouraging growth during the first phase of the Industrial Revolution.

At Catrine, a busy industrial village was planned and built in the years after 1787. On a sheltered holm, a five-storeyed twist mill was erected, having 5,240 spindles driven by large waterwheels and employing 300 persons. Around, a neat village was built (Fig. 44), containing a carding mill, "jeanies" and workers' homes, which, in advance of the still prevalent traditional huts, were stone-built, two-storeyed and slated. The laudatory description in the *Statistical Account* [24] states

[50] *S.A.*, Vol. II, pp. 71 ff.
[51] J. Dobie, *Cuninghame Topographized*, Glasgow, 1876, pp. 372–4.

that after the first five years, the population had reached 1,350. The promoters, Alexander of Ballochmyle and the Dale of Glasgow, tried to ensure the maintenance of decent social standards by building a church and schoolhouse, by providing gardens and pasture for cows, and by police measures, which included gates locked each evening at either end of the main street.

The advance of Kilmarnock between 1760 and 1800 was remarkable. In 1740, serge-making was the chief trade ; but shoe-making was then introduced and prospered. By 1791 there were 56 master shoe-makers employing 408 men. In collaboration, tanning, sheep-skin preparation, glove-making and saddlery became prominent. A woollen company formed in 1766 had greatly fostered the manufacture of carpets, woollen cloth and woollen goods such as duffles, bonnets and caps. Cotton spinning and weaving rapidly increased after 1780, and the new machinery was already being applied to wool before the end of the century. Dyeing, candle-making, tobacco, bar and cast-iron manufactures completed the census of production in 1791.[2]

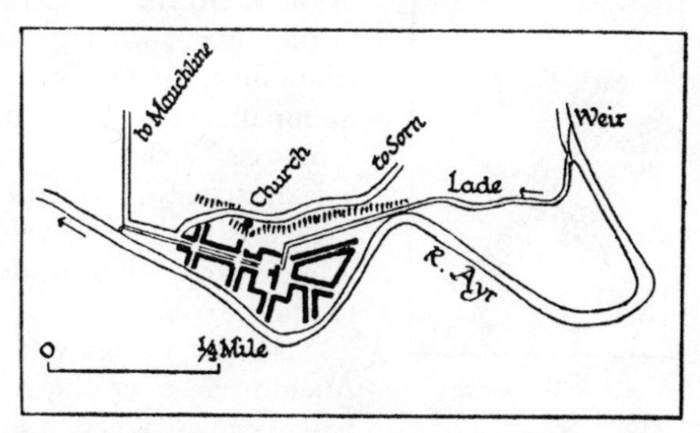

Fig. 44.—The village of Catrine about 1795, based upon the plan in the *Statistical Account of Scotland.*

Other industries which progressed, or were introduced before 1800, included shipbuilding, especially at Saltcoats and Irvine, with its ancillary, rope-making. Iron-works, started at Muirkirk in 1787, and still experimental in 1791, were already being duplicated at Glenbuck, and the rough pottery made at Stevenston, Old Cumnock, Sorn and Coylton, was by 1810 causing wooden cups and platters to disappear.

(v) *Population Changes.* The growth of old villages and the appearance of new, was not entirely due to the progress of the textile industries. Many kirktouns increasingly became the home of colliers, for till 1800 few exclusively mining villages had been founded. But apart from this, the re-fashioning of the countryside, which, in addition to the work of enclosure, was coupled with extensive rebuilding of farmhouses and cottages, caused an increase of masons, carpenters, joiners, wrights, blacksmiths and day-labourers. The growing road traffic demanded ostlers, carriers and carters. The rising standard of life gave more employment to tailors, shoemakers and seamstresses. Many of these tradespeople were doubtless recruited from the children of displaced cottagers as farms were consolidated and enclosed. There must have been a steady drift of families from the countryside to the new or enlarging villages, and thus again the influence of the new roads, in re-grouping the population, became manifest. For Kilwinning, it was stated specifically that from the one or two cottages formerly attached to each farm and now demolished, the inhabitants had moved to Irvine or neighbouring towns to engage in manufactures.[38]

The minister of Sorn [24] described how a village had come into existence at Dalgain (now called Sorn) within the previous 16 years, along the Ayr–Muirkirk road and near the junction of the Galston road. The proprietor of the holm had begun to feu about 1775, and at the time of writing 34 houses had been built containing 191 persons. Giving then an occupational census—3 shopkeepers, 3 innkeepers, 3 masons, 7 shoemakers, 5 weavers, 5 tailors, 4 seamstresses, 7 colliers—he commented "the village is therefore evidently the residence of a large proportion of the tradesmen belonging to the parish."

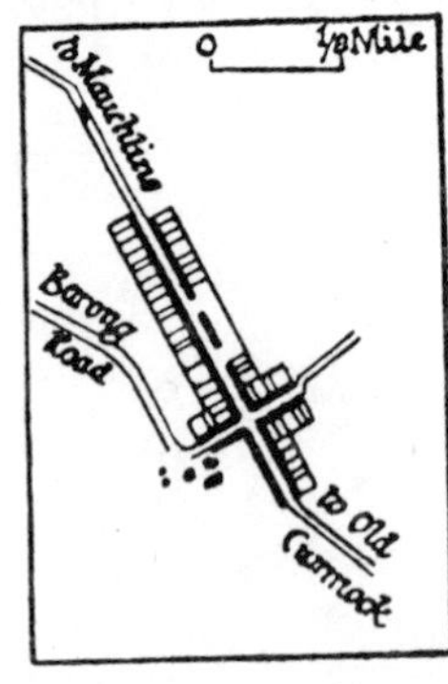

Fig. 45.—The village of Auchinleck, early in the nineteenth century, based upon the Ordnance Survey, Six-Inch Plans (1st ed.), and Thomson's *Atlas of Scotland* (1832).

(Barong Road should read Barony Road.)

A similar village was planned and built at Auchinleck, after the Ayr and Lugar Waters had been bridged to carry the new road between Mauchline and Ayr under the 1766 Act. Till this date, Auchinleck must have been practically isolated when rivers were in flood. On the *Military Survey*, only the kirk, manse and two cottages were shown. But in the 'seventies, the neat double-row of cottages began to extend along the new straight road, each with its vegetable plot in the rear.[52] Here, the rectilinear impress of the surveyor—who dominated road-making and enclosing alike—is especially manifest and serves to distinguish these new villages from the older (Fig. 45).

But it is dangerous to over-simplify an account of population movements between 1760 and 1800. Rural depopulation had begun in the Southern Upland parishes before the Enclosure Movement, and may be connected with the development of cattle-grazing and the decline of cultivation whilst farmland still remained unfenced. This appears to be the cause of much of the decrease indicated on Fig. 42

[52] *S.A.*, Vol. XI, pp. 430 ff.

in the row of southern parishes. It seems also to have contributed largely to the decline between 1755 and 1791 in Kilwinning,[38] Largs [53] and West Kilbride.[41]

The first effects of enclosure, as already stated, were to reduce cottagers and hence the rural population. In Barr [54] and Colmonell,[23] where enclosing had just begun in 1791, this consequence is fully explained ; and hence the decreases shown on Fig. 42 have a double cause—extension of grazing and beginnings of enclosure. But later, as the full benefit of agrarian improvements were being reaped in a higher level of production, the land could support more workers. Cottagers did not return, except in a limited way as married labourers in tied cottages ; but the rise of dairying required more farm-servants—cowmen and dairymaids—who usually lodged at the farmhouse. Thus, in a parish such as Stair, which was among the first to benefit by improved husbandry, and which remained villageless, the population increased by more than a third between 1755 and 1791, and gained a further 20 per cent by the 1801 census. In Symington, too, the minister is positive that enclosures had increased the rural population.[39]

Over the Ayrshire Plain generally, population was increasing at the end of the century. The full effects of enclosure and agrarian improvements, the increase of tradespeople, of the new textile-industries and of coal-working were all reflected in the 10–25 per cent increase recorded in most parishes between Dr. Webster's Census of 1755 and that included in the *Statistical Account* (1791–3). In Beith, Sorn, Loudoun, Maybole and Newtown-upon-Ayr, the increase of manufacturing was especially rapid, and concomitantly the rise in population. But the greatest growth was at Kilmarnock (which was to exceed 10,000 early in the next century) and at Saltcoats (where weaving, coal-mining and coal-exporting were all flourishing after 1785).

[53] *S.A.*, Vol. II, pp. 360, Vol. XVII, pp. 503 ff.
[54] *S.A.*, Vol. XII, pp. 81.

XI

THE DELIMITATION OF MORPHOLOGICAL REGIONS

By D. L. LINTON, M.Sc.

I

Morphological subdivision has been attempted at all sorts of scales. As a class exercise it is quite customary in this country to present students with a sheet of some standard topographic map—our own one-inch series or the U.S.G.S. 1/62,500—and ask them to analyze the landforms of the area shown and suggest a division into morphological units. Equally, and on the continental scale, we recommend them to base their understanding of the regional geography of large areas upon a sound grasp of the pattern and characteristics of the major landform units. Between these extremes of the local and the continental there is, moreover, a sufficiency of examples illustrating our geographical textbooks, monographs and journals, of the morphological subdivision of areas of intermediate size to suggest that the technique is currently accepted as being valid at all scales. Frequently the areas delimited are termed " natural regions " though both their boundaries and the names given to them make it abundantly clear that they are, in fact, morphologically determined.[1]

In fact, it can even be urged that a predilection for morphological subdivision may be considered as a characteristic of British geographical method. If so, it may possibly derive from the close association of the subject in Britain with pedagogy and educational practice and the realization of the high importance in teaching method of a presentation which appeals clearly to the visual memory. Be that as it may, this predilection is one which we seem to share with our American colleagues—if indeed we have not derived it from them—and in which we differ from the French. There must be many in this country who have picked up de Martonne's little volume *Les Régions Géographiques de la France* expecting to find in it a map depicting the areal extent and mutual relations of the fourteen regions so lucidly described in its pages. Their expectations were doomed to disappointment. The genius of French geographers is for the apt and vivid characterization of the essence of a region in words, supported, if need be, by

[1] As for example, W. Fitzgerald, *Africa*, 1934, p. 149, Fig. 18 ; L. D. Stamp, *Asia*, p. 249, Fig. 115, or p. 449, Fig. 252 ; M. Shackleton, *Europe*, p. 46, Fig. 13 ; p. 217, Fig. 54.

illustrative maps and diagrams of carefully selected sample areas. Such characterization is no doubt held to be definition enough, and since the basis of characterization may change from region to region there is logic in avoiding the difficulties of drawing boundaries according to criteria that change as the boundary itself is crossed. Even when, after the lapse of twenty years, de Martonne returns to his subject in the massive " France Physique " of the *Géographie Universelle*, he still shows no tendency to lend definitive regional outlines to his now much more extended morphological descriptions, and the highly ambitious map by which those descriptions are illustrated shows by an elaborate series of cartographic conventions almost every important morphological feature in France, but leaves the question of morphological subdivision untouched. Nor are these two works by de Martonne untypical of those of his countrymen. In all the fifteen tomes of the *Géographie Universelle* there is not a single map in which the morphological regions of any territory are given definition. Privat-Deschanel has a map of the major natural regions of Australia, defined largely on a structural basis, and a few maps of structural units are used by other authors. Only Baulig and de Martonne offer any specifically morphological, as opposed to structural, maps and these are of morphological *features* not regions. It is clear that whatever appeal the delimitation of the latter has for Anglo-Saxons, it has none for the Gallic temperament.

II

The best known, and most successful, attempt at morphological subdivision is that proposed by Fenneman for the United States, and it may be well to recall here what it attempted and what it achieved. Fenneman's was not the first attempt of the kind and many of the actual subdivisions he recognized were adopted ready made from those of earlier workers, notably Powell, 1896, Davis, 1899, Gannett, 1902, Bowman, 1911, Blackwelder, 1912, and Tarr and von Engeln, 1913.[2] The common element in these earlier regional subdivisions is large. All were in fact agreed that the main elements to be given recognition were broadly the Coastal Plain, the Appalachians, the Interior Lowlands, the Ozarks, the Rocky Mountains, the Intermontane region and the Pacific mountains and valleys. There was, moreover, a considerable measure of agreement as to the subdivisions of at least the Appalachian and Intermontane regions, though divergence of

[2] For a useful summary statement of these early attempts illustrated by maps at 1/50,000,000, see W. L. G. Joerg, " Natural Regions of North America," *Ann. Assoc. Amer. Geog.*, Vol. IV, 1914, pp. 55–83.

opinion is apparent as to the status of these subdivisions, i.e. whether they are first or second order regions. Similar doubt existed whether the New England region was entitled to independent status and as to the separation of the High Plains, which Powell preferred to call plateaux, from the rest of the Interior Lowlands.

It is in the Ozark and Great Lakes regions, however, that least agreement is seen. In the former case the divergence is chiefly a matter of where boundaries should be drawn; in the latter it is more fundamental. Powell's *Lake Region* is accepted by Gannett alone among the later writers, its uniformity giving way in Davis's hands to a variety based on structure. Davis recognized in the Adirondacks and the uplands south of Lake Superior outlying portions of the Laurentian region of Canada, and in the low upland of Lower Michigan a structural analogue of the Appalachian plateaux. Here is a disagreement arising from a basic divergence of view as to the criteria upon which recognition of the existence of a separate morphological region should depend. Thus, when Fenneman took up the matter at the Princeton meeting of the Association of American Geographers in December 1913, though much agreement existed there were also divergences of three kinds to be resolved, regarding the very existence of some regions, the status of others and the boundaries of all.

Fenneman's objective and the basis of his method can best be revealed by a quotation from his first presentation of the subject [3]:

> In a broad way the division of the United States into provinces serves two purposes; first, for the discussion and explanation of the features of the country; second, as a basis for the plotting and discussion of social, industrial, historical and other data of distinctly human concern. This second purpose is distinctively geographic. The first has its geographic side also, but is quite as much geologic as geographic. This paper is written in the conviction that these two purposes are consistent and that in serving the first, the physical criteria should be so handled as to serve the distinctly geographic purpose. The ever-present considerations common to both purposes are topography and soils. They are, on the one hand, the end product and record of physiographic history, and, on the other, the beginning of geographic development. These therefore are the most fundamental criteria in determining physiographic provinces.

Earlier attempts at subdivision had mostly been concerned with the first of the two purposes named by Fenneman and boundaries sketched on small-scale maps had been adequate,[4] but any attempt to use a

[3] N. M. Fenneman, "Physiographic Boundaries within the United States," *Ann. Assoc. Amer. Geog.*, Vol. IV, 1914, pp. 84–134, esp. p. 85.

[4] Powell, 1/14,000,000; Davis, 1/30,000,000; Bowman, 1/15,000,000; Blackwelder, 1/40,000,000; Tarr and von Engeln, 1/27,000,000.

map of physiographic divisions as a basis for the plotting and discussion of the facts of social geography requires precision of quite a different order. True, Gannett's map had been made for this purpose and reproduced at 1/27,000,000,[5] the boundaries being drawn, not along physical features, but along the county boundaries most nearly approximating to them, so that census statistics could be directly applied to the natural divisions. In this way a difficulty was avoided that later gave Fenneman much concern, namely, the ascription of small areas by local workers to one or other side of a physical boundary that has never been precisely defined. Hence his insistence that boundaries should be based " as far as possible on important contrasts visible in the field " or, where such contrasts fail, upon the most suitable arbitrary lines. His paper devotes 43 of its 50 pages to the precise definition of boundaries. He claims that " unity or similarity of physiographic history " is a formula that almost expresses the basis of his delimitations, but nearly as important is the consideration that each area recognized as a unit can be so treated in description. " Making due allowance for generalization," he says, " it is still true that the ideal limits of a province are such as admit of *the largest possible number of general statements* before details and exceptions are taken up."

Following further discussions at their Chicago meeting in December, 1914, the Association of American Geographers formed a committee of five, with Fenneman as chairman and including two members from the United States Geological Survey, " to devise a systematic division of the United States." The report of this committee took the form of the now well-known map at the scale 1/7,000,000, which first appeared in 1916, accompanied by a revised description by Fenneman of the provincial boundaries.[6] Map and description were reprinted in 1921, and in 1928 a third edition was prepared after revision of the map and text and consultation with the physiographic committee of the United States Geological Survey. The changes were not large or vital, and the map and its accompanying nomenclature of three orders of morphological regions—*major divisions*, *provinces* and *sections* —has long since become a standard document not only of American geography, but of agriculturists, pedologists and the like, and all Departments and Bureaux concerned with regional variation of surface distributions.

[5] W. F. Willcox, " A Discussion of Area and Population," *12th Census of the United States, Bull. No. 149*, 1902, map p. 10.

[6] N. M. Fenneman, " Physiographic Divisions of the United States," *Ann. Amer. Assoc. Geog.*, Vol. VI, pp. 19–98, Plate I.

III

Naturally the American units are all large. The " major divisions " are of sub-continental dimensions, and the largest provinces are not much smaller (Central Lowland 585,000 sq. miles ; High Plains 532,000 sq. miles). Even among the sections the largest exceed 100,000 sq. miles. These figures offer little hope that the American methods will have much relevance to British problems. Even the smallest units are all large by British standards. Only three of the " provinces " are smaller than England (Ozark Plateaux, 40,000 sq. miles ; Ouachita Mountains, 15,000 sq. miles ; Adirondack Mountains, 10,000 sq. miles). These may be compared with Scotland, 30,000, and Wales, 8,000 sq. miles. Some half a dozen " sections " are still smaller, ranging in area from 2,000 to 6,000 sq. miles.

It is inevitable that any subdivision of the whole United States should be coarse-grained. Not only is the terrain to be divided very large (3,026,000 sq. miles), but the method employed is essentially that of seeking the morphological Highest Common Factor of as wide an area as possible. In the major divisions that H.C.F. is limited to a few basic characteristics, usually those that reflect the general rock structure or structural history. For the provinces the H.C.F. is generally higher, embracing more considerations, and for the sections it tends to be higher still. At each stage an increased measure of homogeneity is found, for the simple reason that each stage of the subdivision has proceeded by the separation of contrasted areas. Thus the Appalachian Highland is first marked off from the Interior Lowlands on the one side and the Coastal Plain on the other. The crystalline Older Appalachians of Davis must then be distinguished from the Newer Appalachians, and in Fenneman's scheme, the opportunity is taken at the same time to separate the Piedmont Plateau Province from the Blue Ridge Province, and the Valley and Ridge Province from that of the Appalachian Plateaux. Finally at the section level the Triassic Lowlands are marked off from the rest of the Piedmont Province, the Catskills from the other Appalachian Plateau sections. We say " finally " because the process may not be carried further. Not because of any lack of variety within the sections : enough of that remains in even the most uniform of them to permit the further separation of unlike portions. But it happens that when this is effected the portions turn out to be individual landform features of precisely the size and kind that have been utilized in the earlier stages of the process to provide boundary lines to major divisions and provinces on the assumption that they could be regarded as simple unitary features. Clearly if they were so treated at one level they cannot now be regarded as complex.

The point is not, however, one merely of scale and degree of generalization. Fenneman states categorically "all lines are necessarily generalized, owing to the scale of the base map," but goes on to make clear that "the determination as to whether a certain point near a boundary lies in one division or the other must rest ultimately on the definition and description of the several units and not on the accuracy of the map maker." [7] It is rather that if account is going to be taken of the internal complexities of the features that comprise sections there must be a change in the criteria used in classification. Consideration of an actual case may here lend point to the argument.

The Black Hills Section (13*c*) is a small section of obvious unity marked off by the strongest of topographic contrasts from the surrounding areas of the High Plains. Not the least element in that contrast is the topographic variety of the Hills when compared with the monotony of feature of the Plains. The section embraces a rough eastern area of granite and schist of mountainous relief, a massive western plateau of virtually unbroken Carboniferous limestone, the remarkable encircling Red Valley and Dakota Hogback, and sundry peripheral cuestas beyond. Viewed as a whole the Black Hills are manifestly a unit—a flat-topped elliptical blister rising decisively from the surrounding plains and with its top worn sufficiently away, in places, to expose its variegated core. But subdivide it further and that unity is lost. The Red Valley is clearly a unity of a sort, but one that has nothing in common with that possessed by the section as a whole.

That, it seems to the writer, is a logical statement of the position from the point of view of the American classification. But if we transpose to a British context the situation changes, and the Black Hills would loom altogether larger and very differently. The Black Hills section (some 6,000 sq. miles) covers an area rather larger than all Yorkshire and nearly as large as Wales, and has a relative relief of about 4,000 ft. with summits rising more than 7,000 ft. above the sea. Its exposed crystalline rocks cover an area about equal to that of the comparable Lower Palaeozoic outcrops of the Lake District, while the limestone plateau, with an extent comparable to the whole Pennine upland between Stainmoor and the Tyne, is without parallel in Britain. The Dakota hogback runs a course not very dissimilar to that of the combined North and South Downs round the Weald and is considerably bolder than our chalk cuesta ever becomes. Viewed through British eyes such features assume a new stature and are emphatically to be

[7] N. M. Fenneman, *op. cit.*, 1916, p. 26, note 6.

regarded not as unitary morphological features but as minor morphological regions, each with its own characteristics of relief and form.

But there lies the rub. Relief and form are not the criteria by which the American subdivision was effected. By subjecting the Black Hills section to a British rather than an American view we have revealed the precise point at which a change of criterion is required if morphological subdivision is to continue. When we divide sections into their component parts we reach the stage at which it ceases to be true that physiographic history is a unifying factor.

Once more an example is called for, but this time it must be British rather than American. The chalk plateau of East Kent and the clay lowland of the Kentish Weald have for the last 30 million years or more had virtually identical physiographic histories. Both participated in the mid-Tertiary uplift of the Weald, and experienced the full course of the erosion cycle that was initiated by that uplift and concluded by the production of a peneplane which was submerged beneath the waters of the Pliocene Sea. Both shared in the later uplift which led to the initiation of a new consequent drainage on the thin cover of marine Pliocene sediments, and the successive downward movements of base-level which led to the incision of that drainage and its superposition on the rocks below. Yet how different in pattern, elevation and aspect are the results of these parallel histories even the most cursory glance at sheets 51/84 and 61/14 of the 1/25,000 map will show. Virtual identity of process and stage is here associated with strongly contrasted forms since the latter are dominated by lithology. A lithological unit, in other words, is not a physiographic region.

An analogy from the field of chemistry may be helpful. The smallest subdivision of a substance that can exist and still retain the physical and chemical properties of that substance is the molecule. The smallest physiographic subdivision of a land mass for which description is simplified by reference to its morphological evolution is the section. Subdivision of the chemical molecule yields atoms of the constituent elements whose properties are unrelated each to the other or either to the compound whole. Subdivision of the physiographic section yields unit areas of the constituent morphological elements, closely related to the subjacent geological formations, but unrelated to each other or to the compound whole.

IV

Reluctantly, therefore, we conclude that we cannot break down the physiographic sections of several thousand square miles apiece and fit them for use in these small islands. But it is well, nevertheless, to attempt the application of the American method to the British case. In this we cannot do better than make a beginning with the Highland and Lowland zones to whose contrast, as Sir Cyril Fox has unforgettably demonstrated, the Personality of Britain owes so much. It is clear from the outset that both are nothing less than insular fragments of " major subdivisions " of the continent; but it is also clear, on closer examination, that both are composite. The former includes, in the north-west, part of the " Atlantic Highlands " that stretch from the

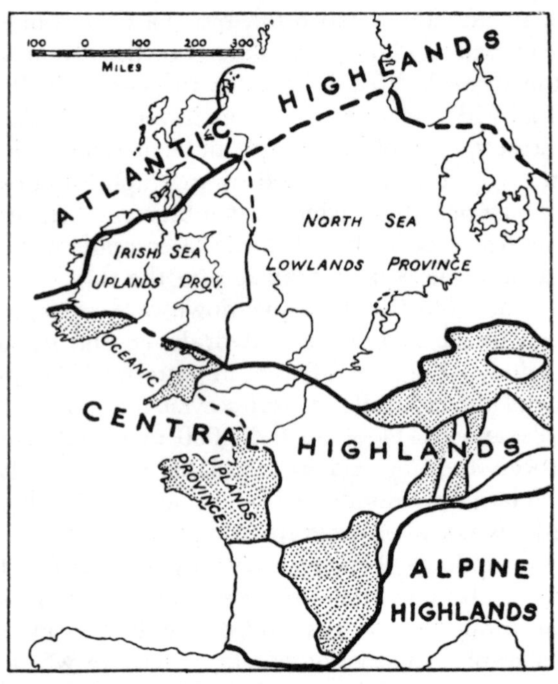

FIG. 46.—The major physiographic divisions of Western Europe and their component physiographic provinces.

North Cape to Connemara (Fig. 46), and correspond obviously to the structural division of the continent termed " *Palæo-Europa* " by Stille,[8] while its southernmost portions in Ireland, Wales and Southwest England are, with Brittany, disconnected sections of the " Oceanic Uplands Province " of that major division of the continent which comprises the old massifs and intervening basins of Central and Western Europe (the Central European Highlands). This latter division is essentially related to Stille's structural division "*Meso-Europa*," and includes both

[8] H. Stille, *Grundfragen der Vergleichenden Tektonik*, Berlin, 1924, pp. 231–4.

upland and lowland provinces. Among the latter is the "Anglo-Parisian Lowland Province" embracing the Paris Basin in France and the Weald and Wessex regions in the English Lowland. The latter is thus seen to be traversed by a major divisional boundary near or south of the line of the Kennet-Thames. The structural significance of this line has received rather full discussion elsewhere by Professor Wooldridge and the present writer, and its course as a major tectonic boundary defined.[9] For present purposes that boundary need not be adhered to. Following Fenneman's excellent advice,[10] having decided that a particular boundary is needed, we "choose the boldest and most consistent line available"—in this case the boundary between the Chalk and the overlying Tertiaries.

The remaining portion of the Lowland Zone is clearly part of the Great European Plain and corresponds to the western portion of Stille's structural unit *Ur-Europa.* Its two curving cuestas and outer, inner and intermediate lowlands presented W. M. Davis with his classic example of a belted plain and the drainage of cuestas, but its inland boundary—a boundary of provincial rank at least—is not easy to draw consistently. The eastern continuation of that boundary, it is worth noting, divides Denmark and Scania from the rest of Scandinavia, and these lowland areas together with the Saxon lowland of north-west Germany, the deltaic lowlands of the Rhine and Maas, Belgian and French Flanders, and the English Plain north of the Thames, may be grouped as the "North Sea Lowlands Province" of the Great European Plain.

In this brief sketch we have attributed the southernmost portion of the Highland Zone of these islands to the Central Highlands Division of Europe, and the north-western portion, including the Scottish Highlands and comparable portions of north-west Ireland, to the Atlantic Highlands Division. Between these two lies a large area which also belongs to the Highland Zone. It is in a rough way triangular and has the Irish Sea at its heart, and Bristol, Aberdeen, and the mouth of the Shannon at its vertices. Because of its situation straddling the Irish Sea it may perhaps be termed the "Brito-Hibernian" or "Irish Sea Uplands Province." It is an area of mingled upland and lowland relief and for its fuller description obviously requires to be broken into some half-dozen sections. The

[9] S. W. Wooldridge and D. L. Linton, "Some Episodes in the Structural Evolution of S.E. England considered in Relation to the Concealed Boundary of Meso-Europe," *Proc. Geol. Assoc.*, Vol. XLIX, 1938, pp. 264–91; *Structure, Surface and Drainage in South-East England*, Publication 10 of the Institute of British Geographers, London, 1939, pp. 4–7, and Fig. 3.

[10] N. M. Fenneman, *op. cit.*, 1916, p. 25, note 7.

uplands are hills and plateaux rather than mountains, and differ fundamentally from both the bold, heavily glaciated, crystalline highlands of the north-west, and the strongly grained but well-subdued worn-down highlands of the south-west. Its lowlands and some of its uplands are largely underlain by relatively little disturbed Upper Palæozoic rocks (sometimes covered by Trias) resting on a strongly-folded Lower Palæozoic basement. Where the latter is exposed in up-domed or up-faulted areas the Caledonian trend is conspicuous, but it is not significant for the province as a whole. Beneath the lowlands opened up by erosion of the less resistant rocks during the last one or two erosion cycles, or exposed upon the lowland margins, are all but two of the British coalfields. Coalfields are almost its leading characteristic. The exceptions are the concealed coalfields of Kent and Lincolnshire, deeply buried beneath the gently-dipping strata of the English section of the North Sea Lowlands Province, and they provide one of the indications that structurally it is with the latter region that the Brito-Hibernian Province has its affinities. But topographically it is unique in Europe—the distinctive British contribution to European relief.

It thus emerges that though the application of the American method to the British case can do nothing to help us define morphological regions of the smaller sizes to which our insular experience naturally inclines us, it is nevertheless well worth while. It focuses attention on larger relationships and continental affinities and brings to light, what the present writer at any rate would not otherwise have apprehended, the fact that the core of the British area is a physiographic province unique in Europe. It is largely our previous failure to grasp either its uniqueness or its unity that has made all our attempts at the physiographic description of our own islands so meagre and fragmentary.

V

Gratifying as this recognition of a larger unity may be it leaves untouched the question of small regions such as those to which, in western Europe, traditional usage has attached *noms de pays*. Scotland is rich in such names, Carrick, the Merse, Lothian and the Mearns being examples from the lowland side of the Highland line, and Lorne, Cowall, Moray and Formartine examples from the other. All these and yet others have most marked morphological characteristics: by what means may they be delimited and defined?

A little earlier we made use of an illustrative analogy from the field of chemistry: it may help here if we push it a little further. Physiographic sections were then likened to chemical molecules since subdivision resolves both into their constituent elements—atoms on the one hand and relief units like the Portland Hills or North Downs on the other. Now it is noteworthy that it is to relief units of just this order of magnitude that *noms de pays* are applied, and moreover, that common experience tells us that such " atomic " units are themselves complex. What is obviously needed to match our " morphological chemistry " is a " physiographic subatomic physics " in which the characters of " atoms " may be described in terms of the ultimate particles of which they are constituted. We need, in a word, " morphological electrons."

Such, however, are not far to seek. The ultimate units of relief are *flats* and *slopes*, small units of surface that are visibly either the one thing or the other, and are not susceptible of subdivision on the basis of form. These are the " electrons " and " protons " of which landscapes are built. They are more varied in kind than their physical analogues, since slopes may be gentle or steep, concave or convex, smooth or rugged ; while flats, even when they fully deserve the name, differ in respect of absolute altitude, extent and relation to the water-table. With these variations go considerable variations in aspect, exposure to sun, wind and rain, and thus in micro-climate, as well as in soil character and profile. Flats and slopes thus offer a certain limited variety of habitats to growing plants and other organisms which ecologists have long recognized under the generic name of *sites*. The operative words here are " limited variety." If in any area the physiographic conditions exist for the production of a particular slope form they are usually rather widespread and examples of that form are likely to be repeated fairly commonly over the area. A small area of uniform physiographic history, coupled with uniformity of lithology of the subjacent rock, and uniformity of degree of relief, may be expected to be characterized by repetition of a few characteristic sites over its whole extent. If one passes out of the area of uniformity of relief, rock and physical evolution, no matter which of the variables be changed, the change will be marked by the disappearance of some characteristic sites and the appearance of some that are new. Such change marks the crossing of a regional boundary, and site-recognition thus offers a delicate criterion by which physiographic regions of the smallest order can be recognized.

The concept of " site " has been in use for more than a quarter of a century by foresters, and it was a forester, Bourne, who in 1931 first clearly enunciated the principle that regions may be delimited by their characteristic site assemblages.[11] The present writer owes

[11] R. Bourne, *Regional Survey*, Oxford Forestry Memoirs, No. 13, 1931, pp. 16–18.

his introduction to these ideas to Professor Wooldridge, who almost alone among British geographers was quick to appreciate their significance for our subject. That significance has two aspects. On the one hand flats and slopes are nothing but the visible expressions of the erosional and depositional processes working with given materials in relation to particular base levels, and can therefore be given generalized expression in terms of structure, process, and stage. On the other hand, sites offer units of relatively uniform environment for the development of soils, natural vegetation or cultivated crops. As much as 18 years ago, Wooldridge pointed out how " those elements of shape which function as controls in human geography " in any landscape, its flats and slopes, are properly to be comprehended only in terms of the denudation chronology of the region,[12] while more recently he has shown how the concept of site may be used to advantage in the explanation of the distribution of grassland and arable land in the North Riding of Yorkshire.[13] Yet if these concepts have remained little used by geographers they have entered deeply into the discussions of pedologists. G. R. Clarke, for example, in his book on the practical aspects of soil study devotes no less than 40 pages to the consideration of " soil-site " [14] and the chief weakness of his discussion is perhaps that his " Item II, Age of Site," and " Item IV, Topography," are divorced from each other and from " Item I, Locality of Site (Region, etc.)," when all are clearly aspects of the manner and stage of dissection of the area. There is evidently room for co-operation of geomorphologists with soil scientists in relating the soil-sites of a region to its morphological evolution. It is, in fact, the present contention that the equation of flats and slopes, the ultimate units of relief, with sites, the units of environment [15] permits the morphologist to make a direct contribution to the work of the pedologist, ecologist and practical farmer, and that the principle of regional delimitation on the basis of site analysis allows geographers to attempt correlations in generalized regional terms which will offer explanations of the detailed variations of soil, vegetation and crops on the ground at the field-by-field scale.

The present writer has made two attempts to apply Bourne's methods, and the second of these two was specifically directed to the end just

[12] S. W. Wooldridge, " The Cycle of Erosion and the Representation of Relief," *Scot. Geog. Mag.*, Vol. XLVIII, 1932, pp. 30–6.

[13] S. W. Wooldridge, *The Land of Britain,* Part 51, " Yorkshire " (North Riding), London, Geographical Publications Ltd., 1945, pp. 388 et seq.)

[14] G. R. Clarke, *The Study of the Soil in the Field,* Oxford, 1938, pp. 13–53.

[15] " A unit of land suitable for a single system of utilization may be termed a site," G. R. Clarke, op. cit., p. 13.

outlined.[16] To frame a description of the physical background of land utilization in the counties of Peebles and Selkirk that would have some reality when tested by farmers and shepherds against the conditions they experience on their own land, the area of the two counties was divided into morphological units on the basis of recognition of their recurrent sites. It is not claimed that detailed site-analysis was rigorously effected for the whole area, but it was carried sufficiently far to permit its extension on the basis of the ascertained morphological conditions. Within the area of the two counties, some 620 sq. miles, 21 regions or parts of regions are represented, giving an average size of some 30 sq. miles apiece. This figure serves as an indication of the size of region over which a characteristic assemblage of sites can be recognized in strongly accidented upland country.

The earlier attempt referred to above was concerned with a much larger area, the whole Central Lowlands of Scotland totalling several thousand square miles, and manifestly, detailed site-analysis in the field could not be pursued over the whole terrain. The resulting map may, therefore, be regarded only as a reconnaissance survey of regions based on site recognition. Nevertheless the results have considerable point in the present context, and the map, which was used at the Cambridge Meeting of the British Association in 1938 to illustrate a contribution to a discussion on " Some Aspects of the Regional Concept " but never published, is therefore reproduced here as Fig. 47. In an area of about 4,500 sq. miles it shows about 130 minor subdivisions so that the mean area of the latter is about 35 sq. miles, a figure of the same order of magnitude as that noted for Peebles and Selkirk.

The large number of subdivisions is the inevitable result of their small average size, and this in turn of the refined character of any delimitation based upon site-analysis. But it emerges also that these minor units may be grouped together into larger wholes by associating like with like and keeping unlikes apart. Of these larger entities some 19 or 20 cover Central Scotland and it is immediately apparent that they comprise such familiar features as Lothian or the Ochil Hills. In other words they are morphological regions of precisely the size we have been seeking—the *pays* of traditional usage, and the " atoms " of our " morphological chemistry." By taking minor regions defined by their characteristic sites and grouping these, we arrive at a series of morphological units which can themselves be combined to yield sections of physiographic provinces, and so we may pass by two further stages of aggregation to the undivided continents.

[16] D. L. Linton and Catherine P. Snodgrass, *The Land of Britain*, Part 24, " Peeblesshire and Selkirkshire," London, 1946, pp. 407–18.

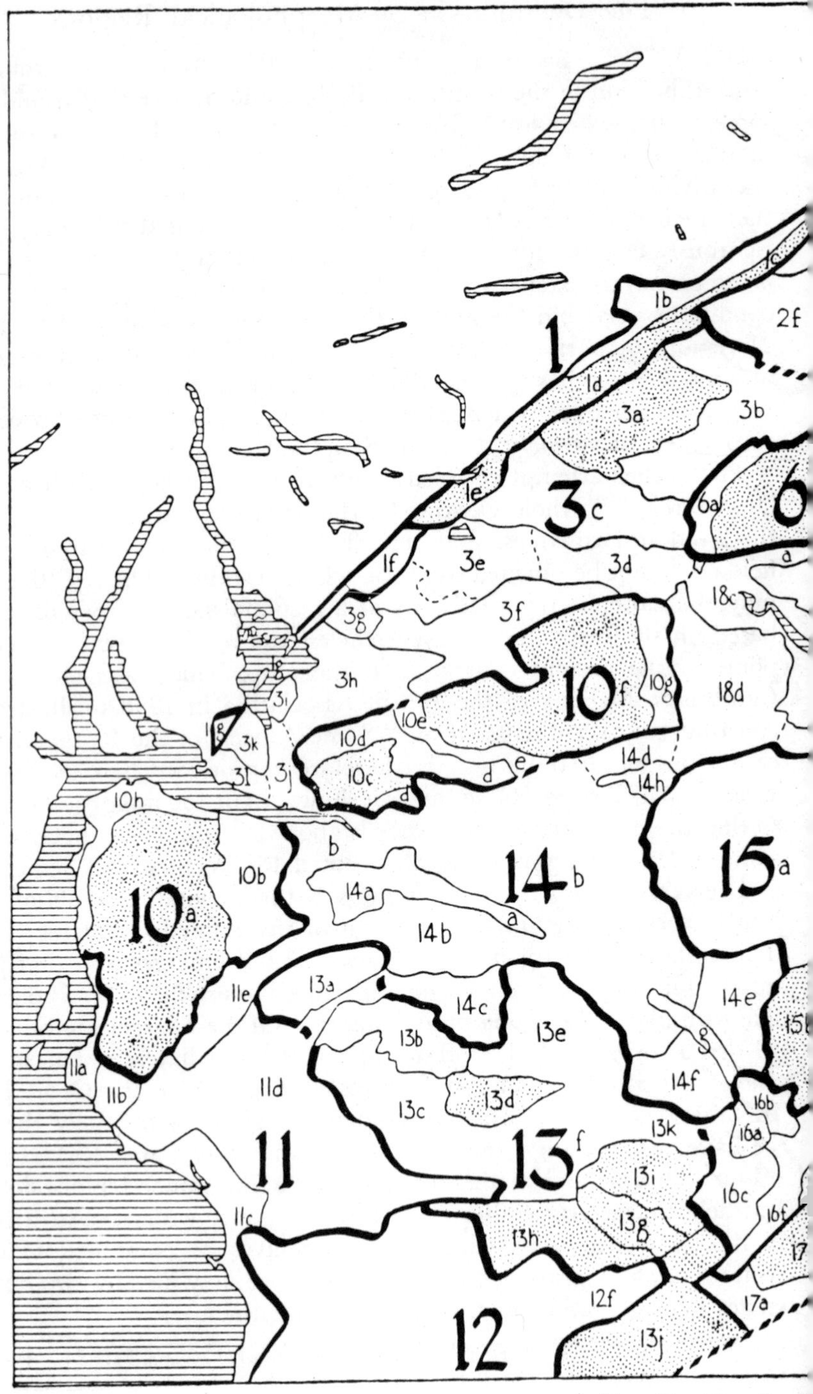

FIG. 47.—The Regi

Central Scotland.

FIG. 47.—The regions of Central Scotland.

(1) *The Highland Border Zone.*

a. Ridges of Angus.
b. Grampian Border Furrow.
c. Strathearn Ridges.
d. Ruchill (Conglomerate) Hills.
e. Menteith Hills.
f. Low Ridges of Gartmore.
g. Balmaha Ridge and Killeter.

(2) *Synclinal Belt of Strathmore.*

a. Alyth Bench (300–400 ft.).
b. Isla Flats and Terraces (fluvio-glacial and alluvial).
c. Cairnleith—Balbeggie Benchlands (300–400 ft.).
d. Logiealmond Bench (600–800 ft.).
e. Gask Ridge.
f. Strathearn—the Pow and Almond lowlands.

(3) *Synclinal Belt of Menteith and Lennox.*

a. Braes of Ardoch.
b. Muir of Orchil and Strath Allan.
c. Teith Valley Drift Area.
d. Vale of Menteith—the Carse.
e. Vale of Menteith—the Moss.
f. Benchlands of Buchlyvie Muir (550–700 ft.).
g. Buchanan Upland.
h. Lennox Lowland.
i, k. Lennox Uplands (cf. 3*a* and 3*g*).
j. Leven Valley.
l. Cardross Lowland.

(4) *Sidlaw Hills.*

a. Central Sidlaws (Craigowl 1,433 ft.).
b. Black Hill Cuesta (King's Seat 1,285 ft.).
c. Braes of the Carse and Moncreiff Hill.
d. Low Hills of Dundee Region.

(5) *The Carse of Gowrie.*

a. The Carse.
b. Lower Earn Flats.
c. Errol Bench.
d. Longforgan Bench.

(6) *The Ochil Hills.*

a. Sherrifsmuir.
b. High or Western Ochils.
c. Middle Ochils.
d. Fifeshire or Eastern Ochils.
e. Newport Upland.

(7) *Interior Lowland of Fife.*

a. Crook of Devon Hills.
b. Lowland of Kinross.
c. Leven Flats.
d. Howe of Fife.
e. Strath Eden.

(8) *Fife Basalt Cuesta.*

a. Saline and Cleish Hills.
b. Benarty.
c. Lomond Hills.
d. East Fifeshire Uplands.

(9) *Coastal Fringe of Fife and Angus.*

a. Dunfermline Low Hills.
b. Lochgelly Plain.
c. Volcanic Ridges of Southernmost Fife.
d. East Fife Drift Plain.
e. Ore and Leven Valley Flats and Coastal Benches.
f. Tents Muir.
g. Angus Coastal Lowlands.
h. Buddon Ness.

(10) *Western Basalt Plateaux.*

a. Renfrew Plateau.
b. Kilmacolm Upland.
c. Kilpatrick Hills and
d. Bordering Benches.
e. Strathblane.
f. Campsie Plateau.
g. Carron Upland Benches.
h. Northwestern Scarp Zone of Renfrew.

(11) *North Ayrshire Lowland.*

a. Clyde Benches.
b. Knockewart Hills.
c. Ayrshire Raised Benches.
d. Cunningham Lowlands.
e. Lochwinnoch Depression.

(12) *South Ayrshire.*

a. Mauchline Area.
b. Kyle Lowland.
c. Uplands of Carrick.
d. Uplands of Ochiltree.
e. Airds Moss.
f. Muirkirk Depression.
g. Cumnock Uplands.

(13) *Clyde Plateaux.*

a. Beith Plateau.
b. Ballagioch Hills.
c. Cunninghame Uplands.
d. Ellrig Hills.
e. Rotten Calder Uplands.
f. Strathavon Depression.
g. Nutberry Hills.
h. Distinkhorn Hills.
i. Kype Moors.
j. Cairn Table Hills.
l. Neilston Trough.

(14) *Clyde Lowlands.*

a. Renfrew Flats and Clyde Floodplain.
b. Lower Clyde Drift Plain.
c. Mearns Lowland.
d. Kilsyth Trench.
e. Carluke Benches.
f. Stonehouse Benches.
g. Clyde Trench.

(15) *Forth-Clyde Uplands.*

a. Slammannan Plateau.
b. Wilsontown Upland.
c. Upper Pentland Benches.
d. Northern Pentlands (volcanic).
e. Southern Pentlands (Up. O.R.S.).

(16) *Middle Clyde Lowland.*

a. Dillar Hill group.
b. Clyde Gorges.
c. Broken Cross Moor.
d. Carnwath Drift Plain.
e. Carmichael Hills.
f. Douglas Valley.
g. Medwyn Valley.
h. Thankerton Lowland.

(17) *Synclinal Zone of Biggar.*

a. Glenspin Moors.
b. Roberton Hills.
c. Tinto Hills.
d. Biggar Lowland.
e. Biggar Volcanic Hills.
f. Dolphinton-Linton Lowland.
g. Conglomerate Hills of Kirkurd.

(18) *Inner Forth Lowlands.*

a. Devon Valley.
b. Clackmannan—West Fife Drift Plain.
c. Carses of Stirling and Kinnell.
d. Stirlingshire Drift Plain.
e. Bo'ness Upland.
f. Linlithgow Depression.
g. Torphichen Hills.
h. Avon Lowland.
i. Almond Drift Plain.
j. Lower Pentland Benches, etc.
k. Edinburgh Queensferry Hills.

(19) *Outer Forth Lowlands.*

a. Coastal Benches.
b. Lower Esk Lowland.
c. Upper Esk Benches.
d. Roman Camp Upland.
e. East Lothian Upland Bench.
f. Hills of Lammermuir Border.
g. Tyne Drift Plain.
h. Garleton Hills.
i. Cockburnspath Lowland.

So the chain is logically completed. Nature offers us two inescapable morphological unities and two only ; at the one extreme the indivisible flat or slope, at the other, the undivided continent. By the aggregation of the former and the subdivision of the latter we may bridge the gap between and develop a related series of intermediate units. At the one end of the series are units formed by the repetition over and over again of a small number of flats or slopes : such units possess a high degree of homogeneity in all their morphological attributes—form, rock structure and evolutionary history. Each succeeding member of the series shows increasing diversity. First uniformity of form is lost though unity of formation may remain. That too goes as we pass to the physiographic section with its diversity of rock structure and unity of erosion history : this in its turn gives way to a unity limited to major relief type and structural evolution. Finally the series ends with one of the continental land masses—a unique entity with order in its diversity, but of which no general statement can be made regarding its form, structure or evolution that will apply to all its parts.

To this hierarchy of morphological units it is most desirable that an agreed nomenclature should be applied. The larger units need not detain us. The American names have the sanction of thirty years' use and the writer can think of no reason why they should not become equally acceptable on this side of the Atlantic. Our problem is thus reduced to that of naming the two grades of morphological unit that intervene between the site and the section. As it happens a useful nomenclature lies ready to hand. In 1933 J. F. Unstead, who had for many years previously advocated the grouping of minor regions in a related series, developed his views in the Herbertson Memorial Lecture delivered to the Geographical Association.[17] In the opening paragraphs of that lecture he outlines his method of " synthetic " regional delimitation. From ultimate units which he terms " features " and may be seen from his chosen example to be nothing other than the flats and slopes of the morphologist, he derives " regions of the lowest order " which he terms *stows*, and by aggregation of these " regions of the second order of magnitude " which he terms *tracts*. As examples of tracts he cites the North Downs and South Downs, each being composed of a number of stows, some being valley stows (the transverse river valleys with their characteristic flood-plain, bluff and terrace " features "), and the others plateau stows (the intervening plateau blocks scored by combes and dry valleys). Unstead's units are identical in magnitude and character with those that would be rather more precisely defined by site-analysis, and his terms would seem admirably suited to our present purpose. Two qualifications, however, have to be made. Although Unstead says that " relief, structure and soils, either singly or in combination, are generally the distinguishing criteria of a tract " he clearly envisages that other factors may on occasion be decisive. London and " other conurbations " he would rank as stows. Further his stows and tracts are basal units in a structure whose higher members are determined increasingly by climate, vegetation and land-use and so come increasingly to resemble portions of Passarge's " Landschaftsgürtel." These qualifications, however, are not fatal to the suggestion here made. Unstead's scheme, taken as a whole, has not met with wide acceptance, possibly because of the imprecision introduced by variations in the basis of classification from relief to climate and climate to human occupancy at different levels in the hierarchy. But it remains true that the names given to his units of the two lowest orders were selected to designate what, in almost every instance he cites,[18] are actually morphological divisions, and are

[17] J. F. Unstead, " A System of Regional Geography," *Geography*, Vol. XVIII, 1933, pp. 175–87.

[18] See for example the application of Unstead's method in his *Systematic Regional Geography*, Vol. I, " The British Isles," London, 1935, pp. 292, especially the maps on the front end-paper and pp. 79, 99, 115 and 133. In Wales, Scotland and Ireland, the method breaks down since the identification of stows, on which the synthetic process depends, is not rigorously pursued. It is worth emphasizing, moreover, that Unstead's method of assembling tracts into tract groups, as exemplified in Chap. X and the map on p. 169, is not one that can lead to the recognition of the physiographic section as the unit of the next higher grade. The essential unity of such areas as Sussex or Lincolnshire at the section level is suppressed in his scheme.

admirably fitted to take their place in a purely morphological classification. And it is fortunate that if *stow* is a useful revival from the archaic past (O.E. *stōw*—a place) with the merit of being without associations, the previous associations of the term *tract* are very much what they would be if it became the acceptable term for the first subdivision of the physiographic section. Such expressions as " the chalk tract of Salisbury Plain " or " the lowland tract of Buchan " accord well with ordinary usage, and, it is suggested, do no violence to Unstead's intentions.

Whether these suggestions for the adoption of a particular nomenclature for a graded series of morphological units—site, stow, tract, section, province and continental subdivision [19]—be accepted or not, the general thesis and conclusions of this essay remain unaffected. Morphological delimitation must proceed from the consideration of one or other of the two natural units of the lands, the site and the continent, by the recognition of the characteristic groupings of the one and the characteristic subdivisions of the other, carrying both processes to the point where their results converge. Such a task, logically conceived and systematically executed, would provide an interpretative statement of the regional variations of surface form well calculated to be of value at all scales, not only in academic studies but equally in practical affairs.

[19] It will be noticed that the equivocal term *region* finds no place in this nomenclature.

XII

SOME ASPECTS OF THE RURAL SETTLEMENT OF NEW ENGLAND IN COLONIAL TIMES

By F. GRAVE MORRIS, M.A.

That the first settlers in New England in the seventeenth century reproduced many of the features of the English village as it existed at that time is well known. Less attention has been given to a comparison of the progress of that settlement with that of the Anglo-Saxons in Britain, for despite certain differences arising from the more advanced state of civilization which obtained in the seventeenth century, there appear to be fundamental similarities in the process by which the settlement was effected in both instances.

The conditions of settlement appear comparable in three particulars. Firstly, the settlers were in both cases primarily agricultural, practising a form of mixed husbandry which combined the growing of cereals with the raising of cattle. Secondly, both Old and New England were mainly forested and the amount of land open to immediate settlement and cultivation was very limited. Thirdly, both Anglo-Saxons and Puritans found themselves in a hostile land which enhanced the need for mutual protection. The first two conditions made it necessary to establish the earliest settlements in places where the labour of clearing the land for cultivation could be reduced to a minimum and where pasture and forage was available for the animals. The third condition made some form of defensive nucleated settlement almost imperative if the new immigrants were not to be destroyed by the hostile natives. Both in the choice of sites for settlement and in the formation of nucleated villages each invasion shows identical characteristics. It may not perhaps be too much to apply to the earliest settlements of New England the term " primary " as it has been applied to those of Anglo-Saxon England.

But as population increased new settlements had to be established. Sometimes they resulted from fission, a new settlement being established within the bounds of an older one ; sometimes they were established in areas well beyond those of the settled communities. On the one hand, the gradual disappearance of relatively open land rendered essential the clearing of patches in the forests, more likely to be the work of individuals than of communities. On the other, the decrease of the hostile native population made the formation of nucleated defensive settlements less necessary. Thus a scattered form of settlement consisting of isolated farms and hamlets replaced the earlier nucleated villages. To this type of settlement, less well recognized but nevertheless clearly apparent in New England, may be given the term, applied to Anglo-Saxon England, " secondary " settlement.

It seems to be taken for granted by many writers, both historians and geographers, that throughout the colonization of New England the characteristic form of settlement was the nucleated village, but recent research has shown that this is not wholly true.[1] This appears to be due partly to a confusion of thought about the meaning of the words " town " and " township " in New England and partly to a lack of adequate geographical studies of the form of the later settlements as compared with their political and legal organization.[2]

Any discussion of the settlement of New England must begin with an examination of the meanings of the terms " town " and " township." The word " town " is in fact the Old English " tun " which does not denote anything in the nature of a modern town but only a rural agricultural settlement, a village. The term " township " refers primarily to the land belonging to the community occupying it. Both terms denote legal conceptions in New England and do not, in themselves, give any indication of the physical nature of the settlement. The idea of a " town " is a body of men organized for certain political, economic and even religious purposes, involving no doubt a certain common centre, the courthouse and the church, but in itself it does not indicate nucleated settlement. The terms " town " and " township " later became interchangeable so that when one reads of a new town being established it may only mean the laying out of a new township or area of land perhaps some six miles square. The people who settled in that area formed a legal corporate body with certain political rights and duties and also a religious body. But there is no indication that the settlement took the form of a nucleated village rather than of scattered farms and hamlets.

In drawing any comparison between the two colonizations it must be admitted that the Puritan settlers of New England showed a consciousness of the need of political and religious solidarity in a way that

[1] See, for instance, Edna Scofield, " The Origin of settlement patterns in rural New England," *Geog. Rev.*, Vol. XXVIII (1938), pp. 652–663 ; G. T. Trewartha, " Types of rural settlement in colonial America," *Geog. Rev.*, Vol. XXXVI (1946), pp. 568–596.

[2] Thus the Atlas of the Historical Geography of the United States creates a false impression in its maps of the growth of towns in colonial times in indicating the " towns " of New England as comparable with the towns of Virginia which were much more like towns in the modern sense.

their less educated and cultured Anglo-Saxon forbears probably lacked, and that this consciousness operated strongly in the direction of nucleated settlements. It may, however, be observed that the earliest Anglo-Saxon settlements seem to have been the work of either clans or organized groups of warriors. But it may reasonably be urged that without this incentive the very conditions of colonization made the establishment of such settlements highly probable and desirable. As the conditions changed so, it appears, did the form of settlement.

The earliest settlements in New England, whether at Plymouth by the immigrants of the *Mayflower* in 1620 or those around Boston from 1630 onwards, were indeed nucleated settlements. To this end many causes contributed. In the first place there was the consciousness of the need for definite political organization, of which the famous Plymouth Compact is an outstanding example, while the whole history of Massachusetts and Connecticut displays the same spirit. Settlement was controlled as far as possible and involved the creation in advance of a responsible political organization, the " town," and the laying out of definite tracts of land. Moreover to the need of political was added that of religious organization. The earliest settlements were organized as " churches " and indeed at first " church " and " town " were almost synonymous. Yet while these factors contributed to the establishment of nucleated settlements they did not compel it as later history shows. But in an unknown country, surrounded by treacherous enemies well acquainted with every track through the forest and able to fall without warning upon isolated settlers, dispersion would have been self-destruction. Defence against the Indians was one of the most important reasons for compact settlements among the Puritans as no doubt it was among the Anglo-Saxons against the Romanized Britons. Although the Plymouth settlement was not at first fortified the threat of the Narragansett Indians caused the settlers in November, 1621, to surround their houses with a palisade and to create a military organization for their defence.[3] That the need for compact settlement in face of the Indians was realized by the earliest settlers round Boston Bay, in the " towns " of Boston, Roxbury, Watertown, Medford and Dorchester is likely enough, yet as early as 1635 the General Court of the Massachusetts Bay Colony found it necessary to issue an order forbidding the erection of any dwelling houses further than half a mile from the " meeting-house " in order to ensure greater safety, so strong already was the tendency for these pioneers to scatter their farm-houses. After the purchase by Massachusetts in 1677 of a large portion of the district of Maine the General Court laid down definite

[3] Bradford, *History of Plymouth Plantation*, p. 134.

conditions for the renewal of settlement, among which was the requirement that the villages should be compact ones.[4] The influence of the Indian menace remained strong throughout the seventeenth century and the erection of " garrison towns " recalls the establishment of " burhs " under Alfred and Edward the Elder as defences, not indeed against the natives of the land but against the invading Danes.

That the Indian menace contributed towards the establishment of nucleated villages without however making it the rule is clear from all the available evidence. That the economic system which the earliest settlers brought with them contributed to the same end is also well established. Putting aside the instance of Plymouth, where a form of simple communism was attempted without success, it is clear that the earliest " towns " were laid out exactly as the contemporary English village ; the houses and " house-lots " around the centre with its " meeting-house," the arable fields divided into strips and the " meadow " or pasture apportioned among the members of the community. The maps of early New England villages, such as Watertown in 1636 or Wethersfield in 1640, are exact replicas in substance of the mediæval English village.[5] Such a system suited both the prevailing mixed husbandry and the nature of the land. In Virginia, where a single staple crop dominated the economy of the colonists, such a form of land holding was never established. The earliest settlers in New England thought in terms of the English village and sought to reproduce it in the new country.

Now the form of husbandry and of land holding which they established controlled their choice of site. The earliest, such as those around Boston Bay or the so-called " river towns " of Connecticut, lay on rivers and streams, partly, no doubt, because of their value in some instances as means of communication but also because on their banks lay the valuable water-meadows so essential in those times to the rearing of any considerable numbers of cattle. These riverside settlements recall at once the earlier riverside villages of Wessex and of many a southern and eastern stream in England and the value attached to their " meadows " in Domesday Book. Such " meadows," together with numerous Indian clearings which are also frequently mentioned in New England documents, had the further advantage of demanding little or no clearing before they could be put to use, thus saving the immense labour and enormous time required for clearing the forest. Thus a parallel is again established between these earliest of New

[4] L. K. Mathews, *The Expansion of New England*, p. 29.
[5] P. W. Bidwell and J. I. Falconer, *History of Agriculture in the Northern United States, 1620–1860*, pp. 50, 51.

England settlements and the older English villages situated on gravel patches which may well have been less heavily wooded than the surrounding land. With a similar form of husbandry the sites of primary settlement were identical in fundamental character.

Such compact and carefully organized villages continued to be formed throughout the seventeenth century, but along with these there was an increasing tendency towards more scattered settlements. A passage in Bradford's *History of Plymouth Plantation* under the year 1632 is so illuminating as to be worth quoting in full as an illustration of a process that no doubt occurred in all the earliest settlements in some degree or other.

For now as their stocks increased, and ye increse vendible, ther was no longer any holding them togeather, but now they must of necessitie goe to their great lots ; they could not other wise keep their katle ; and having oxen growne, they must have land for plowing & tillage. And no man now thought he could live, except he had catle and a great deale of ground to keep them ; all striving to increase their stocks. By which means they were scatered all over ye bay, quickly, and ye towne, in which they lived compactly till now, was left very thine, and in a short time allmost desolate. And if this had been all, it had been less, though to much ; but ye church must also be devided, and those yt had lived so long togeather in Christian and comfortable fellowship must now part and suffer many divissions. Firstly, those that lived on their lots on ye other side of ye bay (called Duxberie) they could not long bring their wives & children to ye publick worship & church meetings here, but with such burthen, as, growing to some competente number, they sued to be dismissed and become a body of them selves ; and so they were dismiste (about this time), though very unwillingly. But to touch this sadd matter, and handle things together that fell out afterward. To prevent any further scatering from this place, and weakning of ye same, it was thought best to give out some good farms to spetiall persons, yt would promise to live at Plimoth, and lickly to be helpfull to ye church or comonewelth, and so tye ye lands to Plimoth as farmes for the same ; and ther they might keepe their catle and tillage by some servants, and retaine their dwellings here. And so some spetiall lands were granted at a place generall, called Greens Harbor, wher no allotments had been in ye former division, a plase very weell meadowed, and fitt to keep & rear catle, good store. But alass ! this remedy proved worse then ye disease ; for within a few years those that had thus gott footing ther rente them selves away, partly by force, and partly wearing ye rest with importunitie and pleas of necessitie, so as they must either suffer them to goe, or live in continuall opposition and contention. And others still, as yey conceived them selves straitened, or to want accomodation, break away under one pretence or other, thinking their owne conceived necessitie, and the example of others, a warrente sufficente for them. And this, I fear, will be ye ruine of New-England, at least of ye churches of God ther, & will provock ye Lords displeasure against them." [6]

[6] Bradford, op. cit., pp. 361–3.

Despite the writer's fears, the process continued and several new " towns " such as Scituate, Marshfield and Duxbery were formed.

Mention has already been made of the tendency which appeared as early as 1635 among the " towns " round Boston to establish separate farm-houses outside the original settlements. The occupation of Brookline throws light on part of this process. This was originally occupied as " outland " for Boston and Cambridge as it contained good soil, timber, marshland and meadow. The farmers of the earlier settlements at first pastured their cattle here in the summer, bringing them into the " towns " in winter without establishing permanent settlements which appear, however, to have been made by 1640. One wonders indeed if this is not an exact parallel in all essentials to the first settlements in the " dens " and " fields " of the Weald or the " shields " of Northern England.

As such settlements increased they were properly organized and separately incorporated as " towns " under the colonial governments. Sometimes, as in the " river towns " of Connecticut, a process comparable to the fission in mediæval English villages appears. Thus the settlers of Hartford moved across the river and became the " town " of East Hartford while about 1662 some settlers from Windsor did the same and formed the " town " of East Windsor, a process similar to the formation of English double villages such as Worth Matravers and Langton Matravers.

While it is true that the colonial governments in the Seventeenth Century endeavoured to keep control of settlement and to organize it in advance, they could not prevent the pioneers from moving out of the original settlements either into some part of the as yet unoccupied lands of the " township " or else into areas outside any organized " township." Many instances could be quoted of " towns " being formed within an older " township " by some group of settlers, who had left the original nucleated settlement or else by settlers who, entering the " township," found the land around the original settlement already fully occupied. The pioneers who went far from the older settlements incurred the wrath of such men as the Puritan divine, Cotton Mather, who in 1694 wrote :

> Again, Do our Old People, any of them Go Out from the Institutions of God, Swarming into New Settlements, where they and their Untaught Families are like to Perish for Lack of Vision ? They that have done so, heretofore, have to their Cost found, that they were got unto the Wrong side of the Hedge, in their doing so. Think, here Should this be done any more ? [7]

[7] Quoted by F. J. Turner, *The Frontier in American History*, p. 64.

These secondary settlements, either within or without the bounds of the established townships, were the work of individuals and in many cases must have involved the laborious task of clearing the woodland step by step. But, wherever possible, the labour of clearing the wooded uplands and cultivating their inferior soil was evaded. Thus Worcester County (Mass.) was passed over for 60 years while settlers pressed up the better valley lands of the Connecticut and Housatonic rivers.[8]

In the eighteenth century the tendency towards more scattered settlement increased. After the Peace of Utrecht in 1713 the Indian menace, which had been enhanced by the French wars, began to decline until the outbreak of the Seven Years War in 1756, although the danger of sporadic Indian raids was never entirely absent. Moreover, the policy adopted by the Colonial governments changed from one of carefully organized control to one which gave rise to land speculation. Not one but whole groups of "townships" as yet unsettled were surveyed and sold to speculators who in turn sold to pioneer farmers. Though the Colonial governments usually specified that a school and church should be erected within a certain time, it is clear that this condition was not always fulfilled in some of the outlying settlements and the absence of churches and schools in some parts of Maine and New Hampshire was unfavourably recorded in 1754.[9] The old corporate spirit characteristic of the earlier settlements was losing its grip among the hardy, individualistic pioneers.

In these frontier settlements, as well as in the hamlets formed within the older townships and later incorporated as separate "towns," the elaborate system of land tenure characteristic of the old English village and the earlier settlements of New England was never established. Pioneers carved out of the forest their own farm lands and erected their own farm-houses. Under such conditions the isolated farm appeared even though it formed part of an incorporated "town." The process is well illustrated in the following quotation regarding Western Connecticut in 1780:

> I saw, for the first time, what I have since observed a hundred times; for in fact, whatever mountains I have climbed, whatever forests I have traversed, whatever bypaths I have followed, I have never travelled three miles without meeting with a new settlement, either beginning to take form or already in cultivation. The following is the manner of proceeding in these improvements or new settlements. Any man who is able to procure a capital of five or six hundred livres of our money, or about twenty five pounds sterling, and who has strength and inclination to work may go into the woods and

[8] L. K. Mathews, op. cit., p. 103.

[9] L. K. Mathews, op. cit., p. 104.

purchase a portion of one hundred and fifty to two hundred acres of land, which seldom costs him more than a dollar or four shillings and sixpence an acre, a small part of which only he pays in ready money. There he conducts a cow, some pigs, or a full sow, and two indifferent horses which do not cost him more than four guineas each. To these precautions he adds that of having a provision of flour and cider. Provided with this first capital he begins by felling all the smaller trees, and some strong branches of the large ones ; these he makes use of as fences to the first field he wishes to clear ; he next boldly attacks those immense oaks, or pines, which one would take for the ancient lords of the territory he is usurping ; he strips them of their bark, or lays them open all round with his axe. These trees, mortally wounded, are the next spring robbed of their honours ; their leaves no longer spring, their branches fall, and their trunk becomes a hideous skeleton. This trunk still seems to brave the efforts of the new colonist, but where there are the smallest chinks or crevices, it is surrounded by fire, and the flames consume what the iron was unable to destroy. But it is enough for the small trees to be felled, and the great ones to lose their sap. This object completed, the ground is cleared ; the air and the sun begin to operate upon that earth which is wholly formed of rotten vegetables, and teems with the latent principles of production. The grass grows rapidly ; there is pasturage for the cattle the very first year ; after which they are left to increase, or fresh ones are brought, and they are employed in tilling a piece of ground which yields the enormous increase of twenty or thirty fold. The next year the same course is repeated ; when, at the end of two years, the planter had wherewithal to subsist and even to send some articles to market : at the end of four or five years, he completes the payment of his land and finds himself a comfortable planter. Then his dwelling, which at first was no better than a large hut formed by a square of the trunks of trees, placed one upon another, with the intervals filled by mud, changes into a handsome wooden house, where he contrives more convenient, and certainly much cleaner apartments than those in the greatest part of our small towns.[10]

Such, *mutatis mutandis*, must have been the process of clearing the woodlands of England.

That scattered settlement was characteristic of many of these frontier "towns" is clearly shown both by eighteenth-century maps,[11] and also by written accounts. Thus the author just quoted writes : "For what is called in America a "town" or "township" is only a certain number of houses dispersed over a great space, but which belongs to the same incorporation . . . the centre or headquarters of these towns is the meeting house or church. This church sometimes stands single, and is sometimes surrounded by four or five houses only." [12]

Within such scattered townships nuclei must later have appeared

[10] F. J. Chastellux, *Travels in North America in the Years 1780, 1781 and 1782*, p. 34. Quoted by Bidwell and Falconer, op. cit., pp. 76, 77.

[11] See maps in Scofield, op. cit.

[12] Chastellux, op. cit., Vol. I, p. 20, quoted by Trewartha, op. cit., p. 580.

at cross-roads, bridges and fords, but such nucleations were not part of the original settlement. Even today there are within a very few miles of Boston "townships" with no very obvious centre, which recall not the nucleated village at the foot of the Sussex Downs but rather the parish of scattered farm and hamlet still found within the Weald.

Such was the evolution of settlement in New England in the seventeenth and eighteenth centuries and the parallels which it appears to offer to the earlier settlement of England. If the process in the former case was more rapid it was due to the more rapid increase of population and perhaps to somewhat better means of taming the wilderness. The history of New England is fully recorded and through it we may perhaps see something of the less well recorded history of the settlement of much of our own land, and the process by which its woods were cleared, its fields cultivated and its villages, hamlets and farms established.

XIII

A CENTURY OF COAL TRANSPORT—SCOTLAND 1742–1842

By ANDREW C. O'DELL, M.Sc.

The hundred years considered in this essay, while now remote, saw the establishment of much of the industry of Scotland as we know it today and its legacy of industrial and urban growth is responsible for many of the problems which now confront the country. During this period was laid down the framework of modern mineral transport, for it was in these years that the railway supplanted the road and its packhorse and the canal and its barge. Much of the impetus for the development came from a desire to exploit the industrial resources, the chief being coal, of central Scotland.

The financial year 1742–3 saw the commencement of a long series of manuscript Customs Accounts now preserved in the Register House, Edinburgh. A century later was marked by a milestone in the social history of the coalfields, for it was in 1842 that the Act was passed which forbade the employment of women and children in mines. Between these unconnected incidents was a century of fundamental change in the social and industrial evolution of Scotland. While the population of Scotland increased by some 40 per cent., of more significance was the concentration of the population in the coalfields with the rise of large towns such as Coatbridge. This development could not have come about without the improvement of land communications. Coal was in ever increasing demand for both domestic use, in preference to peat and wood, and for industrial use, particularly in the iron, chemical, glass and textile manufactures. It is not possible to obtain full statistics of the increasing industrial use but records have come from the past which illuminate the change. The first coke-produced pig-iron in Scotland was made by the tapping of the first Carron furnace on January 1, 1760, and from this date the output of iron rose, at first slowly, and then with great rapidity as the blackband ores and the hot-blast process were used: 1788 Scotland produced 7,000 tons of pig-iron (of which 1,400 tons by charcoal), 1796, 16,086 tons (900 tons by charcoal), 1806, 22,840 tons, 1828, 36,500 tons, 1839, 196,566 tons and 1843, 311,000 tons.[1] The main centre for the iron

[1] R. Meade, *Coal and Iron Industries of the U.K.*, 1882, pp. 731–6 and 830.

industry was the parish of Old Monkland, and here the total consumption of coal was 36,000 tons in 1794, but 530,400 tons for blast furnaces alone in 1839.[2] There was, in addition, a considerable use for foundry work after Henry Cort successfully utilized it in 1784 for forging iron. The other trades did not consume so much as the iron industry but the amounts were not negligible, for in 1808 Robert Bald estimated that while the pig-iron producers used 160,000 tons the glass-works used 25,000 tons and the distilleries 53,000 tons.[3]

With a growing home market, as well as an expanding export, it became necessary to go further afield and deeper after the coal. Both these trends created difficulties in the way of transport. So long as the coal was wrought on the *in-gaun-ee* (i.e. in-going or adit) system the movement of coal could be performed on the backs of women and girl *bearers* who formed a team with their men relatives who were the hewers. This system was continued even after shafts were sunk although it meant much toil and danger for the bearers (see Fig. 48). To speed the lifting of the coal the horse-gin was introduced, but even then most of the coal was still carried to the shaft-foot by the female *bearers*. The first gear erected in Scotland for the raising of the coal entirely by steam-power was in 1792, but the horse-gin method persisted in the Midlothian field until 1844.[4] The increasing lift was responsible for the elimination of the horse-gin in favour of steam-power. About 1830 an underground tramway was introduced in the Midlothian field and in place of *bearers* female *pushers* were used to move the coal in wooden hutches running on rails, or sliding on a corduroy road, to the shaft foot. In other fields underground wagonways had been introduced much earlier, e.g. at Bo'ness before 1754. The use of female labour became a matter of reproach to the public conscience from 1793 and in 1839 a Commission was appointed to survey the employment of women and children in mines and the publication of this report revealed so scandalous a situation that an Act was passed prohibiting their use. In some cases Court proceedings had to be instituted to prevent the women descending the mines since the loss of wages was a serious matter to the mining families. The century also saw a change in the status of the miners. Until 1775 the men and their families were bound to the pit from birth and this state of serfdom was only broken by an Act, 15 Geo. III, c. 22, of 1775, which had to be re-inforced by another Act, 39 Geo. III, c. 56, in 1799.

Having reached the surface the coal had to be carried to the

[2] *New Statistical Account*, VI, p. 948.
[3] R. Bald, *A General View of the Coal Trade of Scotland*, 2nd Edition, 1812, p. 96.
[4] P. M'Neill, *Tranent and its Surroundings*, 2nd Edition, 1884, p. 30.

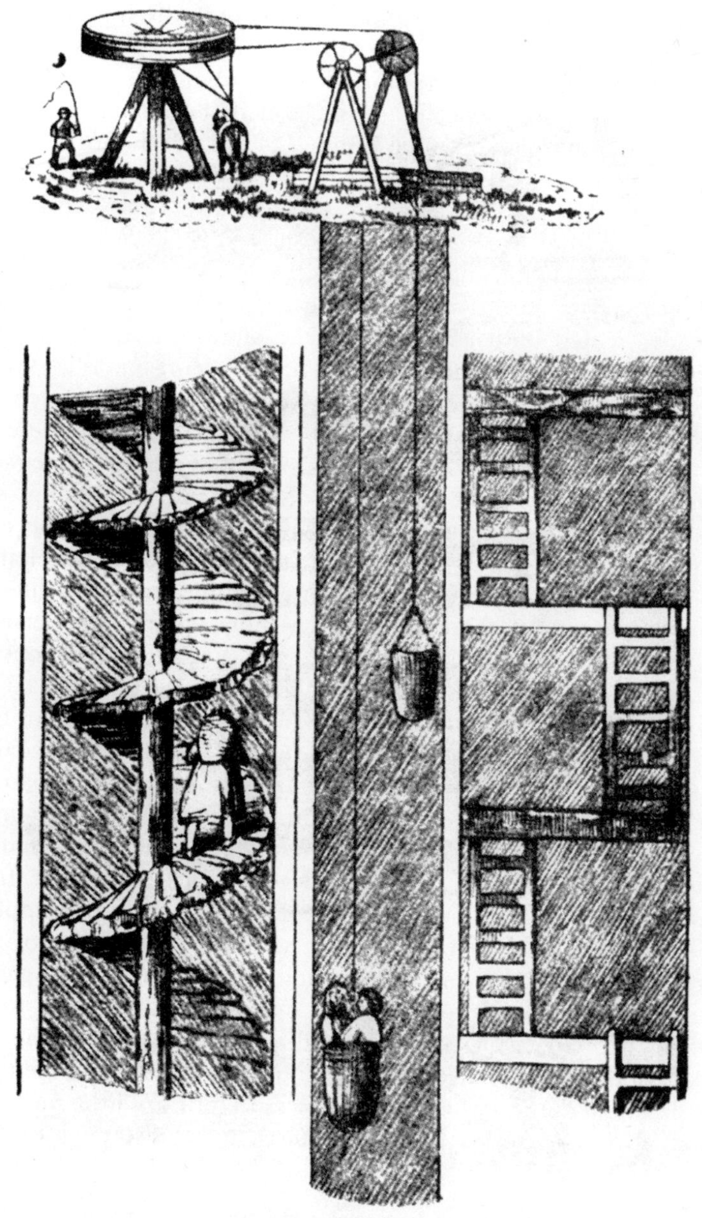

Fig. 48.—Raising coal to the surface (reproduced from P. M'Neill, *Tranent and its Surroundings*).

Fig. 49.—Slype and hutch drawing by women and boys (reproduced from P. M'Neill, *Tranent and its Surroundings*).

consuming centres. A number of the early pits were close to the centres of population, e.g. at Govan near Glasgow, and carts could conveniently carry the coal to the houses and to Broomielaw Quay for shipment, and even so late as 1808 most of the coal used in the city of Glasgow came from collieries only four miles distant.[5] Of necessity many pits were away from the burghs which had been established because of their suitable situation for agricultural markets and for commerce and a longer haul was necessary. In the days before good roads the coal was carried from the pit-mouth to the consumer, or the shipment point, either by creels on the backs of men or by packhorse, but, with the increase in the tonnage carried it was essential to get a speedier and a more commodious method. As early as 1722 the York Buildings Company had laid wooden rails downhill from the collieries at Tranent to the small port of Cockenzie, where the coal was shipped coastwise or to Sweden in exchange for high-grade iron billets. The route of this tramway is still partly followed by a railway. Other early tramways in to the Forth basin were the Carron Railway from Kinnaird Colliery to the ironworks laid in about 1766, the Elgin line constructed about 1768, the Alloa Railway constructed in 1768 and the Halbeath Wagon Road opened in 1783. It will be noticed that all but one of these primitive lines were from the coal districts to the seaboard and, simple in construction as they were and worked by either gravity or horse-power, they greatly facilitated the coal trade and permitted the movement of coal from the interior coastwise or abroad. The Halbeath Wagon Road is the best example as it was constructed by Sampson Garsyne Lloyd and Cornelius Lloyd, British merchants in Amsterdam, to develop the export trade of pits at Halbeath which they had purchased.[6] Unlike the Northumberland Tyne the Forth is not deeply incised and these railways terminated on quays, and not on staiths, alongside the vessels ; a factor which added to the shipping costs.

In the west of Scotland there were various projects for tramways, e.g. in the preamble to the first Act of the Monkland Navigation, but it would seem as if most of these remained as projects. The western area was at first more interested in canals than in railways as a method of reduction in the transport charges on coal—an interest perhaps aroused by the successful transformation of the Clyde from a fordable stream into a navigation. The first project was for the Forth and Clyde Navigation and powers were obtained in 1768 (8 Geo. III, c. 63) for construction, but owing to shortage of funds the proprietors did not

[5] R. Bald, op. cit., p. 36.

[6] W. Stephen, *The Story of Inverkeithing and Rosyth*, 1938, p. 102.

succeed in opening the canal for through navigation until 1791. Whilst the object of the work was to provide this through routeway the canal, traversing as it did coal districts, was of value for the development of the mineral trade from the opening of the first section between Carronmouth and Port Hamilton, near Glasgow, in November 1777. The Monkland Navigation Co. received its first Act in 1770 for a canal on the line laid down by James Watt. The canal was specifically designed to open up a new coal district to serve Glasgow, for the petition states:

> That the price of coal is daily increasing in Glasgow, to the great detriment of its trade and manufactures, and that there are almost inexhaustible funds of coal in the parishes of Old and New Monklands, which can be wrought at a very moderate expense, but on account of the distance, cannot easily be brought to Glasgow by land carriage.[7]

This canal was invaluable in the development of the region it traversed although at first the revenue was most disappointing to the proprietors. In 1792 it was joined by a cut to the Forth and Clyde Canal and so permitted Airdrie coal to reach the east coast. Some of the collieries alongside the Monkland canal shot the coal from the hutches, at the pithead, over screens, direct into the barges, which was an economical procedure avoiding the losses of the "coal-hills."

After the opening of the above canals there was a lull in construction which was broken by the formation of the Edinburgh and Glasgow Union Canal, opened in 1822, and of the eastern end of the Ardrossan Canal from the Clyde to Johnstone in 1812 (the western portion was never completed owing to the spread of the railway system).

The canals of central Scotland all carried a considerable coal traffic although only the Monkland Navigation attained 0·2 million tons in any one year. Even after the railway era had opened, these canals were of value to the works sited along their banks, for so long as the waterway was maintained the manufactories were able, under stringent protective clauses in the Acts which authorized the railway companies to acquire the canals, to obtain lower rates for transport of minerals and manufactured goods from the railway since otherwise they would send by water. This would not have had so much force unless, as did happen, the railway concerned was a competitor of the railway owning the canal. Useful as the canals were they had a drawback which was experienced in 1794 when a coal-pit was

[7] Quoted E. A. Pratt, *Scottish Canals and Waterways*, 1922, p. 146.

flooded out, with loss of life, by the Monkland Canal bursting its banks.[8]

The disadvantages of canals were various, but the major one which operated in the Glasgow region was that with the exhaustion of the pits situated on the canal banks it was necessary either to run cuts to virgin coal districts or to carry the coal by land to the canal side and there trans-ship it to barges. Land transport with trans-shipment was expensive, but it was tried. The Monkland and Kirkintilloch Railway was made to bring down coal to the canal, as also was its extension, the Ballochney Railway opened in 1828, two years after the M. and K. Railway. The Ballochney Railway was of great value since for the first time the mineral reserves of the high land north-east of Airdrie could be exploited. Other feeding lines of railway were the Hurlet line which joined the Hurlet collieries to Paisley, and so to a canal before 1824; the Wishaw and Coltness Railway which was opened in 1833 and extended the influence of the Monkland and Kirkintilloch Railway into central Lanarkshire; the Slamannan Railway, opened in 1840, which joined the Ballochney Railway to Causewayend on the Edinburgh and Glasgow Union Canal.

Besides these feeding lines railways had also been constructed as independent units to serve coal districts. The first of the public lines was the Kilmarnock and Troon Railway which received its Act in 1808 and was opened in 1810. This line was laid out to promote the coal export trade of Troon. The Garnkirk and Glasgow Railway was the first line constructed in Scotland which directly rivalled a canal, and it took traffic from the Monklands district to St. Rollox which was then in the outskirts of Glasgow. It became very popular for coal, as well as passenger, traffic, and the tonnage carried rose from 114,144 tons in 1832 to 254,010 tons in 1840. By means of reducing tolls and improving the works the Navigation Company was able for many years to continue to thrive despite this rivalry, but the canal ultimately fell into disuse and the works have recently been abandoned by statutory powers.

Elsewhere in Scotland there was an improvement of lines of communication leading in from the ports and the principal works in this connection were the Caledonian Canal, which never carried so much traffic as was hoped, the Inverurie Canal opened in 1807, the Dundee and Newtyle Railway, opened 1831, the Dundee and Arbroath Railway and the Arbroath and Forfar Railway both opened in 1839. All these improved links carried some coal into the interior along with other

[8] A. Miller, *Rise and Progress of Coatbridge*, 1864, p. 3.

imports. The Aberdeen and Inverurie Canal from its opening encouraged the use of coal instead of wood and peat. The traffic gradually increased as the table below shows.

Coal Traffic on Inverurie Canal [9]

1808	4,335 bolls
1809	5,521 ,,
1810	6,192 ,,
1832	1,888 tons
1833	2,184 ,,
1834	2,144 ,,
1835	2,659 ,,
1836	3,199 ,,
1837	3,968 ,,
1838	4,759 ,,

The opening of the railways provided new depots for the coalmasters, and since these were outside the previous group of suppliers it was impossible for the various attempts by a combination of suppliers to fix prices of coal to succeed for long.

For 1831 Cleland gave a view of the coal trade of Glasgow from which it is clear that improved communications had widened the area of supply from the time of Bald's account, when most came from a four-mile radius.

Supply of Coal to Glasgow, 1831 [10]

	tons
Collieries Govan to Polloc and Mountvernon	268,497
From 13 collieries by the Monkland canal	194,223
Suggested one-third added to this as barges carry more than dues paid	64,741
By Monkland and Kirkintilloch Railway, including 2,148 tons of coke	24,088
Cannel coal for gas works	9,050
Cannel coal for general use	450
	561,049
Of above estimated exported or used bunkers	124,000
Retained for domestic and industrial use	437,049

Often simultaneous with the improvement of land communications was that of the ports to permit the movement of coal from an adjacent coal district. Without giving a tedious list of particular works it is interesting to notice some of the constructions required to render the ports suitable for the larger tonnage of coal handled, the iron hulls of the steamships and the increasing size of vessels.

The harbours along the exposed shore of Midlothian were sited at

[9] Based on *N.S.A.*, Vol. XII, pp. 68–9, and W. Thom, *History of Aberdeen*, Vol. II, 1811, pp. 176–9.

[10] J. Cleland, *Enumeration of the Inhabitants of Glasgow*, 2nd Edition, 1832, p. 199.

small creeks and with the increasing size of vessels it proved difficult to enlarge these, and the nearby major port of Leith, although with a longer land haul, could handle the coal traffic more expeditiously. The West Lothian-Stirlingshire district has two ports, Borrowstounness and Grangemouth. The older port was Bo'ness and this port in 1843 consisted of two piers which provided shelter from easterly gales. A considerable drawback to the port was the liability to silting owing to the great mudbanks deposited in this reach of the Forth and attempts were made to counteract the trouble by the use of a sluicing pond which allowed a rapid flow of water across the basin at low tide. Grangemouth first became a port on the cutting of the Forth and Clyde Canal and at first the works were simple. The close of our period saw the construction of the first wet dock and it was only with the opening of this in 1843 that the port became significant. Previously it had just been a creek at the mouth of the Carron, and the vessels, the largest which could be admitted being of 90 tons burden, lay on the mud except at high water.

Along the north bank of the Forth was a series of minor ports. In the west was Alloa which for many years had as its staple the shipment of coal both coastwise and foreign. The coal was drawn from the immediate hinterland and the trade was early encouraged by the opening of the Alloa Railway so that by 1791 56,000 tons were shipped of which most went coastwise. The works were simple, consisting of a commodious quay alongside the river. At Charlestown, further downstream, were extensive limekilns built in 1777 and the wagonway which fed these with coal from near Dunfermline also facilitated the coal trade so that by 1792 a great tonnage, by the standards of the time, was exported. The harbour dried out at low water but vessels were able to lie safely on the mud then exposed. Inverkeithing made many attempts to retain its shipment of coal but was handicapped by the small size of basin which could be made and by the presence of a reef across the harbour mouth. While the works carried out kept the trade alive until after 1840, in which year 1000 tons a week were exported, the coal trade ultimately ceased as vessels grew still larger. Further east to West Wemyss the coast has a series of minor ports such as Dysart and Burntisland which carried on a coal trade. All these ports were exposed to south-easterly gales and all suffered from heavy seas at some time in their history. Methil, now the greatest coal port along the coast, could only be created after our period when engineering skill and financial resources were greater.

On the west coast the Clyde provided a waterfront near the coal deposits, but it was useless until the river was deepened after 1750.

The Forth and Clyde and the Paisley Canals brought coal to the lower Clyde before the days of the railways. South of Ardrossan the coalfield comes to, or near, the sea and the ports of Ardrossan, Irvine, Troon,

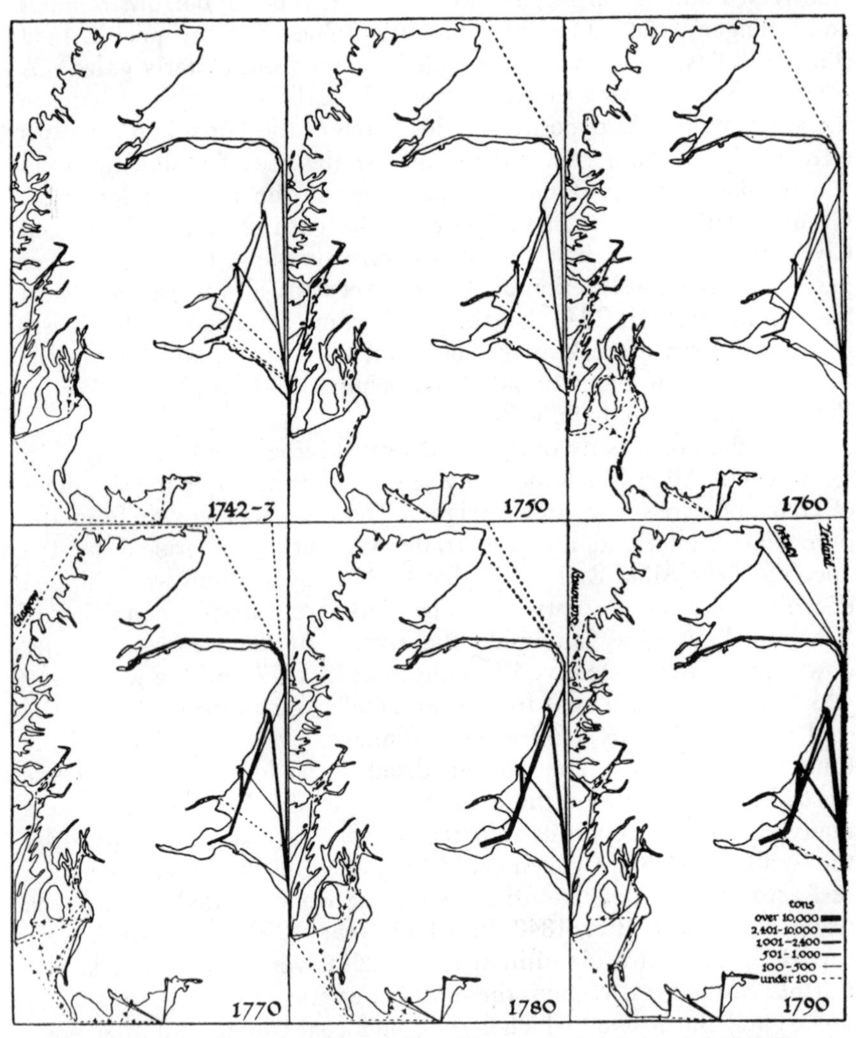

Fig. 50.—Coastwise coal arriving at Scottish ports during the period 1742–90.

The tonnage derived from Forth and N.E. England shipment points grouped for each of those areas (based on MS. Customs Accounts, Register House, Edinburgh).

Ayr and Girvan have had an interest for long in the coal shipment trade. Ardrossan was created as a harbour in 1806 by Lord Eglinton in connection with his canal scheme, but did not become important until the railway to it was opened in 1836. Troon was a creation of the Duke of Portland who from 1806 commenced the works which culminated in a wet dock cut out of solid rock in 1842. The port was fed with coal from the Kilmarnock district and in 1839 130,500 tons were brought by the Kilmarnock and Troon Railway for export. Irvine, Ayr and Girvan are all at the mouths of rivers and all were subject to bars formed when the rivers came down in spate with a heavy load of sand and gravel. The sheltered mouths of the rivers gave sites for earlier port development than the more projecting sites of Ardrossan and Troon.

While it is impossible to present a picture in detail of the movement of coal about Scotland during the century Fig. 50 shows one facet—viz. the growth of the seaborne coal trade as revealed by the manuscript Customs Accounts. The increasing trade of the shipment ports in most, if not all cases, implies an improvement in the land communications of the immediate hinterland. The carriage of coal by sea from the Forth ports was burdened by duty of 3*s.* 8*d.* per ton if it went north of Redhead and south of Dunbar. English coal reached the Scottish market without this charge which was a factor in favour of its use along the north-eastern seaboard. Some of the coal which went coastwise undoubtedly was later exported both legally and otherwise.[11] After the end of the records of the series of the Customs Accounts used for Fig. 50 the movement of coal by sea grew still further and the coastwise shipments in Scotland rose from 455,000 tons in 1825 to 557,000 in 1835 and 724,000 tons in 1840. At the later date the most important shipping points in decreasing order were Irvine, Bo'ness and Glasgow, all of which sent out over one hundred thousand tons.[12]

For the close of the century under consideration the situation of the fuel supply of Scotland is summarized in Fig. 51. The coalfields and areas reached by railway or short road haulage burnt coal as did the coastal districts. From the importing ports the spread inland varied directly with the improvement of the land communications as is seen if the districts near Aberdeen and Peterhead be compared.

In conclusion, the century here considered saw the change from human to tramway carrying of coal in the pits, the change from cart and packhorse to canal and from canal to railway in the journey from the pithead and the change from the minor creeks to the better endowed or equipped ports for trans-shipment. It was these changes which laid the foundations for the coal traffic of the following century

[11] V. E. Clark, *The Port of Aberdeen*, 1921, p. 101.
[12] R. Meade, op. cit., pp. 274–5.

and the routes laid out in response to prevailing geographical controls of a century or more ago still tend to be followed today.

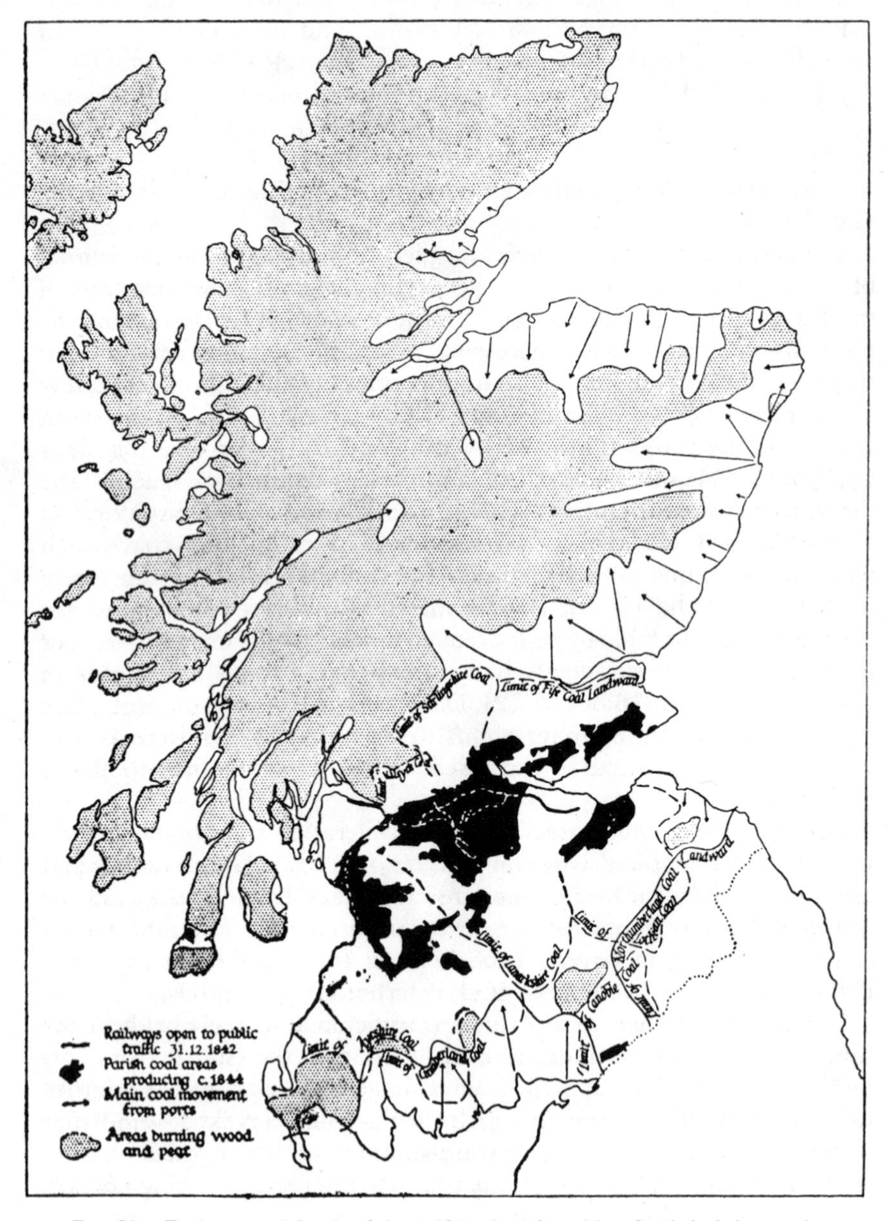

FIG. 51.—Fuel areas of Scotland in *c.* 1844 (based on New Statistical Account).

XIV

GEOGRAPHICAL FACTORS IN THE EXPLOITATION OF MINERALS

By N. J. G. POUNDS, M.A., Ph.D.

The existence of a mineral deposit, locked up in the crust of the earth and waiting on man for its discovery and utilization, is one only of several factors in its exploitation. Its value is conditional. It may be of such overwhelming importance that difficulties and high cost of mining become of relatively small importance. If it consists of the more common minerals, the costs of working, the amount of capital required, the size of the ore body and the likelihood of the complete amortization of the equipment—for old mine gear has very slight value—by the time that the reserves are exhausted, assume a greater importance. Human nature is fallible and human knowledge limited; the variables and the unknowns in this equation are so many and so great that in no occupation is there such risk as in mining, nor such possibility of profit; "so expensive in its operations and so uncertain in its success," as William Pryce expressed it, "that few or none are carried on at the risk of one or two persons." [1] The dangers and uncertainties of gold-mining—such is the value of gold—have not been too great to attract thousands to California or Klondike. But there has been no iron rush, at least, not since the end of the Bronze Age, and within no measurable space of time are all the known ironfields likely to be worked, least of all worked out. It may be said, however, that a deposit containing appreciable quantities of the precious metals will always have value, wherever it is situated, while reserves of copper, lead, zinc, tin and others of the more important non-ferrous minerals will be studied with a view to future development. The exploitation of iron ores lying in remote areas is likely to remain for many years highly improbable, while beds of clay, limestone or building stone, are so common that they arouse little interest and scarcely add to the value of the land in which they lie.

The factors which help to give value to minerals, determining in varying degrees the extent to which they are likely to be developed or whether they will remain undisturbed, may be summarized thus:

(i) *Geological.* These include not merely the grade of the ore, which may determine whether or not it can stand the cost of transport; the

[1] W. Pryce, *Mineralogia Cornubiensis*, 1778.

presence of small quantities of subsidiary minerals, which may, on the one hand, like silver in lead-zinc or gold in copper, give a greatly enhanced value to the ore, or, on the other, like titanium or phosphorus in iron ore, increase the difficulties of smelting and refining, and diminish the value of the ore. The shape of the ore-body, whether massive or bedded, made up of lodes or stringers, or consisting of the impregnation of a country rock by mineralizing vapours, all influence the method and cost of working. Open-cast mining is sometimes possible on shallow bedded deposits of a medium or low grade. Many of the porphyry copper ores of the United States and Chile, the copper sulphides of Rio Tinto and the Kimberley and Pretoria diamond pipes are examples. Further geological factors are the strength of the country rock, determining the extent to which timbering is necessary, the hardness of the rock for cutting and boring, the presence of quartz-bearing minerals, which may make precautions against silicosis necessary. An example is known of mild poisoning from the occurrence in a lode of the arsenic-iron sulphide, Lollingite.

(ii) *Geographical Factors.* Transport and communications constitute the most important factors of a specifically geographical nature. Climate and water supply, problems of feeding the miners and maintaining health ; mountain sickness and malaria ; water in the mines and the temperature gradient underground ; supply of fuel and power and recruitment of labour, are others.

(iii) *Economic Factors.* The world consumption of the minerals produced, the introduction of valorization and restriction schemes, the price levels attained, the possibility of the introduction of alternative materials and of the discovery and development of richer or more easily worked deposits are all of importance. A problem of great interest arises in connection with asbestos and china clay, both of which are obtained in more than one quality in fixed or only slightly varying proportions, command different prices and are normally destined for different uses, though interchangeable to some extent, according to market conditions. Technical developments, lastly, may bring vast reserves of metal within the sphere of economic working. The cyanide process has thus replaced the mercury amalgam method of separating gold. The introduction of the flotation process in ore-dressing has lowered the grade of workable copper ore from some 2½ per cent of metal to 0·75 per cent, thus bringing within the sphere of operation the great reserves of the mountain states of the U.S.A. Many of the low-grade ore producers of the Canadian Shield owe their success entirely to this process.

The factors which have been outlined in these paragraphs do not lend themselves to mathematical analysis, and, while they are necessarily taken into consideration in mine valuation, there must remain a wide margin of error in any estimate based on them. In general it may be said that a rich, easily worked ore-body can compensate for difficulties of transport and for an adverse physical environment. This is not only true of the precious metals but also of those non-ferrous ores in regular demand. The gold fields of the West Australian Desert, the silver-lead of Broken Hill (N.S.W.), pitchblende of the Great Bear Lake region, and copper and tin in the high Andes and iron ore in Lappland are all examples of mining carried on in harsh climates, where the supply of food and materials and even of water is difficult. On the other hand, poor ores may be worked if other circumstances are favourable, the low grade iron ores of the English Scarplands, for example, and the even poorer ores between Hannover and Braunschweig, which the Germans with government support have recently been able to utilize.

The following essay is concerned primarily with the second group, the geographical factors in the development of the world's mineral resources, but it should be clear that no judgment which underrates or ignores the rest can have any validity whatever.

The appearance of capriciousness in the distribution of mineral resources through the crust of the earth is modified by the knowledge of the periods in the earth's history when the mineralizing processes were most active. The bedded minerals are associated primarily with three geological periods ; coal with the upper Carboniferous ; rock-salt, potash, gypsum, anhydrite and similar minerals mainly, though not entirely, with the Trias, and bedded iron ores, in Europe at least, mainly with the Jurassic. The formation of the lode minerals and of those ore-bodies derived in one way or another from mineralizing vapours and solutions has been, in general, contemporary with the great periods of mountain building. Tertiary ores are infrequent in Europe ; common in the Americas. In consequence, the chief mineralized regions are found, in Europe at least, in areas of Caledonian and Hercynian folding ; in the United States, mainly in the Tertiary and Laramide, but also in earlier mountain systems.

Mining has always been carried on more in upland and mountain districts than in the plains, on the fringes of civilization rather than at its centre. To Aristotle the miners were " men of the hills " ; to the Welsh they were *gwyr y mynyddau* (men of the hills). The Germans knew a mine as a *Bergwerk*, and miners as *Bergleute*, latinized as " *montifodina* " and " *montani* ".

It might seem that the most important geographical problems confronting the miner, both in early times and today, would arise from the harshness of the climate of upland districts and from the difficulties of transport. These may be eased in the mineral regions of greatest age by the peneplanation of the land mass, which has softened the asperities of its relief. Nevertheless, these oldest rocks occur in the "shields" whose size, extremes of climate and relative unproductivity of anything except minerals have hindered their development.[2] The essentially geographical factors in mining may be grouped under five heads :

(i) *Climatic.* Surface temperature changes are scarcely felt a few feet underground, but drought and frost are of serious importance in the surface operations. They hold up equally the ore-dressing processes, whicn depend in greater or lesser degree on running water. On many occasions and in several places, working has been confined to the winter months. Examples range from the alluvial gold diggings of California to the china-clay pits of Cornwall. The provision of a perennial water supply by means of reservoirs has been possible in Cornwall, but there are many places where such a scheme could not possibly be economic. Climatic factors may slow down other surface processes. In Swedish Lappland, the stock-pile at the iron mines freezes and the ore is found to be frozen into the trucks which carry it to Narvik, so that it has to be thawed out with steam. In Alaska and the Canadian North, placer deposits are frozen hard throughout the year below a depth of a few feet. Here a method has been evolved of injecting water into the deposits before working them. The rarified atmosphere at mines of great altitude induces mountain-sickness, and diminishes the efficiency of labour. The problem of the supply of drinking water becomes acute in mines situated on the desert fringe of Australia, South Africa and North America, and this factor is likely to render improbable the exploitation of any except the most valuable minerals, and these only in a relatively high degree of concentration, in these areas.

(ii) *Transport.* The movement of ores from the mines to the coast or to the consuming centres is likely to be a downhill one, but distance and the irregularity of the terrain more often than not neutralize this advantage. Most of the important mining regions have now been linked with other areas by rail, though it is doubtful in many cases whether the line would have been built if the expectation of traffic

[2] J. J. O'Neill, "The Exploitation and Conservation of Mineral Resources," *Trans. Roy. Soc. Canada*, XXXIV, iii, 1940, pp. 1–14, stresses the contrast between the Pre-Cambrian mineralized areas of Canada and the more recently mineralized areas of the U.S.A. in connection with the problem of absorbing into agriculture or industry the mining population when reserves are exhausted.

derived exclusively from the mines. Recent discoveries in the Orange Free State are, it seems, thought sufficient justification for such development. There must, on the other hand, be an immense number of ore bodies, unfavourably situated with reference to rail or water transport, whose development must remain improbable because the expected gain from the sale of the mineral would not cover reasonable profit and also the amortization of investment in gear for mining and transporting the produce. A similar deposit close to a railway already built might reasonably be worked. The cost of transport of minerals has to be related, not merely to distance, the difficulty of the terrain and to whether or not the line bears other transport, but also to the grade of the mineral. Only in exceptional cases will a low grade iron ore bear the cost of transport over any great distance before smelting. Such ores in Germany were smelted close to the workings at Peine; in Lorraine, much of the ore worked is smelted on or near the ore field, and in England the tendency is for the low-grade bedded iron ores to be smelted close to the workings. The journey from the upper end of Lake Superior to the Pittsburg area must be considered an unusually long one. In matters of this kind, however, the map is an uncertain guide, because the factor of mere distance is only one of several. More important than the extension of the distance of land or water travel is break of bulk. Steel works are more mobile than iron mines. In the U.S.A. the location of furnaces at Gary, Detroit and other lake-side ports illustrates the tendency to avoid, where possible, the transference of the ore from ship to rail. In England, the migrations of the iron and steel industry from the middle ages until the present, are in large measure, responses to changes in the location of the major sources of ore.[3]

If, however, ore is smelted close to the mines, this normally necessitates an import of fuel, and it is largely a matter of comparative costs, whether the ore is smelted near the mine, exported as ore, or partially smelted to a concentrate or matte, if this is technically possible. In many instances, coal forms a return cargo for ore, so that a smelting industry is carried on at each end of the axis of transport. Very broadly, this is what happens in the Great Lakes area; it also happens between the Magnitogorsk iron field and the Kuznetsk coal basin. In the Katanga region, the copper is smelted to a concentrate for export. In North America there tends to be a number of smelters comparatively well-placed for each of the mineral producing areas, though the Sudbury nickel ore is concentrated to a matte and finally smelted at Port Colborne, 300 miles away on Lake Erie. The factors

[3] "Iron and Steel Industry," Cmd. 6811, pp. 12–13.

which determine the pattern which the movements of ore, fuel and concentrates will assume are not, however, necessarily those which would be operative under conditions of free competition, and are not necessarily explicable wholly on economic grounds.

(iii) *Local Factors.* Few mining areas are without some lesser disadvantages, which may be, on balance, offset by corresponding advantages. Stable foundations for mine buildings ; firm rock through which to sink the shaft and to form a roof ; freedom from flooding and from excessive water in the pit, are all factors which contribute to the profitability of mine-working. The closing down of one mine and the cessation of its pumping increases the difficulty of keeping neighbouring mines clear of water, as recent instances in Cornwall have shown. The subterranean temperature gradient may be steep, and mechanical ventilating devices or air-conditioning may become necessary at a comparatively shallow depth. The gradient in the Rand mines is unusually gentle, greatly increasing the ease of working.[4] In general, the gradient steepens when the shaft passes through zones of oxidation in the lodes.

(iv) *Fuel and Power.* Only alluvial mining, and this only on a very small scale, can be carried on without power-driven machinery. The older mining regions, situated generally in mountainous or hilly country, with only small units of production, were often able to use local resources of water power. The ingenuity displayed by the early miner is delightfully shown in the woodcuts which adorn Georg Agricola's *De Re Metallica.* It could be seen until recently in the devices of the Cornish miner to extract the last horse-power from the streams to operate his pumps, winzes and buddles. Such dependence, however, could in many cases be only seasonal, and the units of production were necessarily small. The utilization of larger water-power units has been relatively costly, and, in general, more than the mining interests alone would appear to justify. Canada is, however, particularly well favoured in this respect, and here greater use has been made of large-scale hydro-electric constructions than elsewhere. In certain instances it was thought cheaper to set up the installations required than to open a route through the forest for liquid or solid fuel, even for a mine whose expectation of life was not great. Electric power, whether hydro or thermal, is probably the most satisfactory form of power that can be used on mines, because of its mobility, its rapid acceleration and of the varied uses to which it can be put. Power

[4] It would appear that man's capacity to stand the high temperature and humidity of deep mines has been underestimated. See *Mining Magazine,* Sept. 1940, pp. 131–7, and Jan. 1946, p. 27.

is required, not merely for pumping and winding, but also for air compressors, tool sharpening and grinding equipment, the operation of the many and varied ore-separating plants and other similar purposes. Current is generated on the mine or taken from a public supply in most cases. Power generation on the mine and the utilization of steam power is made possible only by an import of fuel. Only in rare cases is coal found in close juxtaposition to metalliferous mines, a consequence of the fact that upper Carboniferous measures are not highly mineralized. At the present day, the cost of coal is one of the major items in the operation of mines remotely placed from coalfields such as, for example, those of Katanga. The mining history of Africa is, in fact, very closely linked with that of railway development, and exploitation of most of the major mineral areas had to be delayed until railway construction had been completed.[5]

It is, however, in the location of the attendant smelting industry that the cost and availability of fuel is most important. Long-established mining and smelting industries have built up around themselves a nexus of services, a pool of skilled labour and a body of " good will." These may represent so great a capital asset that any movement of the industry to take advantage of cheaper fuel would not be worth while. With a developing mining industry, however, and most of the great mineral regions of the world are in process of development, a degree of planning is usually possible, and the smelting industry can sometimes be located in a position which allows it to profit most or suffer least from the relative location of ore and fuel. Briefly, the smelting operations can be carried on at the mine, at the source of fuel or at some intermediate position, which allows of easy access to both. Relative costs of fuel are to a large extent a function of the cost of transport. A further possibility is for the ore to be partially smelted to a concentrate at the mine before export. A familiar example of the latter is the production of matte, or " blister " copper on the minefields of Katanga and Northern Rhodesia and also of a nickel matte at Sudbury for refining elsewhere. The Central African copper is of a low grade, some 3 to 5 per cent, and the export of ore would clearly be costly in relation to its metal content. The import of sufficient coal for complete smelting would also be expensive. There are many instances, dating mainly from a period when the mechanism of transport was less developed than it is at present, of the movement of ore to changing sources of fuel : the migration of the Swedish ore through the forests to the " bruk " where there was an adequate

[5] S. H. Frankel, *Capital Investment in Africa*, 1938, p. 249. Also Lord Hailey, *An African Survey*, p. 1484.

supply of charcoal [6]; of the Cornish tin in search first of charcoal, and then of coal.[7] Lastly, the movement of refined bauxite to the sources of developed hydro-electric power constitutes the extreme example of the dependence of smelting operations on an abundant supply of cheap and non-transportable power. Alternatively, fuel has sometimes moved towards a changing supply of ore, as in the North Riding of Yorkshire and in South Wales. Unless political considerations intervene, the dominant factor in the pattern of movement of fuel and ore is likely to be the grade of the ore itself and the ratio of coal to ore used in the furnace.

(v) *Labour.* The cost of labour, it has been said,[8] usually accounts for some 60 per cent of the whole cost of operating a mine, and it hardly needs emphasis that a supply of labour is an essential of all except the most rudimentary forms of mining. It is apparent also, from what has been said of the geology of mineralized areas, that metalliferous mines are at most likely to occur in areas of no high fertility and of relatively sparse population. There have been few mining areas opened up during this century or the last where the supply of labour has not been a more or less serious problem. The development of reserves in the Canadian North and in the desert regions of northern Chile, South Africa and Australia are cases in point. "The distinguishing feature of labour problems in Africa," wrote Lord Hailey, "is the inadequacy of the supply of labour." [9] The recruitment of native labour for the mines in all Africa south of the Sahara has become not an occasional necessity, but a regular business. There are on the Rand two organizations, the *Witwatersrand Native Labour Association* and the *Native Recruiting Organization*, which together employ over 200 European and nearly 2,000 native recruiting agents.[10] The problem has been scarcely less serious in the Belgian Congo, and the recruitment of labour was extended for a time even to the mandated territory of Ruanda-Urundi. It has not been easy "to steer a course between the pressure of humanitarian influences and the demands of employers." [11]

Not only is labour likely to be scarce; its quality in areas in which the white population is small is likely to be inadequate.

Labour costs (it has been said) are almost always greater in districts depending heavily upon native labour. Irresponsibility, incapacity to

[6] G. A. Montgomery, *The Rise of Modern Industry in Sweden*, 1939, p. 84.
[7] A. K. Hamilton Jenkin, *The Cornish Miner*, 1927, pp. 177–8.
[8] T. J. Hoover, *The Economics of Mining*, 1933, p. 143.
[9] Lord Hailey, *An African Survey*, 1938, p. 603.
[10] *The Mineral Resources of the Union of South Africa* (Dept. of Mines, 1940).
[11] Lord Hailey, op. cit., pp. 645–6.

use explosives and machinery, lack of judgment and initiative, and need for constant supervision are such marked traits of most of the races furnishing large bodies of cheap labour that these characteristics more than counterbalance any saving in wages.[12] On the other hand, the costs of supervision rise and a highly paid managerial group is required. Mining in Mexico where labour is cheap but notably inefficient, is not cheaper than similar work in the States.[13] It is not uncommon for labour to prefer agriculture to mining work and to undertake the latter only in the last resort. In most areas where coloured labour is employed in mines, men are recruited for short periods only and, although many, especially the Chinese, are capable of becoming comparatively expert, they do not stay long enough to acquire more than the rudiments of mining technique. Least of all can such a temporary and shifting labour corps produce skilled foremen and overseers. It is unwise, however, to generalize. Each body of native labour should be studied in relation to the type of mining work that it is called upon to perform.[14] Chinese labour is thus intrinsically cheap, and the mineral output, particularly of tin, wolfram and antimony of the interior of China is a measure, not so much of the mineral wealth of the country, which has been greatly overestimated, but of the relative cheapness and efficiency of labour.[15] Again, much of the mining of prehistoric times was carried on in deposits that were essentially poor with a labour corps that was large and cheap.[16] If the labour is cheap, it can, under suitable conditions, displace or prevent the introduction of machines. The alluvial tin-fields of Nigeria and Malaya afford examples. An example of the marginal profitability of cheap labour can be seen in the china clay mining industry of Cornwall. A large corps of poorly paid and but slightly skilled labour had been used to break down the clay in the "stope." After the strike of 1911, its wages were increased and it at once became more profitable to instal pumps and hoses and to work the clay by the Californian method of hydraulicing, so reducing very considerably the number of employees.

The working miner of the non-ferrous metalliferous mines of Europe has been, in the past, a very highly skilled workman. The nature

[12] T. J. Hoover, op. cit., p. 143.

[13] S. R. Finlay, *The Cost of Mining*, p. 48. These opinions are confirmed by H. F. Bain in *Min. & Met. Soc. Amer.*, No. 114, 1921.

[14] F. Spearpoint, "The Native African and the Rhodesian Copper Mines," *Journ. Roy. Afr. Soc.*, 1937, Supplement, July, is a useful review of this problem in one area.

[15] W. F. Collins, *Mineral Enterprise in China*, 1918, p. 31, and H. F. Bain, *Ores and Industry in the Far East*, passim.

[16] T. A. Rickard, *Man and Metals.*

of the work in lode deposits, often faulted, and frequently containing a considerable variety of economic minerals, called for no ordinary degree of geological knowledge. With the decline of this mining, particularly in Cornwall, in the nineteenth century, an extremely competent labour corps was scattered to the minefields of newer areas, perpetuating in some degree the traditions of the older. But changes in ore-dressing methods in the twentieth century are tending to make possible the utilization of lower grade deposits. Mining becomes less selective and the onus is transferred from the miner to the assayer and ore-dresser. The miner is ceasing to require the skill of his forefathers, but is beginning to require techniques of a different kind, a knowledge of the internal combustion engine, of turbines and dynamos, rather than of the behaviour of lodes. It seems likely that the old race of skilled miners will gradually be replaced by a very much less skilled labour corps, dependent upon a very few, highly trained geologists and technicians. There are many who view the change with grave misgivings.

From this brief consideration of certain factors in the profitability of mining we turn to an even shorter review of its geographical distribution. De Launay once developed the thesis that in any major unit of the earth's surface, minerals have been exploited in a certain order, which happens, at present, also to be their order of diminishing value. In a newly discovered land mass, the first to be mined would be gold, followed by silver. Then the non-precious metals would come within the range of profitable exploitation, beginning with copper and followed with lead, zinc and iron. One might add tin, which historically would come between silver and copper. It can readily be appreciated that the precious metals would be sought for before the non-precious and that base metals would not be imported from the newly discovered continents to Europe which had an abundance. It is perhaps less apparent that Europe has itself passed through these phases of mineral development. The approaching exhaustion of each has in turn stimulated production in remoter areas. Gold and silver were being worked in the American Cordillera when tin and copper were the most important mineral products of Europe. In Europe the chief copper deposits, those of Rio Tinto and Tharsis, had first been worked in classical times for gold and not seriously for copper until the sixteenth century, when the enterprise of the Fuggers was predominant.[17] Gold-mining was of some small importance in England during the middle ages, but, for a considerable period, tin was

[17] R. Ehrenberg, *Capital and Finance in the Age of the Renaissance*, and W. G. Nash, *The Rio Tinto Mine*, London, 1904.

the chief mineral product until the development of copper production at the end of the sixteenth century. The search for the more valuable base metals has since been pushed further towards the peripheral regions of the settled and developed parts of the world. The gold of Cripple Creek and the silver of the Comstock lode have yielded pride of place to the coppers of Idaho, Montana and Arizona, while in the more settled regions of the East, the copper of Michigan pales before the iron ores of the Great Lakes.

Such a picture of mining development can be accepted only in very general terms. It applies only to a small range of minerals, known since ancient or medieval times; and it takes no account of the "new" minerals, whose value has been discovered only recently. These minerals, especially nickel, chrome, manganese, tungsten, antimony, molybdenum are more local in their occurrence than the "older" minerals, and their exploitation may "rejuvenate" a country, such as Portugal, which had long been abandoned as a serious mineral producer. Nevertheless, the "older" minerals remain of overwhelming importance, and are likely to remain predominant in the near future. Nor does this scheme consider minerals known since ancient times, but little used, such as mercury. Lastly, such a scheme can only be put forward subject to the caprice of nature in concentrating minerals into deposits of workable size. It is evidently necessary, in any case, to take unit areas large enough to have a varied mineral endowment, but not too large to have had a more or less uniform historical development. If there can be said to be cycles in mineral production, Europe can be said to be nearing the end of its cycle; it is definitely in its iron age. South Africa remains in the age of gold, though perhaps on the threshold of a later phase of development.[18] South and Central America, Australia, Central Africa and Asia are progressing into the phase when the mining of the non-ferrous base metals is predominant.

It might be asked whether this trend is reversible, whether any region that has reached the stage in which iron is the dominant mineral can again become a producer of more valuable minerals. No simple answer is possible. A number of minerals are widely scattered through the earth's crust, particularly iron, aluminium, manganese, titanium, and in the foreseeable future no shortage is conceivable provided that the rising costs of production, consequent on the mining of lower grade deposits, can be met. On the other hand, there is a range of minerals which, by the nature of their occurrence, are concentrated in a relatively

[18] The accomplishment of Lord Hailey's prophecy, *African Survey*, 1935, is likely to be postponed by further discoveries of gold.

small number of areas. Tin, lead, zinc, mercury, nickel belong to this group, and copper is but little more widespread. Absolute exhaustion of these is possible, and some European producers are either extinct or have only small reserves—Cornwall, Sweden, Mansfeld, Freiburg, Rio Tinto. It would be rash to assume that a region which has ceased to produce is necessarily worked out, because an ore body can only be exhausted within the scope of contemporary mining technique. Perhaps the Cornish copper is the only significant European mineral deposit that can be said to have been completely exhausted. It is nevertheless true that any region that has been intensively worked and studied for several generations becomes fairly well known, and it is not likely that any major ore bodies remain to be discovered.

It has been said that any considerable advance in mining technique would open up very deep deposits now below the limit of economic mining. The thickness of the mineralized zone of the earth's crust, however, is not great, and, with the exception of gold mines, metalliferous mines, are comparatively shallow. "Most of the mines in the classic metal districts, such as Joachimsthal, the Harz, Linares and Almaden, range from 1,200 to 1,600 ft. . . . and in most places little or no deeper work seems to be in sight."[19] The vertical range of minerals is comparatively small, and, unless there is a clear zonation of minerals in depth, deep mining holds out little prospect. This zonation of minerals has been one of the most clearly marked and influential features of the mineral lodes of the south-west of England, but it is less clearly marked in Europe. It would appear, then, that short of the discovery of fresh deposits in the older mining areas—a far from likely contingency—the major part of the supply of these more localized minerals will be obtained from areas nearer the pioneering fringe of mining, where the greater costs of mining will be met automatically as supply becomes more restricted.

Editors' Note

Dr. Pound's essay offers a general discussion of factors and principles which are further illustrated elsewhere in this volume, notably in the essays of Mr. Beaver and Mr. O'Dell. We may call attention also to the remarks of Dr. Willatts (pp. 297–300) on the mineral aspect of land-use planning. Conditions in Britain present indeed a rather exceptional case ; there are few areas which are not, potentially,

[19] D. F. Hewett, "Cycles in Mineral Production," *Trans. Amer. Inst. Min. and Met. Eng.*, 1929, p. 65.

the source of some mineral product and in particular, the high degree of urbanization puts a premium on a wide range of building materials. The circumstances of the sand and gravel industry have recently been studied in detail. In quantity its products are second only to coal in the country's mineral output and they add very greatly to the value of the land beneath which they occur (cf. p. 241). The required deposits are widely spread. The product is bulky and of low intrinsic value and rarely travels far from its point of production. The cost of loading gravel and transporting it by road to a point two miles distant may involve an addition of 30 per cent to the cost of excavation and dressing. For every additional mile of haul the cost rises by a further 6 per cent; at thirteen miles from the pit the cost is nearly doubled. These facts make clear the magnitude of the problem involved in determining "optimum" land use. Sand and gravel frequently underlie land of high agricultural quality.

The questions here briefly indicated are fully discussed in the Report of the Advisory Committee on Sand and Gravel, Parts 1 and 2, H.M.S.O. 1948, and in *Geog. Journ.*, Vol. CXV, 1950, p. 42.

XV

THE AGRICULTURAL REGIONS OF BELGIUM

By STANLEY W. E. VINCE, M.Sc.(Econ.)

Introduction.

By any standards Belgium is principally urban and industrial. 60 per cent of its 8,386,553 people live in communes of over 5,000 in population, and one quarter are in the conurbations of Greater Brussels, Antwerp, Liége, Ghent, Charleroi and the Borinage.[1] 90 per cent of its exports by value are manufactures, and heavy imports of wheat, maize, barley, vegetables, fruit and wines are necessary.

The balance of agriculture and industry is, however, closer than in Britain.[2] Before the war there were large export surpluses of eggs and market gardening produce, particularly tomatoes and glass-house crops, and the country was virtually self-supporting in beet sugar and meat.

Belgian agriculture is, then, especially worthy of study. A notable feature is the importance of small holdings and a long and strengthening tradition of intensive cultivation of " industrial " crops. This tradition arose naturally from the mediæval textile towns of Flanders, in a region of light sandy soil which, if not inherently fertile, responded gratefully to heavy manuring with urban waste and to intensive cultivation stimulated by urban demand.

Statistics show [3] that 91 per cent of all holdings (1,033,797 out of 1,131,146) are below 5 hectares (about 12 acres), of which three-quarters are smaller than 1 hectare (2·42 acres). Many tiny holdings are worked part-time by industrial and other workers, especially in Flanders and the Borinage.

[1] Greater Brussels, 912,774 ; Antwerp, 492,645 ; Liége, 253,689 ; Ghent, 213,844 ; Charleroi (with the 5 satellite towns of Jumet, Montignies sur Sambre, Gilly, Marchienne au Pont, and Marcinelle), 147,187 ; The Borinage (Mons, Quaregnon, Wasmes, Jemappes, Frameries, Boussu and Hornu), 108,758. Total : 2,128,591. Based on estimates in *Le Moniteur Belge*, April 13, 1939 (Brux.).

[2] The Recensement General of 1930 gives 16·9 per cent. of the working population in Agriculture and Forestry, compared with 5·7 per cent in Britain in 1931.

[3] Statistics of holdings are based on the Recensement Général Agricole, 1929, the official statistics of the Ministry of Agriculture and the most recent available. These give the numbers of holdings of each size category for each of the 41 arrondissements or administrative counties.

Notable features are therefore :

(i) the frequency of intensive cultivation and of industrial crops, merging imperceptibly into horticulture.

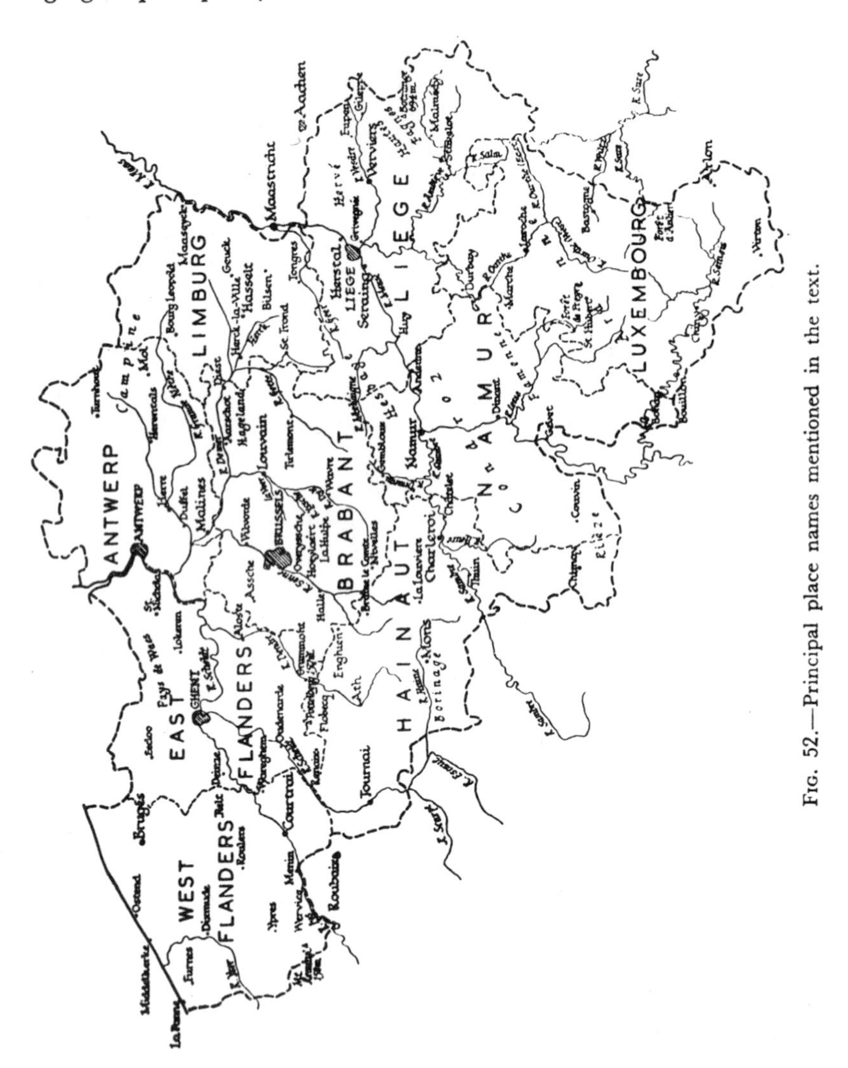

Fig. 52.—Principal place names mentioned in the text.

(ii) The frequent association of industrial workers with part-time agricultural holdings.

(iii) The prevalence of small holdings, resulting in a remarkably high density of rural population especially in Flanders, unique in Western Europe.

(iv) The important export market for certain specialized agricultural produce.[4]

Despite its smallness (one half of Scotland) there are striking regional

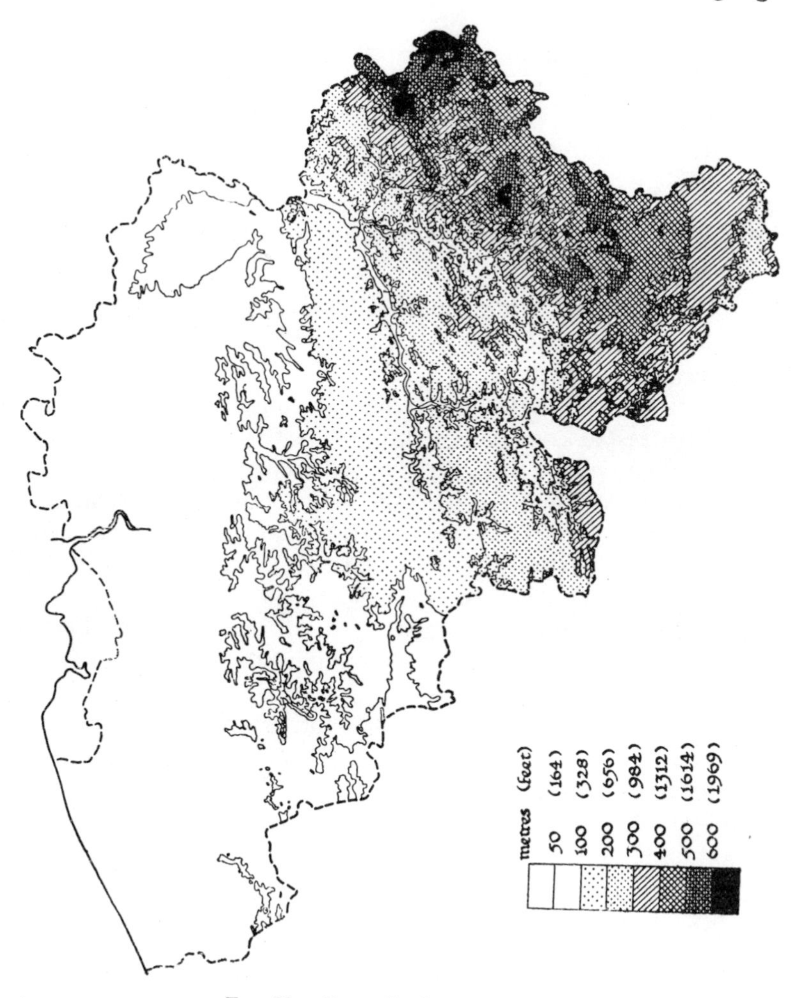

FIG. 53.—Generalized relief features.

While showing the special significance of the 200-metre contour along the line of the Sambre-Meuse in delimiting upland from lowland Belgium, this map also illustrates the function of the 300-metre contour in dividing the High Ardenne from the Pre-Ardenne. The depression of the Famenne and the ridge and vale topography of the Condroz is also shown.

Within Lowland Belgium, the highest part of the limon-covered plateau immediately north of the Sambre-Meuse line (and one of the chief regions of large-scale arable farming) is seen to be relatively undissected, contrasting markedly with the mature erosion further north and north-west by the Dendre and Dyle, tributaries of the Scheldt.

---

[4] Statistics of export trade are obtainable from the *Bulletin mensuel du Commerce avec les Pays étrangers, de l'Union Economique Belgo-Luxembourgeoise*, 1939, Bruxelles, 1940.

variations in agriculture which can be clearly related to the physical background and economic pattern, controlled by geographical factors. The present study attempts to distinguish precisely agricultural regions based upon crops, land-use and farming practice, and to interpret this classification in relation to the geographical background.[5] A short conclusion deals with the effect of the German occupation, in order to test the strength of the geographical factors previously discussed and the stability of this regional classification, during the greatest food crisis in Belgian history.

The Physical Background—Structure and Relief

The valley of the Sambre-Meuse is the approximate boundary between the Palæozoic Ardenne Massif (exceeding 600 and usually 1,000 ft.) and the plains and lower plateaux of central and northern Belgium, formed of gently dipping Cretaceous and Tertiary material, usually mantled with Quaternary limon and fluviatile sands.[6] In the extreme south-east corner of Belgium, the old rocks of the Ardenne disappear beneath Trias and Jurassic sediments dipping successively southwards with a marked scarp and vale topography. This small region is usually known as Belgian Lorraine and may be regarded as the north-eastern edge of the Paris Basin. It drains westwards to the Meuse system via the Semois.

The Ardenne Massif forming the western end of the Rhine Plateau, has typical Hercynian parallel folds trending almost east and west.

The highest moorlands (the Hautes-Fagnes) are developed on the oldest rocks, chiefly Cambrian quartzite and Silurian schists. Beginning near La Roche, they culminate in the inconspicuous summit of Botrange (2,227 ft.) north-east of Malmédy.

Most of the Ardenne is however developed on resistant Devonian rocks—including quartzites, conglomerates, sandstones and schists; this resistant sequence is temporarily interrupted by Middle Devonian limestone (Couvinian) and Upper Devonian shales forming the trough of the *Famenne*, standing at little more than 600 ft. (Figs. 53 and 54) which extends from Chimay near the French border, crossing the Meuse near Givet, to the neighbourhood of Durbuy. It is drained

[5] The classification itself rests primarily upon the 1929 statistical data (chiefly for individual cantons) supplied by the Belgian Ministry. No regional boundary has been drawn primarily to agree with any physical factor.

[6] For the physical background of Belgium reference has been made to *Géographie Universelle*, Vol. II, Paris, 1927. *Belgique, Pays-bas, Luxembourg* (A. Demangeon). For certain aspects of the geology, see L. Dudley Stamp, "The Geology of Belgium," in *Proc. Geo. Assoc.*, Vol. XXXIII, London, 1922. For the evolution of the river system see M. L. Lefèvre, *Notice sur la Carte oro-hydrographique de Belgique à 1:500,000*, Turnhout, 1937.

by the Lesse and the Middle Ourthe, reaching the Meuse near Dinant and at Liége respectively.

North of the Famenne is the undulating *Condroz* region—650 to 1,100 ft.—transitional in physical and agricultural respects to lowland Belgium. Here Devonian sandstones and shales alternate with narrow bands of Carboniferous (Dinantian) limestone—with a north-east trend. Young erosion has produced a striking ridge and vale topography, vales etched in limestone and shale being some 150 ft. below the rounded sandstone ridges. The Condroz is traversed by the Meuse flowing between Dinant and Namur in its rejuvenated and frequently

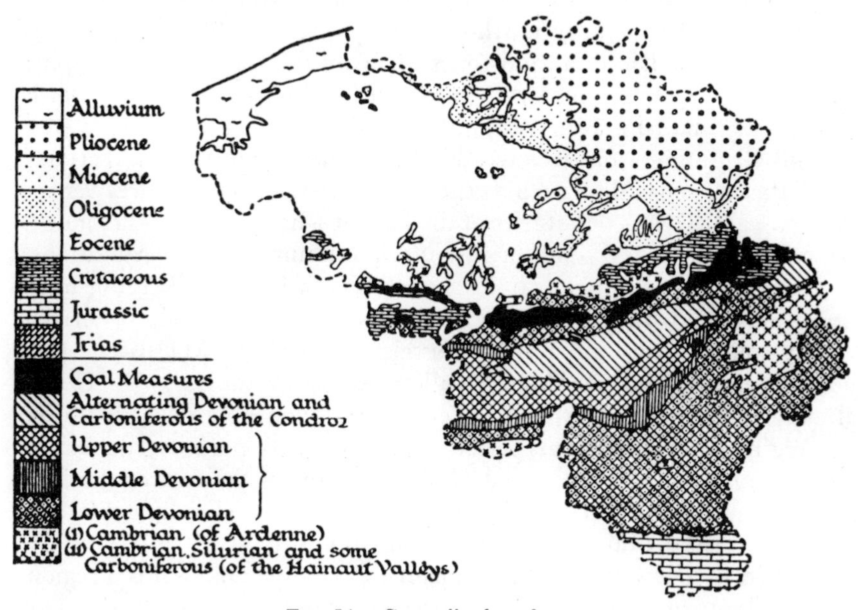

Fig. 54.—Generalized geology.

gorge-like valley, and its eastern limit is approximately marked by the incised lower Ourthe Valley immediately south of Liége. The alternation of Devonian and Carboniferous rocks is masked eastwards by Cretaceous chalk, in turn partly masked by Limon in the distinctive region of *Herve*, which drains steeply north-westwards to the Meuse, via the rivers Vesdre and Geul.

The northern edge of the Condroz is an anticline, in which Lower Devonian and Silurian sandstones form a wooded ridge at 650 to 700 ft., overlooking the well-defined valley of the Sambre-Meuse, between Charleroi and Liége. The valley itself is largely etched in the *Coal*

Measures of the major synclinal extending from Northern France into Germany. Westwards, near Mons, this trough has been partly filled with Chalk, Tertiary sands, and clays, and now drains sluggishly via the Haine into France to join the Escaut.

Lowland Belgium. Beyond the Sambre–Meuse the Palæozoic rocks of the Ardenne dip northwards beneath a succession of Secondary and Tertiary strata. The narrow outcrop of the *Chalk* forms a plateau-like region immediately north of the valley, reaching 550 to 600 ft., and widest immediately west of Liége. It is succeeded by the broad outcrop of *Eocene* sands and clays, and in turn by the narrower outcrop of *Oligocene* and *Miocene* (chiefly sands), forming a band some 10 miles wide from St. Nicolas (East Flanders), through Malines and Hasselt to Maastricht. The entire north-east is covered with coarse *Pliocene* sands (in which Diestian sandstones diversify the relief), over-stepping directly on to Oligocene strata in a south-western extension towards Louvain. Small Pliocene cappings give diversified relief further west around Grammont and Renaix (350 ft.). In the south of this Tertiary zone, the deeper valleys such as the head waters of the Dendre, Senne, Sennette and Dyle—tributaries of the Escaut—have cut into the underlying Palæozoic rocks, resulting in somewhat damper surface conditions.

Although largely unglaciated, large tracts of lowland Belgium have been masked with the largely wind-transported Limon, occurring chiefly south of the line Ypres, Brussels, Hasselt and Maastricht, but not usually beyond the Meuse valley except in the Pays de Herve already mentioned. Although everywhere favourable to agriculture, the régime and settlement pattern depend on its porosity, thickness, and the underlying rock. The Hesbaye limon in the east of the belt rests largely on fissured chalk, and is a well-drained rolling plateau of sparse surface drainage ; the Hainaut limon rests on thin Tertiary sands directly overlying the impermeable Palæozoic floor—it is a region of abundant surface drainage, dissected relief, and much damper agricultural forms.

The Pliocene sands of the north-east have been largely covered with coarser sands and gravels, laid down in post-glacial times as the alluvial fan of the Meuse. West of Maastricht, this detritus forms the sterile Campine plateau reaching some 250 ft. and still a region of heaths, shifting sand-dunes, pine plantations, and boggy hollows, caused by leaching and the formation of hard pan. Towards Antwerp, the sterile cover thins, the height drops to some 40–50 ft., and a higher water table permits cultivation.

Flanders is a low plain of gentle relief (50–150 ft.) developed chiefly on Eocene sands, except west of the line Dixmude, Deinze, Pottelberg

Hill, Brussels and the Upper Senne, where Flanders clay prevails. Its mixed light soils, ranging from sands to sandy loams, have become the most productive, if not the most fertile, in Belgium.

The straight coast line has been formed chiefly of dunes, accumulated along former off-shore sandbanks. The lagoons behind have silted up with marine clays, now dyked and drained to form the flat plain of the *Polders*, nowhere more than 10 ft. above sea level and reaching a maximum width of 12 miles in the lower Yser valley. The "*Dune*" belt between Middelkerke and La Panne is two miles in width and reaches a height of 80 ft.

The Climatic Background

Lowland *temperatures* resemble those of south-east England, with slightly greater extremes, increasing inland.

MEAN MONTHLY TEMPERATURES (° F.)

	Jan.	*Feb.*	*Mar.*	*Apr.*	*May*	*June*	*July*	*Aug.*	*Sept.*	*Oct.*	*Nov.*	*Dec.*	*Range*
Ostend (*Coast*)	37·9	37·9	41·9	46·2	53·9	57·9	61·8	61·8	60·9	52·9	42·9	39·0	23·9
Brussels (*Centre*)	36·1	37·5	41·8	47·8	55·7	60·6	63·6	62·9	58·6	50·7	42·1	37·5	27·2
Gembloux (*Centre*)	35·6	36·7	41·5	46·6	54·5	59·1	63·0	61·3	57·0	50·0	41·2	37·4	27·4
Bourg Léopold (*Campine*)	35·4	36·7	41·7	47·3	55·0	60·8	64·4	63·0	58·4	50·2	41·0	36·7	29·0
Chimay (*Ardenne*)	34·0	36·1	42·6	45·9	55·6	58·8	63·3	61·8	57·2	48·2	37·5	35·1	29·3
Arlon (*Lorraine*)	33·6	39·5	40·5	45·5	52·3	58·4	61·7	61·2	56·3	48·9	40·6	35·0	28·1

Tables based on data in Bulletins of the Institut Royal Météorologique de Belgique (for Brussels and Gembloux). Figures for the other stations taken from the *Encyclopédie Agricole Belge*, Vol. I, p. 46.

The higher inhabited Ardenne valleys have mean winter temperatures scarcely above freezing, and the short summers rarely exceed 60° mean for more than two months. Fig. 55 shows that the mean number of days with frost exceeds 120 annually in the high Ardenne, but is under 60 in the narrow coastal tract open to maritime influence.

In most of the lowlands the average is from 60 to 80 days, but in the Campine the inland situation and night radiation from the porous sands force the average well over 80 days.

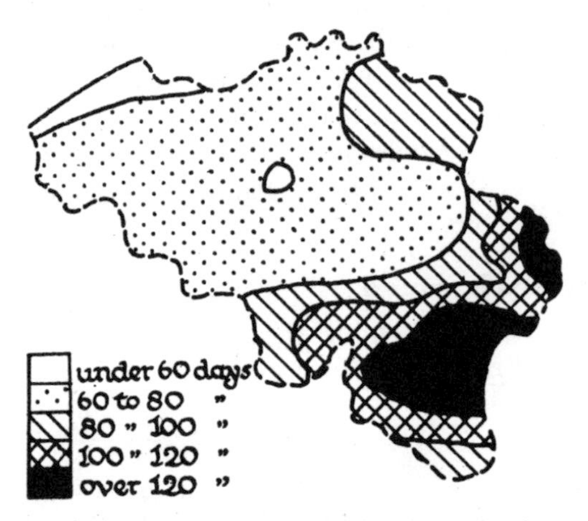

Fig. 55.—Mean annual number of days with frost.

Frost frequency and the length of the growing season is, in general terms, a function both of relief and of continentality of situation. Thus, within the limits of lowland Belgium, the Campine of the north-east is seen to be relatively continental, while along the coast frost frequency is no greater than in East Anglia. The readings for Brussels (which appears to be a relatively frost-free " island ") would appear to be influenced by the effects of local artificial heating.

Days with Frost (Mean) [7]

	Jan.	*Feb.*	*Mar.*	*Apr.*	*May*	*June*	*July*	*Aug.*	*Sept.*	*Oct.*	*Nov.*	*Dec.*	*Total*
Ostend	11	10	6	1	—	—	—	—	—	1	6	9	44
Brussels	15	13	9	4	—	—	—	—	—	1	8	12	62
Gembloux	15	14	12	5	1	—	—	—	—	3	10	13	73
Bourg Léopold	18	16	13	7	1	—	—	—	—	4	12	15	86
Stavelot (*Ardenne*)	19	17	12	6	1	—	—	—	—	5	11	13	84
Virton (*Lorraine*)	19	18	15	10	3	—	—	—	1	7	12	15	100
Arlon (*Lorraine*)	24	21	16	7	1	—	—	—	—	3	12	19	103

Mean Dates of Last and First Frosts

Ostend	3 Apr.	12 Nov.
Brussels . . .	27 Apr.	5 Nov.
Bourg Léopold . .	7 May	22 Oct.
Huy (*Meuse*) . . .	26 Apr.	26 Oct.
Gileppe (*Ardenne*) . .	12 May	12 Oct.
Arlon	28 Apr.	25 Oct.

[7] *Encyclopédie Agricole Belge*, Vol. I, p. 46.

Rainfall distribution is closely related to the relief.[8] The highest ground of the Ardenne and also the districts lying south of the main watersheds (because the cyclonic rain comes chiefly from the south-west) receives over 40 in., and the whole massif exceeds 30 in., except for a minute area in the basin of the Heure south of Charleroi. The coastal plain and much of West Flanders receives between 25·0 and

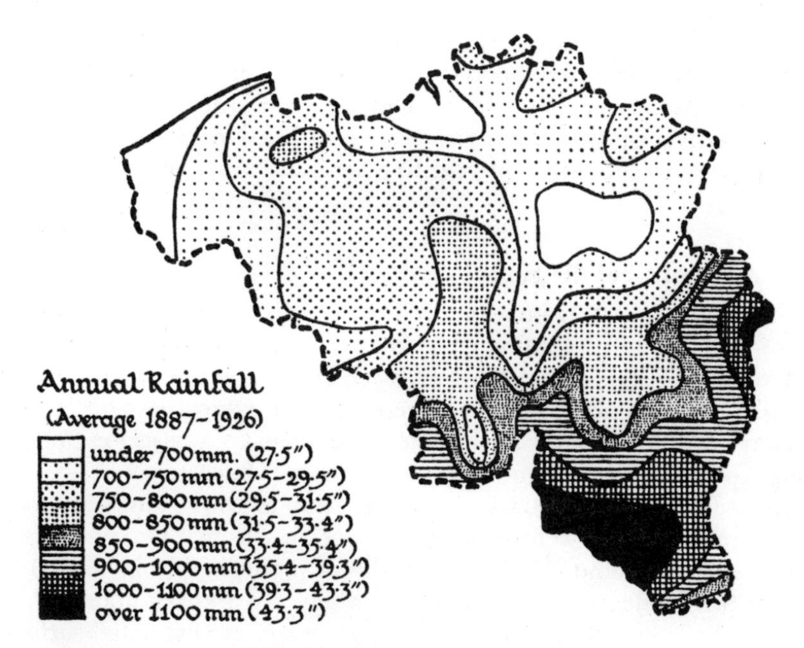

Fig. 56.—Mean annual rainfall.

There is a broad general correlation with relief, but it will be noticed that the highest totals (over 43 in.) tend to occur south and west of the main Ardenne watershed. Most of the cyclonic rain, especially in winter, comes from these directions.

Parts of the Pre-Ardenne have only a moderate rainfall, the Condroz and even the Famenne recording usually below 35 in. Within the Lowlands the eastern interior is drier than the west, although the lowest averages in the whole country occur in the coastal Polders of West Flanders. Near La Panne, for example, the total averages some 25 in.

27·5 in., but inland the total rises rapidly to about 31 in. on the quite modest " heights " between Bruges and Ghent.

The undulating plateaux south of Brussels also have up to 31 in., but the low ground around the Scheldt estuary below Antwerp, and the eastern parts of the limon region of Hesbaye (well towards the interior) receive well below 27 in. The Campine averages 30 in., offset, however, by the porosity of the sands.

[8] Fig. 57 and the figures quoted are based on data given by Emile Vanderlinden, " Sur la distribution de la pluie en Belgique," *Mémoires de l'Institut Royal Météorologique de Belgique*, Vol. II, 1927.

Although everywhere the seasonal distribution may be described as fairly even, the interior shows some tendency to continental features such as a dry cold late winter and spring, and an August peak owing to convectional storms. The coast has the normal autumn and winter maxima of oceanic stations. In the Ardenne with cooler summers and fewer convectional storms most precipitation is also in autumn and winter, much being in the form of snow.

Except, therefore, in the High Ardenne, where the high rainfall and exposed conditions ally themselves with soil to limit agriculture to a grass and pastoral economy, the climate of Belgium can be said to favour cultivation, although the precise utilization is a function also of soils, markets, and pre-existing urban and rural settlement patterns.

The Agricultural Regions

These are based primarily on the detailed *Agricultural Census of 1929*,[9] and the official Atlas of 53 maps,[10] showing densities of crops and of livestock in the 211 cantons. These are the most detailed pre-war statistics available, but, representing typical pre-depression and pre-tariff conditions, have some advantages as a basis for calculation, especially when related to the war-time figures.

The statistics show a marked concentration of wheat and sugar beet on the limon and the centre, where they occupy from 15 to 30 per cent of the exploited land.[11] They are absent or negligible in the sandy regions of Flanders and the Campine, where rye and potatoes replace them and account for from 25 to 40 per cent of the exploited surface. Other regional localizations are easily distinguishable.

By a technique of intersecting " Isopleths "[12] of crop densities, and consideration of types of farming and size of holding, 27 agricultural regions have been distinguished. This sub-division is more detailed than in other geographical accounts which the writer has met with, and is considerably more complex than the eight-fold division used for administrative purposes by the Belgian Ministry.[13]

[9] Ministère de L'Agriculture belge, *Recensement Général Agricole du 31 décembre 1929.*

[10] Ministère de l'Agriculture belge, *Atlas du Recensement Général de 1929* (prepared by the Service d'Études of the Ministry).

[11] The term " exploited land " comprises the whole surface excluding urban areas, rough grazing, and heath, but including woods.

[12] For an example of the technique involved, see Fig. 57.

[13] For a recent discussion and description of the eight traditional regions, see *Bulletin du Ministère de l'Agriculture,* " Partie Economique," No. 2, 1939. A map of these is also reproduced in *Geography,* March, 1946 ; " The Tobacco Industry of Belgium."

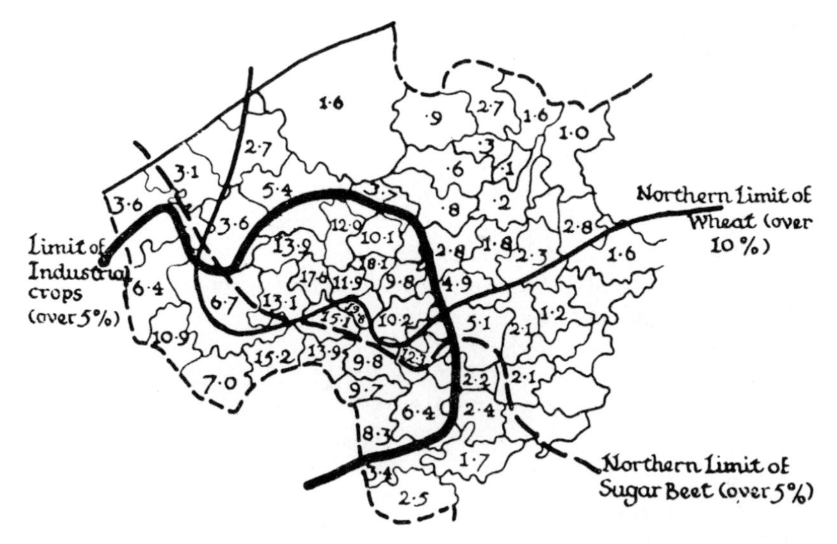

FIG. 57.—The limits of certain crops in West Flanders.

This also serves as an illustration of the method of delimiting agricultural regions. The figures appearing in each canton show the percentage of the exploited land under *industrial crops*. From these, the approximate isopleth of 5 per cent has been drawn, enclosing those cantons with not less than 5 per cent under such crops. From similar data for other crops (in this example *Sugar Beet* and *Wheat*), other isopleths have been drawn. Their intersections suggest regional groupings and even on the limited basis of these three crops, the following regions can already be distinguished:

(1) A region extending along most of the French frontier where wheat, sugar beet and industrial crops are all important.

(2) An inland region where industrial crops attain their greater importance (between 8 and 17 per cent), but where wheat and sugar beet are of minor importance (below 10 and 5 per cent respectively).

(3) Another inland region, and further to the south-east than the preceding region, where industrial crops are unimportant (from nil to 2·5 per cent) but where wheat and sugar beet (and especially the former) are of importance.

The 27 agricultural regions shown in Fig. 58 are based on the study of a large number of such isopleths, supplemented by reference to details of size and type of holding, and in some cases by field study.

1. *The Coastal Regions.*
   (*a*) The Coastal Dune Belt.
   (*b*) The Polders.
2. *The Sand Regions of Flanders.*
   (*a*) The Sandy Plain of East Flanders.
   (*b*) The Roulers Region of Industrial Crops.
   (*c*) Southern Flanders Transitional Region.
3. *The Campine.*
   (*a*) The Dry Campine.
   (*b*) The improved Campine Fringe.
4. *The Limon Regions of the Centre.*
   (*a*) The Ypres-Courtrai Region of Wheat, Beet, and Industrial Crops.
   (*b*) Tournai Wheat and Beet Region.
   (*c*) Southern Hainaut Wheat and Beet Region.
   (*d*) The Central Region of Wheat and Beet.

(*e*) Eastern Hesbaye Region of Wheat, Beet, and Fruit.
(*f*) The Central Mixed Farming Region.

5. *Market Gardening Regions.*
   (*a*) The Brussels–Hoeylaert Region.
   (*b*) Malines–Lierre Region.
   (*c*) The Antwerp Region.
   (*d*) The Liége Region.
   (*e*) Charleroi and the Borinage.
6. *The Condroz Mixed Farming Region—the Pre-Ardennes.*
7. *The Herve Fruit and Dairying Region.*
8. *The Famenne–Chimay Pastoral Region.*
9. *Regions of the Ardenne.*
   (*a*) The High Ardenne.
   (*b*) The Ardenne Forest and Cattle Region.
   (*c*) The Bastogne area.
   (*d*) The Semois valley.
10. *Belgian Lorraine.*
    (*a*) The Arlon Mixed Farming Region.
    (*b*) The Virton Forest and Mixed Farming Region.

The boundaries of these regions are shown on Fig. 58.

1(*a*). *The Coastal Dune Belt.* The discontinuous zone of dunes [14] varies in width from a few yards to some two miles. Where relatively stable, marram grass and shrubs occur, and in a few areas, pines have been cultivated to fix the sand.

Agriculturally negative, coastal holiday resorts and ports have fostered a limited amount of market gardening in the "pannes" or hollows on the older inland belt between Ostend and La Panne; here a reasonably high water table and the weathering of shells have permitted the coarse sandy soils to be cultivated for potatoes, vegetables, and small quantities of rye and turnips. Liberal manuring (often with fish waste) is essential, and the very small holdings are frequently worked as part time or seasonal variants to fishing or catering.

This region is too small for cantonal statistics to be applicable and has been delimited from personal observation and large-scale maps.

1(*b*). *The Polders.* This important fattening region, with 40 per cent of its surface [15] in permanent grass, has a concentration of beef cattle and substantial interest in wheat, as well as beans and other fodder crops (35 per cent). The heavy grey or black soils derive from marine

[14] Cut by the mouth of the Yser, the harbour and docks of Ostend, the Bruges Canal at Zeebrugge, and the mouth of the Zwin on the Dutch border.

[15] The term "surface" will be used normally to imply "exploited" land, i.e. all land except urban areas and heathland and waste land unused agriculturally.

clays and, despite their fertility and high lime status, are late and cold in spring and difficult to work. There are limited amounts of sugar-beet, flax and potatoes, but rye, early potatoes and market gardening are negligible.

In aspect and agriculture the Polders contrast with the sandy interior

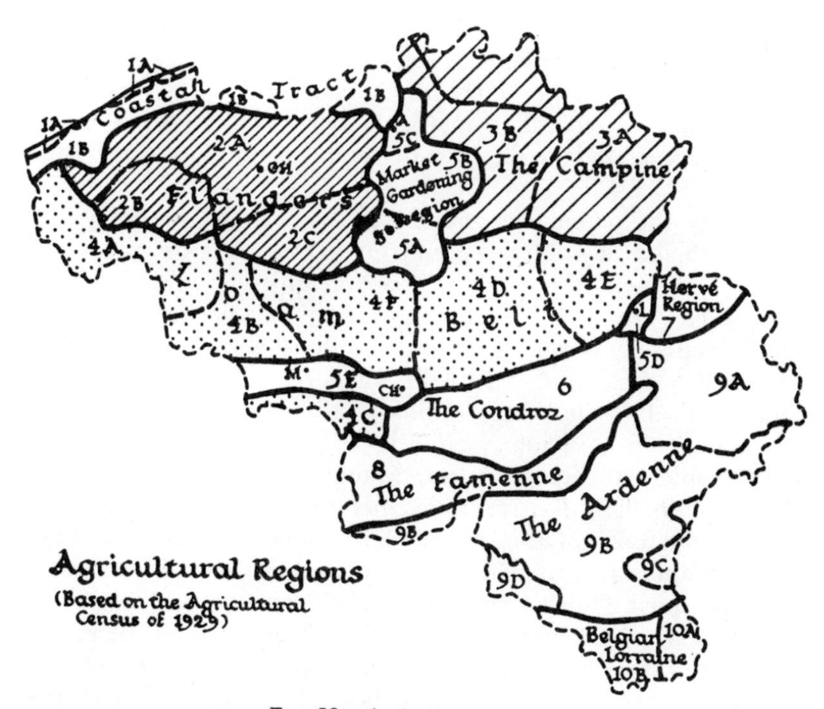

FIG. 58.—Agricultural regions.

The 27 regions shown here may be grouped into 10 broad types. They have been delimited strictly on agricultural criteria—chiefly the crop and livestock distributions recorded in the census of 1929, supplemented by reference to the types of holdings.

Excluding the five Market Gardening regions (5A–5E) which exist partly and most obviously for economic reasons, there is a marked correlation with soil, relief, and climatic factors. The three Flanders regions (2A–2C) tend to be sandy, and are essentially the home of the small-scale intensive cultivator—except for the coastal plain of the Polders, with its heavier soils. The broad belt of the limon or Loam Regions (4A–4F) includes the major arable farming districts, and between these and the forested pastoral areas of the High Ardenne are the three transitional regions of the Pre-Ardenne.

The war-time statistics of 1941–5 show comparatively little change in agricultural distributions and provide striking proof of the validity and strength of the geographical factors behind this classification. It is for this reason that the dot distribution maps are all for 1943, the year of greatest deviation from the pre-war norm.

of Flanders. Their flat treeless aspect is broken only by occasional lines of poplar or willow along a road or one of the innumerable drains. The typical unit is the small grass farm of from 5 to 20 hectares, and some large farms of 50 or more hectares occur. After Flanders, with its dense rural housing and patchwork cultivation, the Polders appear bleak, with houses and roads frequently lining banks of dykes above flood level. The lower Scheldt alluvium below Antwerp and north of Eecloo [16] also belongs to the Polders.

2(*a*). *The Sandy Plain of East Flanders.* Here, *par excellence*, is small-scale intensive cultivation, a minute sub-division of holdings, and possibly the most concentrated agricultural population in Europe. The arable land (80 per cent) grows chiefly rye (25 per cent), potatoes (15 per cent), clover, trefoils and market gardening crops, with but limited amounts of other cereals, sugar-beet, horse beans, or industrial crops. Arable-fed cattle (especially dairy cows), pigs, and poultry are more concentrated than in the Polders, and the eastern region around Lokeren and Dendermonde supplies milk to Antwerp and Brussels.

Physically the region has gentle but varied relief reaching 150 ft. in the south, based on Eocene and Quaternary marine sands. The Flanders Clay near the surface, however, gives a satisfactorily high water table. The driest district—the Houtland, between Bruges and Ghent—retains some woodland and the illusion is heightened by the presence of hedges (timber actually totals only 5 per cent). East of Ghent, in the Pays de Waes, fertility increases, soils tending to sandy loams with much alluvium. Here there is more cultivation of fruit, and vegetables are especially important around Ghent. The chief industrial crop, flax, has survived chiefly in the Lokeren district.

The productivity of this region dates from the mediæval period. Urban demand, manuring, and the early development of a commercial acquisitive spirit fostered this intensive Flemish agriculture. Now, few farms exceed 20 hectares and some 76 per cent of all full-time holdings are below 5 hectares.[17] Many small plots are cultivated by industrial workers. 73 per cent of holdings are rented and owner cultivation is most typical of the very smallest holdings.

This intensive utilization is ultimately due to the light workable soils, permitting elaborate intercropping, easily absorbing manures and warming readily in spring. Early cultivation is also fostered by the *relatively* frost free, semi-maritime situation.

South of the approximate line Oudenarde–Alost the sands pass into sandy loams heralding the transition to the Central Loam Regions, and the agricultural pattern changes.

[16] The comparatively new polders along the lower Scheldt north-east of Kieldrecht were reclaimed only some 50 years ago ; *vide* Van der Vaeren, *Encyclopédie Agricole Belge*, Vol. I, p. 140.

[17] From the Recensement of 1929. In the arrondissements of Ghent, Termonde and St. Nicholas, of the total of 25,520 holdings exceeding 1 hectare, 19,454 were below 5 hectares. In comparing holdings for the various Regions only those over 1 hectare have been considered.

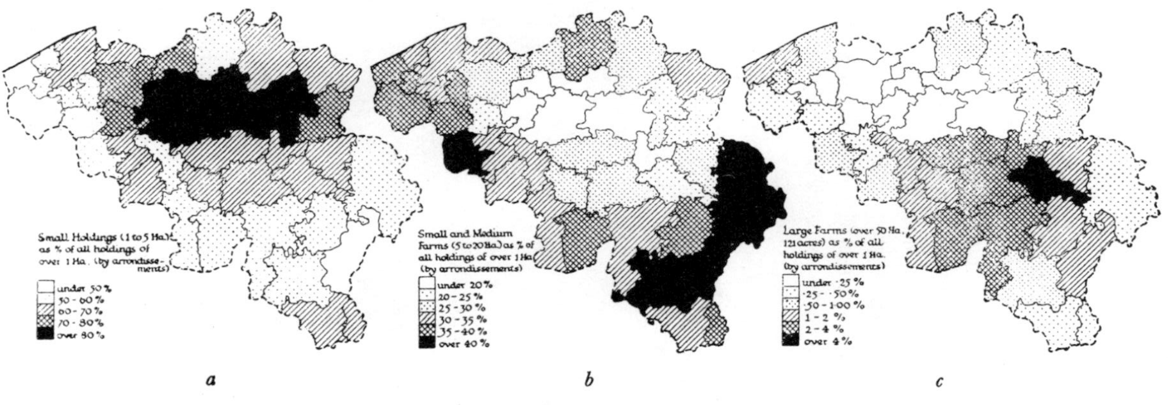

FIG. 59.—The distribution of holdings.

(*a*) *Small-Holdings.* Most of those holdings smaller than 1 hectare (2·42 acres) are part-time, and have been excluded from consideration in the maps of agricultural holdings. Although the proportion of small-holdings is everywhere high, judged by British standards, they are dominant especially in the light-soiled areas of East Flanders, around Brussels, and on the southern and western outskirts of the Campine, where the proportion is over 70 per cent. This reflects the tendency towards the intensive cultivation of vegetables and fruit, and contrasts with the loam tract and even more with the High Ardenne where there is little scope for such cultivation.

(*b*) *Small and Medium Farms.* In the High Ardenne as well as in some of the lighter areas of the limon belt, as in West Flanders and Hainaut, the dominant unit is the small farm of 5–20 hectares (13–50 acres). In the Ardenne such farms are concerned with rearing beef cattle. Elsewhere they tend chiefly to intensive cultivation of industrial crops such as flax, chicory, tobacco and sugar beet in Flanders, and towards market gardening and fruit in the Antwerp districts.

(*c*) *Large Holdings.* The large farm (exceeding 50 hectares or 140 acres) is most typical of the Loam (limon) districts of the centre, especially of Hesbaye where they are concerned with the cultivation of sugar beet and wheat. In the Condroz of the Pre-Ardenne, large mixed farms are also important. Although even in these regions the proportions of such farms rarely exceeds 3 or 4 per cent of the total holdings over 1 hectare, it should be remembered that this represents from 40 to 45 per cent of the farmland.

The sandy, intensively worked land of Flanders and the western Campine fringe are seen to have virtually no large farms of this category. The heavy soil Polders are seen to have some large farms, chiefly held by large-scale graziers.

2(*b*). *The West Flanders Sands.* This region, centred on Roulers, is damper. Flanders Clay is exposed in the Lys and Mandel valleys and the sandy interfluves have some tendency towards limon.

Three-quarters of the surface is arable and the Flemish tradition of small-scale intensive cultivation is continued though with fewer tiny holdings [18] than in East Flanders. Rye and potatoes remain dominant, but cattle are less important. The essential feature is however the importance of *industrial crops*—especially *flax*, *chicory* and some *tobacco*,[19] which together equal rye in importance. It is notable that wheat and sugar-beet (indicative of heavier soils) are of some importance.

About 80 per cent of Belgian chicory for coffee and at least 40 per cent of the flax is grown in this small region. The specialization on such crops, and formerly also on hemp and colza (which have now virtually disappeared through foreign competition) is due both to physical and economic-historical factors. Chicory grown for its roots thrives on a loamy sand with slight alkaline reaction, and a deep well-drained soil permits free root development and harvest pulling. Production has fallen seriously since the nineteen-twenties from 10,000 hectares in 1910 to 7,000 in 1933, and the formerly large exports to Germany, France and Switzerland have long been declining. In West Flanders preliminary drying of the roots is usually done in *tourailles* on the holdings—special coke-fired kilns. Flax is tolerant of heavier soils but prefers well-manured sandy loams on which it frequently follows chicory. Flax, chicory and tobacco all make heavy demands for skilled labour in cultivation, harvesting, and preliminary farm treatment. They are therefore attractive and remunerative crops for small cultivators with large families [20] and specialized marketing and distributive channels have developed which discourage their spread to other regions. Flax cultivation is also influenced by the linen centres of West Flanders (chiefly Tielt, Roulers, Courtrai and Wareghem) and of Ghent, and the lime-free waters of the Lys are still widely used for retting.[21]

2(*c*). *The Southern Flanders Transitional Region.* From near Oudenarde this region of gentle but diversified relief reaches 350 ft. near Grammont, and continues eastwards through Alost to the western

[18] 59 per cent. of holdings are under 5 hectares, and Figs. 59A and 59B show that small farms are more typical of the area. 64 per cent of all holdings are rented.

[19] For a study of the distribution and geographical basis of tobacco cultivation see "The Tobacco Industry of Belgium." S. W. E. Vince, in *Geography*, March, 1946.

[20] Larger growers frequently farm out the harvesting and cultivation of tobacco and chicory to a peasant family, who take an agreed proportion of the returns.

[21] Much flax is now retted in tanks and reservoirs filled from the river rather than in the river itself.

outskirts of Brussels. It is drained north-eastwards to the Scheldt by the Dendre and its tributaries. Its soils and agriculture are transitional between the Flanders sands and the loams of Brabant.

Wheat slightly exceeds *rye* in importance, and with *potatoes* these account for 40 per cent of the exploited land. *Grass* is more important than in Flanders proper, amounting to 33 per cent, and apples and pear *orchards* are also widespread (6 per cent), making this district the third in Belgium for fruit.

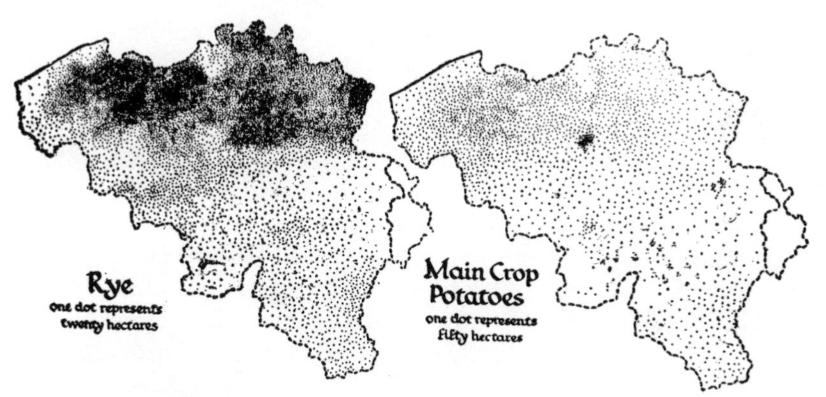

FIG. 60.—Rye. FIG. 61.—Main-crop potatoes.

Rye being primarily a crop of sandy or slightly acid soils, it occurs in Belgium chiefly on the sands of Flanders and of the Campine, where it may amount to 20 per cent of the surface. It is noticeably absent from the limon tracts from West Flanders to Hesbaye, though it reappears to a limited extent in the dissected limon region of Eastern Hainaut. It is also noticeably absent from the heavy lands of the Polders.

The southern limit of rye concentration on the sandy lands of the north is one of the best-defined agricultural "frontiers" in Belgium and corresponds closely with the transition from sand to loam.

Like rye, main-crop potatoes are found chiefly on the lighter and free-drained sandy soils of the north, especially in Flanders and the improved western parts of the Campine. The higher and more sterile Eastern Campine plateau is, however, unimportant. There are, however, other dense concentrations around the chief urban areas, especially Brussels, Liége and Charleroi, where in conjunction with other market garden crops they justify the separation of several market gardening regions. The potato concentrations around Antwerp and Ghent merge into the main producing areas. Within the Ardenne, cultivation is unimportant, but increases south of the main watershed—i.e. in the Arlon, Semois, and Bastogne areas.

Small holdings prevail [22] but cattle are fewer than in Flanders, though the nearness of Brussels and Ghent results in a high proportion of dairy cows. Sugar-beet is negligible, and industrial crops unimportant except for *hops* in the Alost and Assche districts towards the east. There the permanent hop poles are a striking feature of the landscape. Sheep are an important sideline.

3(*a*). *Dry Campine.* Probably at least a third of this comparatively negative region—the north-eastern section of the Campine adjoining the Dutch border—is heath, birch scrub, bare sand-dunes and marshy hollows. Considerable areas are used for military purposes. This

[22] Particularly in the Alost area towards the east of the region. Here 85 per cent of holdings are below 5 hectares, the majority being rented.

region is the highest and bleakest part of the Campine and ends abruptly on the east in the steep bluff overlooking the Meuse valley between Maastricht and Maaseik.[23] Heights of 260 ft. occur in the bleak Lanaekersheide between Genck and Bourg Leopold. The Turnhout and Arendonck district is lower, but has a similar land-use pattern. There is plentiful *wood and plantation* (30 per cent—much of it recent afforestation with Scots pine and other conifers). The original forests of dry oak—birch—heath type were largely cleared

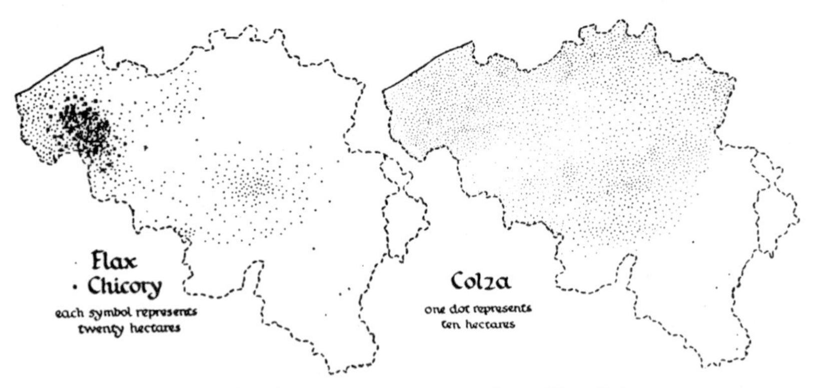

Fig. 62.—Flax and chicory. Fig. 63.—Colza.

Flax is more widely distributed, but occurs chiefly on rather heavier soils throughout West Flanders (largely on sandy limon) and extending in the south on to the limon zone proper around Tournai. There is a smaller producing area between Brussels and Namur on the central limon tract, and also in the fertile Pays de Waes of East Flanders. Its localization is also limited by such factors as local skill and facilities for retting, and by the heavy demands of the linen industry of Courtrai, Roulers, and other West Flanders towns.

Chicory cultivation is virtually confined to the well-drained West Flanders sands region, where soils are more loamy and inherently fertile than in East Flanders. Also important as a localizing factor is the skill involved in cultivation, and in treatment of the roots on the farms, especially in the drying process.

Colza is one of the few crops whose distribution in 1943 was radically different from that of 1929. From being in former times an important industrial crop in West Flanders, it was by 1929 almost non-existent, its area totalling in that year only 32 hectares for the whole country. In 1942 a compulsory quota was allotted to all growers in suitable districts, and in the following year its acreage was some 25,000 hectares. One of the principal uses of Colza seed oil during the occupation was in the preparation of edible fats, and by 1946 its acreage had slumped to only 700 acres. Although generally distributed in 1943 throughout lowland Belgium, and in the Condroz, it was most important on the light soils of Flanders and the improved Campine, but also in Hesbaye.

by the seventeenth century. Most of the modern timber production is for pit-props.

The small-scale cultivation has traditionally been of typically sandy land crops of rye and potatoes, based largely on the manure of store-fed cattle and heath-run sheep. *Rye* (17 per cent), *oats* (8 per cent) and *potatoes* (4·6 per cent)[24] remain dominant, but progress has been towards a grass and dairy economy, with intensive use of floodable alluvial valleys such as the Zwartbeek. Sheep are now unimportant.

[23] The Meuse Valley below Maastricht has not been separately distinguished.
[24] These proportions relate to "exploited" land.

Colliery development near Genck and industrial development along the Albert and other canals is stimulating dairying and market gardening.

Small holdings prevail, especially around Hasselt, where 83 per cent are below 5 hectares and some 60 per cent are owner-cultivated.[25] There is an abrupt transition south of the line Hasselt–Maastricht to the rich loams of Hesbaye.

3(*b*). *The improved Campine fringe.* This much lower region with a higher water table and nearness to Antwerp and Brussels, has an older and closer settlement pattern and more intensive agricultural forms with an increasing proportion of industrial workers interested in part-time agriculture.[26] 80 per cent of the holdings are below 5 hectares, with a high proportion owner-cultivated.

Although the sandy soils result in a high proportion of *wood and plantation* (20 per cent as against 30 per cent in Dry Campine) two-thirds of the surface is cultivated. *Rye and potatoes* remain dominant (30 per cent) but there are appreciable amounts of wheat, market gardening and early potatoes, as well as of *fruit.* Most of the cattle are *dairy cows* (71·5 per cent) and sheep are virtually absent.

Soil and agriculture improve westwards and the boundary with the market gardening and vegetable region of Antwerp, Malines and Brussels is transitional. *Green peas* become very important in the Aarschot area of the south, and strawberries, cauliflowers and witloof (bleached chicory, grown for its leaves) are widespread.

In the south is the " Hageland " around Diest and Aarschot, sometimes regarded as transitional to the Central Limon belt. It stands higher than the rest of the region and the rolling surface (sometimes influenced by the Diestian sandstone of the Pliocene) shows some sandy limon on interfluves, though the well-defined valleys are sandy, with dairying based on typically Campine cultivation. The southern limits have been drawn where rye, potatoes, and market crops are superseded by wheat and sugar beet on the larger farms of the limon areas. This division roughly follows the line from Louvain to Glabbeek.

4(*a*). *The Ypres–Courtrai Region of Wheat, Beet, and Industrial Crops.* Like other limon regions of Belgium, the extreme west of Flanders adjoining the French border has *wheat, sugar-beet* and *oats* as important crops (38 per cent). Unlike the others, however, it grows also important

[25] In Maaseik arrondissement, 4,508 out of a total of 7,441 holdings.

[26] It is also essentially a small holdings region, with over 80 per cent under 5 hectares, and most of the rest below 30 hectares. Rather more than half are owner-cultivated.

industrial crops (10 per cent), *potatoes* (9 per cent) and some *rye* (4·5 per cent). The typical unit is the small farm of from 5 to 20 hectares.[27]

This unusual variety derives partly from the presence of lighter loams than in Brabant, but also owes much to economic factors bound up with traditional Flemish intensive cultivation.

The regional limits pass south-eastwards through Furnes to Ypres, Courtrai, Renaix (Ronse) and thence westwards just north of Tournai. One quarter is *grass*, much being in the wide valleys of the Lys and

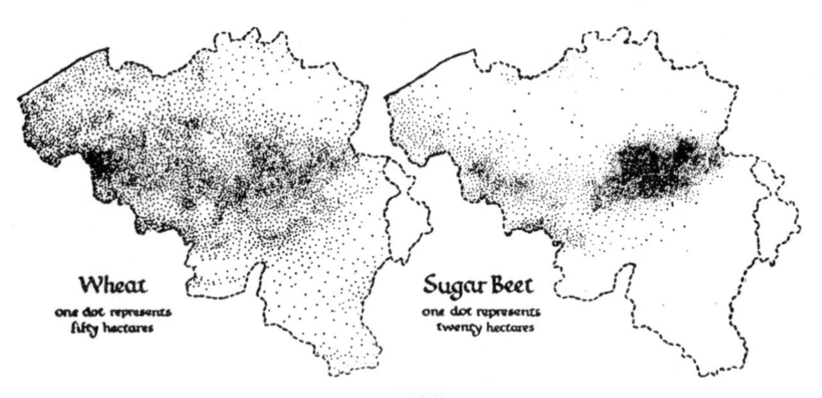

Fig. 64.—Wheat. Fig. 65.—Sugar beet.

Although wheat had increased greatly in all districts by 1943, it is seen to be still primarily the grain crop of the limon areas. Among the negative regions are still the Campine and particularly its unimproved eastern section, the East Flanders Sands, the higher parts of the Ardenne, and the Herve grass region immediately east of Liége. The considerable war-time cultivation in the Condroz (largely spring wheat) is clearly shown. The Charleroi, Borinage, and Haine valley districts appear as negative regions.

Even at the height of the food production drive the cultivation of sugar beet remained virtually tied to the limon tracts, especially of Hesbaye and the east, where the influence of the Tirlemont sugar factory and other plants in that district was an important localizing factor. There is a fair production in the western limon tracts along the French border, but the crop is absent from the damper regions of the Haine and the Borinage (where also small-holdings prevail) and it is of only limited importance in the damper dissected limon tracts around Soignies and Ath. The small production north of Antwerp tends to occur towards the better-drained margins of polder land, and is not entirely a war-time development.

Upper Escaut (Scheldt). Wheat is especially important in the Templeuve area almost on the French frontier between Tournai and Roubaix, and *tobacco* (4·5 per cent) is concentrated in and around the Lys valley (south of the Cassel-Kemmel ridge). *Flax*, while generally distributed, is concentrated on the heavier land in the valleys where also soft water is available for retting. *Chicory* is grown near Menin, but its importance nowhere approaches that of the West Flanders sands region.

4(*b*). *The Tournai Wheat and Beet Region.* The rolling limon-covered plateau south and east of Tournai has been deeply dissected by rivers

[27] 35–40 per cent of all holdings are from 5 to 20 hectares, and only some 50 per cent are below 5 hectares.

of the Escaut and Upper Dendre system. Wooded gravelly Pliocene cappings south of Renaix stand well above the general level and reach 400 ft.

Typical limon crops of *wheat* (16 per cent), *sugar-beet* (6 per cent) and *oats* (12 per cent) comprise one-third of the surface, and larger farms prevail, with important livestock interests, especially fodder beet and trefoils, chiefly for beef cattle and horses. *Grass* and orchards account for a third of the surface. The heavy loams explain the small acreages of industrial crops (2 per cent), rye, potatoes and market gardening.[28]

East of Ath this region merges into a damper mixed farming region (4(*f*)), while along the line Peruwelz-Mons a wooded sandy edge overlooking the Haine valley marks the transition to the market gardening region of the Borinage.

4(*c*). *Southern Hainaut Wheat and Beet Region.* Resembling the Tournai region in land use, this region is separated from the main limon tract by the marshy basin of the Haine. It is based mainly on chalk, with inliers of Devonian sandstones around Thuin in the south-east, and the plateau surface (reaching 600 ft.) is trenched by several small north-flowing tributaries of the Haine.

Compared with Tournaisis, although wheat and sugar-beet remain dominant, the chief differences are the higher proportion of *barley* (4 per cent) and *grass* and orchards (38 per cent against 29 per cent, although only 3 per cent are orchards).

The *sheep* density, almost the highest in Belgium, partly explains the grass and trefoil acreage.

4(*d*). *The Central Region of Wheat and Beet.* This, the principal wheat and sugar-beet area of Belgium, is essentially a region of large and medium sized farms.[29] It is 80 per cent arable, of which more than half is wheat, sugar-beet and oats. The limited grassland is chiefly in valleys such as the Mehaigne (a tributary of the Meuse as Huy), Ormeau (Gembloux) and the north-east flowing Gette below Tirlemont. The prevailing limon cover is well-drained but fairly heavy. The rolling plateau surface rises from 180 ft. in the north-west near Louvain to 660 ft. at Petit Waret, north of the Meuse near Andenne, and is dissected by several well-defined but wide valleys with a south-west to north-east trend.

There are fair quantities of fodder crops such as beet and trefoils,

[28] But local specialities include asparagus and witloof (chicory grown for its leaves).

[29] Only 60 per cent of holdings are below 5 hectares, and some 3 per cent are above 50 hectares, giving the highest proportion of large farms in Belgium (see Figs. 59A and 59C).

but fewer cattle than in either Flanders or the Campine. Barley and potatoes are not extensively grown and industrial crops are also unimportant, except for flax in the Gembloux district. Crops tend to be more varied on the smaller holdings.

Settlements are mostly nucleated, especially on the higher south-eastern section of the plateau.

The southern limits approximate to the physical line of the Meuse valley from Châtelet (east of Charleroi) to Huy. The northern limit passes from Louvain through Glabbeek and Léau to St. Trond ; beyond this, wheat and sugar-beet are abruptly superseded by the rye, potatoes and market crops of the Improved Campine, with smaller farms and a dispersed settlement pattern.

Among the many sugar factories is the large modern plant at Tirlemont, with nearly half the Belgian output.

4(*e*). *Eastern Hesbaye Region of Wheat, Beet and Fruit.* Physically, this resembles the previous region in being a limon-covered plateau sloping northward from a maximum height of some 620 ft. near Liége. The higher southern part is relatively undissected, forming part of " Dry Hesbaye " in which the limon rests directly upon fissured chalk. Near St. Trond and Tongres further north the limon overlies Oligocene clays ; this damper region, traversed by a succession of valleys tributary to the River Demer (Molenbeek, Herek, Yenne, and Geer) is " Damp Hesbaye "—less continuously arable, but with even greater orchard development.

This region differs agriculturally from the central limon region in the importance of *orchard fruit* (11 per cent).[30] *Wheat* and *sugar-beet* are the chief arable crops, with oats. Medium-sized farms prevail (Fig. 59B) and broadly speaking the largest holdings are in the higher, drier, south-east where the villages are markedly nucleated.

Beyond the line Maastricht–Bilsen–Herck la ville–Diest, wheat, beet and large holdings give place fairly suddenly to the Dry Campine, with a fringe of richer cultivation (though essentially rye and potatoes) in the Demer Valley near Hasselt.

4(*f*). *The Central Mixed Farming Region.* Damper soils, dissected topography, a high proportion of permanent grass (41 per cent of which only 3·6 per cent are grass orchards) and fewer large farms, all announce the more pastoral economy of this region in which horses, cattle and sheep are of special importance.

From the physical standpoint it is the upper basins of the Dendre

[30] Chiefly apples, pears, cherries, and currants. There is a big development of co-operative marketing, canning, and fruit juice manufacture, and an important Fruit Research Station at St. Trond (see also Fig. 68).

and Sennette, in which the Palæozoic floor is partly exposed. Limon occurs on the interfluves, and especially on the less-dissected plateau of Nivelles towards the south-east of the area.

Wheat remains important on the heavier soils (14 per cent) but *sugar-beet* is very limited compared with other limon areas. The remaining arable land includes *oats* (11 per cent), *fodder crops* such as beet (5·5 per cent) and trefoils (6·2 per cent) and small quantities of *potatoes* (3·6 per cent). Tobacco is important in the Dendre basin around Flobecq in the north-west.

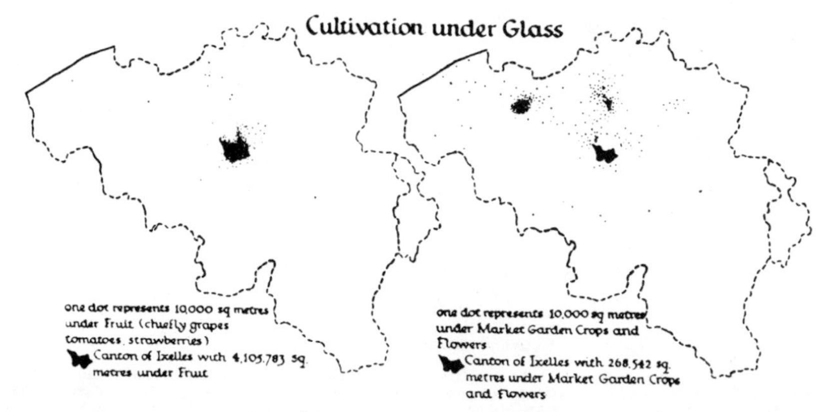

FIG. 66.—Cultivation under glass.

This, the most intensive form of horticulture, has developed primarily around three of the largest urban markets, Brussels, Antwerp and Ghent, and is an apt illustration of the innermost of Jonasson's "concentric zones."

Around Antwerp, Ghent and Malines the crops raised in this way are chiefly early vegetables and fruit, including lettuce, witloof (forced chicory grown for its leaves), celery, flowers, tomatoes, cucumbers, and strawberries. These latter are most important on the lightest soils of all—east of Antwerp.

The Brussels concentration is dominated by grape cultivation, entirely under glass, and occurring around Hoeylaert, some 10 miles south-east of the City, and within the canton of Ixelles. Here occurs some 85 per cent of the considerable Belgian production of high quality table grapes; originating to supply the local market, in pre-war years over half the output was exported, much of it to England.

5(*a*). *The Brussels–Hoeylaert Region.* This region is characterized by intensive open-air vegetable cultivation (mostly on small holdings) and a high proportion of glass devoted chiefly to fruit (grapes, cucumbers and tomatoes).

Physically transitional between the Limon plateau and the northern sandy regions, its soils range from sands to medium loams, but are everywhere easily worked. It is undulating, standing between 180 and 400 ft. and has been deeply dissected by north-east flowing streams such as the Senne, Woluwe, La Vaer, Issche and Lashe—its south-eastern limit being the deep valley of the Dyle between Wavre and Louvain.

Open-air market crops (20 per cent of the surface) increase towards Brussels. Specialities include *witloof* (chicory grown for its leaves) amounting to half the area, *green peas, flowers*, and *nursery plants*. Open-air fruit is chiefly west of Brussels, especially strawberries, tomatoes and orchard fruit.

Glasshouses for grapes abound in the remarkable district of Hoeylaert, Overyssche and La Hulpe beyond the Forest of Soignes. Here occurs 80 per cent of Belgian grape production, with a large export surplus.[31] The typical specialized grape holding is less than 1 hectare

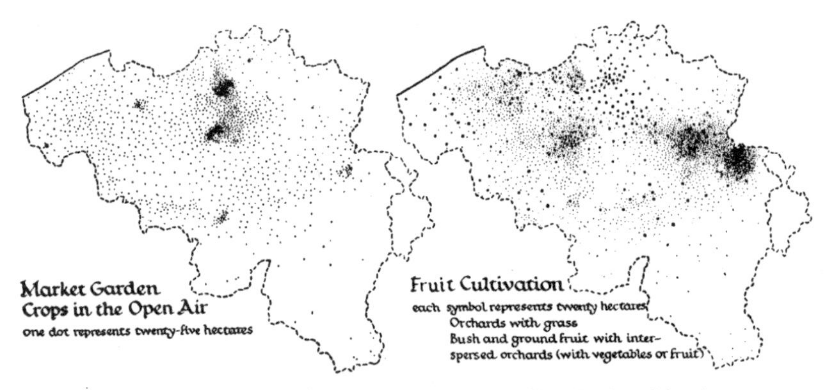

FIG. 67.—Open-air market-garden crops. FIG. 68.—Fruit cultivation.

Such crops are related to (*a*) local urban demand, (*b*) suitable light soils capable of intensive cultivation and intercropping, and (*c*) a relatively long growing season free from late frosts. The map shows four main concentrations. The greatest is the Antwerp–Malines–Brussels region of which the greater part has a light, free-drained soil, and is close to the two largest conurbations of Belgium. Around Brussels one of the specialities is witloof; elsewhere, peas and cauliflowers. Others are related to Liége, to the Charleroi–Borinage industrial region, and there is a more dispersed but in aggregate very large production in Western Flanders and around Ghent. This latter area is related not only to the dense urban and semi-urban population of Flanders, but to the maritime situation and the very light soils which warm rapidly in spring.

A distinction may be made between the *Grass Orchards*, which are usually also grazed by cattle and some sheep, and the *intercropped orchards* where bush and ground fruit occur, as well as vegetables, merging thus into market gardening.

Grass orchards are generally distributed throughout lowland Belgium, except in the Campine, but they are of the greatest importance in the Pays de Herve (east of Liége in the Pre-Ardenne), Eastern Hesbaye (between St. Trond, Hasselt, and Liége), the district of Assche and Hal in the southern transitional belt of Flanders, and throughout eastern Flanders.

Intercropped orchards are also generally distributed in all of the foregoing lowland regions, except in the Pays de Herve, and they are most concentrated in the Assche–Hal district. It will be seen that the improved Western Campine has also a considerable production, especially immediately east of Antwerp, and within the whole triangle Malines–Louvain–Aarschot—much of this merging into market-gardening.

in extent, and as in the Lea Valley the concentration suggests a glass-covered landscape.

General farm crops include *wheat and rye* (20 per cent) especially towards Louvain. The limited permanent grass is confined to damper valley bottoms, e.g. La Rede in Anderlecht west of Brussels. Dairy cattle predominate. Main-crop potatoes are important, with " earlies " nearer the city.

5(*b*). *The Malines–Lierre Region.* Physically and agriculturally dis-

[31] For a more detailed account of the Belgian grape industry, see the article entitled " Viticulture in Belgium," S. W. E. Vince, in the *Geog. Journ.*, Mar.–Apr. 1946.

tinct from the previous region, this low sandy plain centred on Malines extends on the west from the Scheldt between Termonde and Tamise to the valley of the Grande Nèthe on the east, where a transitional boundary with the Improved Campine region can be drawn. Its elevation is from 30 to 60 ft.

Intensive open-air horticulture is important but not dominant (6 per cent as against 20 per cent in the Brussels region). *Potatoes* account for no less than 24 per cent of the surface, however, and " earlies " predominate around Duffel, which is the chief " early " district of Belgium. Other crops include green peas, cauliflowers, and asparagus ; witloof is relatively unimportant.

Rye is widespread (19 per cent) and the dense concentration of dairy cattle providing milk for Brussels and Antwerp is based essentially on fodder crops, since only 20 per cent of the surface is permanent grass. Some 90 per cent of holdings are below 5 hectares, being mostly rented, and large farms are fewer than in the previous region.

5(*c*). *The Antwerp Region.* This small region resembles the Improved Campine, but greater market accessibility has fostered more horticulture and market gardening of which a limited but appreciable amount is under glass.

The timber has been completely cleared.

It differs from the plain of Malines–Lierre in the smaller proportion of potatoes (especially of earlies) and increased attention to flowers and fruit (especially strawberries and tomatoes). There is also more permanent grass (43 per cent compared with 20 per cent) and somewhat higher cattle densities, two-thirds being dairy cows.

The holdings are predominantly small.

5(*d*). *The Liége Region.* This small area around Liége occupies the south-east slopes of the Meuse valley between Seraing and Herstal, and the river terraces of Grivegnée at the junction of Meuse and Ourthe. *Vegetables* (much in part time holdings) amount to 19 per cent of the surface, the rest being chiefly *fruit* (12 per cent) and grass (20 per cent). Dairy cows, pigs and poultry, are of great importance, the latter density being the highest in Belgium.

5(*e*). *Charleroi and the Borinage.* From the industrial district of Charleroi on the east, this region passes westwards through La Louvière into the Borinage, including the marshy valley of the lower Haine.

It is essentially a *milk* and *market-gardening* region, with considerable vegetable production on the part time holdings of colliery and other industrial workers. Grassland (40 per cent of the surface) is especially important in the Haine basin west of Mons. The dense dairy cattle population is primarily for the local milk market.

6. *The Condroz Mixed Farming Region—The Pre-Ardenne.* Extending from the Meuse on the west to the lower Ourthe on the east, the plateau of the Condroz is transitional to the Ardenne proper, standing between 600 and 1,000 ft. with a rainfall of some 30 to 35 in.

The rolling landscape has much timber, especially on the exposed ridges of Devonian sandstones and slates. Farmland is chiefly in the calcareous vales where also remnants of limon are found. Its essentially mixed farming shows increasing tendencies towards grass and dairying. About half the farmland (36 per cent) is *grassland* (with limited orchard development) and beef slightly exceeds dairy cattle.

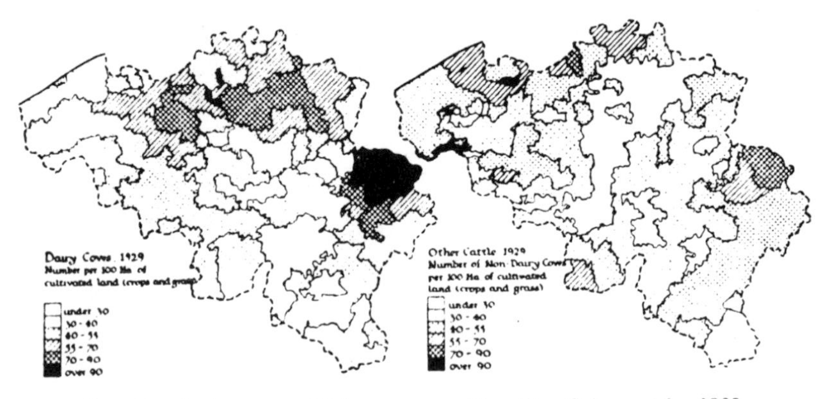

FIG. 69.—Dairy cows, 1929. FIG. 70.—Other cattle, 1929.

Dairy Cows. By far the densest concentration of dairy cattle is in the grassy Pays de Herve, where the highly dissected landscape, partly masked with heavy limon and "chalon" soils, favours small grass farms. This region has over 100 cows to the 100 hectares of exploited land—the highest density in Belgium. Other important regions are East Flanders and the south-western fringe of the Campine. Here production is related partly to small-holdings and also to the local demand of the Ghent, Brussels and Antwerp conurbations.

The main limon tracts with their larger arable holdings have a much lower density, usually below 35 to the 100 hectares, and the Ardenne region in general has a low density, its remoteness from markets fostering beef rather than milk production. The Polders are also primarily a beef-cattle district.

Other Cattle (mainly beef animals). The Pays de Herve, with from 50 to 80 animals per 100 hectares of exploited land, is very important for beef, as is also the Polders region, both within the main coastal area and also in the detached areas along the lower Scheldt. This latter producing area extends also on to the adjacent fringes of the Flanders sands region.

The High Ardenne also has a fairly high density, but most of the lowlands have a much lower density of about 35 beasts to the 100 hectares.

Arable land grows chiefly *oats* (13 per cent) and fodder crops (especially trefoils, 5·3 per cent) but wheat is not unimportant (4·6 per cent), and there is a little sugar-beet.

The region is essentially one of large tenanted farms, and much land has long been marginal for cultivation because of low beef and cereal prices. Moreover, in common with other parts of upland Belgium, there has been some labour shortage due to declining population. The trend towards grass and dairying recalls many British features. Holdings average nearly 50 hectares, almost the largest in Belgium.

7. *The Herve Fruit and Dairying Region.* This small, highly-dissected region of predominantly damp, heavy soils, occupies the north-western flanks of the Hautes Fagnes, between 600 and 1,000 ft., and has a rainfall of from 30 to 40 in. One of its most widespread soils is the "Chalon" or residual clay derived from underlying chalk; heavy limon of the Hesbaye type also occurs.

Over 80 per cent of the area is *permanent grass*, of which some 24 per cent is grass orchard (chiefly apples). The typical small farms are dominated by dairying,[32] over two-thirds of the dense cattle population being dairy cows, and the concentration is the densest in Belgium. Much liquid milk is sold in nearby Liége, considerable butter is also produced, and pigs (reared partly on dairy by-products) are more important than elsewhere in Belgium. Sheep are almost absent but poultry are numerous. The restricted arable land grows oats, a little wheat, and but limited fodder crops in this essentially grass region.

8. *The Famenne–Chimay Pastoral Region.* Developed on the relatively soft Middle and Upper Devonian shales and limestones of the Famenne Depression, this region has heavy soils, contrasting strikingly with the Condroz. Although standing at only from 600 to 700 ft., rainfall is as high as 40 in. owing to the nearness of the High Ardenne, and its small-farm economy is directed chiefly to the rearing of beef and dairy cattle. There is considerable timber (38 per cent) and *grassland* (orchards relatively restricted) accounts for 41 per cent or two-thirds of the farmland. The calcareous corridor of Corwin and Chimay (a western extension west of the Givet salient) forms a remarkable strip of cultivated land three miles wide, surrounded by forests. Here there is more arable land and even a little wheat, but throughout the region as a whole arable land is mostly for oats and fodder crops.

9(*a*). *The High Ardenne.* This extreme north-east corner of the Ardenne Massif comprises the high moors and forests of the Hautes Fagnes (over 2,000 ft.) developed largely on Cambrian and Silurian quartzites, together with the upland basins of the Amblève and Salm, centred on Stavelot, and draining north-west to the Ourthe and Meuse at Liége.

Half of the surface is timber, chiefly deciduous forest of beech, oak and oak-birch, with coniferous plantations of spruce and pine. There is also much heather and grass moor and peat bog.

Farmland[33] is chiefly in the Amblève basin which itself stands at over 1,000 ft., with a high rainfall and short growing season. It is almost entirely grassland (40·5 per cent) raising dairy cows and young

[32] No less than 50 per cent of holdings are between 5 and 20 hectares.

[33] Essentially in small farms, over two-thirds of which are owned by their cultivators.

store beef cattle, and concentrated on butter-making owing to the limited local market for liquid milk. The extremely limited arable land grows oats, rye and fodder crops.

9(*b*). *The Ardenne Forest and Cattle Region.* The central and western Ardenne is lower and less dissected than the previous region, standing between 1,200 and 1,600 ft., but reaching 1,900 ft. in the Forêt de Freyr near St. Hubert. It is developed largely on lower Devonian sandstones and slates. In climate (especially as regards rainfall) it closely resembles the High Ardenne.

Agriculturally it differs from the High Ardenne in the higher proportion of timber (53 per cent). The farmland is almost equally divided between grass and arable, the latter being chiefly oats, trefoils and rye. Even allowing for the greater forest cover cattle densities are lower than in the higher region (37 as against 57 per 100 hectares of improved land). Most are beef animals.

The wooded Rièze upland along the French frontier west of the Givet salient belongs essentially to this region.

9(*c*). *The Bastogne Area.* This district centred on Bastogne adjoining the Luxembourg frontier, although elevated, has a south-eastern aspect, a slightly lower rainfall, and drains to the Wiltz and Sure towards the Moselle.

Only one-third is forested, arable land exceeds permanent grass, and some cereals are grown. The prevailing small farm economy is primarily beef cattle rearing, and the arable land grows *trefoils* (14·5 per cent), *oats* (13 per cent) and other fodder crops. The proportion of dairy cows (though not the density) is the lowest in Belgium. Siliceous, but damp soils derived from the Devonian encourage *rye* cultivation (3·5 per cent) and the small amount (1·5 per cent) is almost the only *wheat* grown within the Massif.

The ridge of the Forêt d'Anlier to the south separates it from Belgian Lorraine.

The essential differences from the main Ardenne region are clear from the cantonal statistics for Bastogne, Fauvillers, and Sibret, but the limits of the region have been largely drawn from topographic maps.

9(*d*). *The Semois Valley.* Below Chiny the rejuvenated Semois with its deeply incised meanders south of the main Ardenne watershed, has a relatively warm and sheltered summer, although the rainfall is almost the highest in Belgium—47 in. On the steep wooded valley sides the proportion of timber rises to 64 per cent.

Arable land slightly exceeds permanent grass and, as everywhere in upland Belgium, is chiefly oats and fodder crops for the rearing of beef cattle. The characteristic crop, however, is tobacco, which, although only 1 per cent of the total surface of the two cantons concerned, is immensely important in the valley itself, and on most farms is the chief source of income. The output (the Semois is the third tobacco region of Belgium) comes from small specialized holdings on the alluvial floor and lower slopes, where the *toubatière* system of continuous cropping prevails, often for as long as ten years. Drying sheds or *séchoirs* are characteristic landscape features below Bouillon, and especially round Bohan, and have largely superseded the practice of hanging the leaves from the eaves of the houses. This specialized cultivation is favoured by the deep winding valley, sheltered from wind, with a relatively long growing season. The risk of late frosts from temperature inversion is said to be offset by summer night mist which maintains night temperatures and substantially reduces the diurnal range.[34]

10(*a*). *The Arlon Mixed Farming Region.* In this region on the southern side of the Ardenne watershed the former forest cover has been reduced to about one quarter, remaining chiefly on the higher sandier tracts associated with the two parallel escarpments of Habergy and of the Luxembourg sandstone. The climate is extreme, the rainfall fairly heavy and, with the prevailing clay and marl soils on the lower ground, grassland almost equals arable. There is a little *wheat* (nearly 3 per cent), *rye* and *potatoes*, but these are overshadowed by oats and other fodder crops. Dairy cattle, pigs and sheep are more important than in the Ardenne. Grass orchards also reappear. There are more small holdings than in the Ardenne (see Figs. 59A and 59B).

10(*b*). *The Virton Forest and Mixed Farming Region.* In climate and situation resembling the previous region, this western part of Belgian Lorraine is very dissected, has more residual woodland (amounting to almost half the surface), and arable is subsidiary to grassland. The utilization of arable land resembles that of Arlon, but there is more rye. The boundary between the two regions is transitional and has been drawn along the cantonal boundaries.

War-time Changes in Agricultural Distributions[35]

The effect of the war was to increase arable land by some 20 per cent, from 873,764 hectares in 1941 to 1,064,510 hectares in 1943, at the expense of both permanent grass and wasteland. Cantonal figures are not available prior to 1941, but this is a suitable base year, as the 1940

[34] Carlsbourg in the lower Semois has a mean daily range of only 17·68° F. for the period April to October, comparing with 20·64° for Virton (Upper Semois), and 20·20° for Bourg Léopold (Campine). Its range is actually slightly less than that of Ypres in maritime West Flanders—18·87°.

[35] Based on the May 15th returns of the Ministry of Agriculture for 1941–5 inclusive.

invasion and evacuation of many cultivators seriously interfered with agriculture. For most crops the production peak came in 1943.

As might be expected, the cultivation of *food grains* (i.e. wheat, rye, barley) and *sugar-beet* was greatly extended, while *fodder crops* (oats, roots, clovers, trefoils) decreased. This reflects the limited land (and labour) available for agricultural expansion, particularly in view of the forced labour emigration to Germany. Certain typically Flemish industrial crops were severely " axed "—especially flax and chicory, but tobacco showed a steady increase to 7,300 hectares in 1944. Colza was also revived for margarine manufacture, and a compulsory quota for all cultivators brought it to 30,000 hectares in 1944, compared with a quite negligible figure—31 hectares—in 1941.

There were catastrophic falls of most livestock, but especially *beef cattle* (from 1,159,232 in 1941 to 652,296 in 1943), *pigs* and *poultry.* (The 3,281,000 fowls of 1941 had dwindled to 2,058,000 in 1943). These were obvious reactions to the cutting off of imported feeding stuffs, the reduction in home fodder crops, and the diversion of home-produced rye, barley, and potatoes into the Belgian loaf. The tendency to retain breeding and milking stock is however shown by the maintenance of numbers of *dairy cows* (825,418 in 1941, 834,876 in 1943), and *sheep* definitely increased, suggesting a more intensive use of grass and grass orchards.

With regional variations, *wheat* increased everywhere, especially in the Improved Campine and in Flanders, and although showing a large *percentage* increase the Ardenne and Dry Campine remained negative (see Fig. 64). The decline in *cattle* was most serious in arable or mixed farming areas where the food drive reduced the fodder crop area. Thus in Flanders and on the Limon numbers were halved, while in grass regions such as Herve and the Polders the decline was only some 14–18 per cent. The overall increase of 30 per cent in *sheep* was chiefly due to rises of from 70–100 per cent in the Polders, Herve, and Flanders, and some arable regions recorded sharp declines.

The dot distribution maps are all for 1943, the year of greatest deviation from the pre-war norm. Comparison with 1929 conditions on which the regional classification is based is also assisted by the regional tables for 1943 and 1945. It will be seen that, broadly speaking, the wartime changes do not invalidate the regional classification described in this article, and a detailed inspection of 1944 and 1945 statistics suggest some move back towards pre-war " normality."

It must be concluded that the 27 regions described represent reasonably stable features of Belgian agricultural geography—features which have their roots in the physical and geographical background.

1929

Regions	Wheat	Rye	Oats	Barley	Industrial Crops	Sugar Beet	Maincrop Potatoes	Early Potatoes	Horse Beans	Trefoils	Other Fodder Crops	Open-air Market Gardening	Glasshouses *	Fruit	Grasslands and Orchards	Woods	Total Cattle	Beef Cattle	Dairy Cows	% of Dairy Cows	Sheep	Pigs	Goats	Horses	Poultry
	% of "Exploited" Land																Livestock per 100 Hectares of "Exploited" Land								
1(a) The Coastal Dune Belt	—	—	—	—	—	—	—	—	—	—	—	—	—	—	—	—	—	—	—	—	—	—	—	—	—
(b) The Polders	**11·6**	3·7	**13·0**	4·5	3·2	3·5	5·6	·7	**4·8**	2·7	5·4	·7	—	·8	**39·5**	·5	80	43	37	40·5	9	40	6	14	989
2(a) Sandy East Flanders	2·8	**25·1**	7·5	·5	2·5	—	**13·8**	1·0	—	6·7	4·9	1·7	—	4·9	24·3	5·2	110	48	**62**	56·4	13	86	6	14	1,241
(b) The Roulers Region	5·0	**14·3**	**10·8**	·1	**14·1**	1·3	**14·4**	1·5	—	1·8	6·0	1·6	—	2·4	27·0	·8	76	33	43	57·6	8	63	22	13	1,884
(c) Southern Flanders	**16·0**	**12·2**	5·0	1·1	1·1	·2	**10·0**	—	—	5·6	7·9	1·7	—	**6·0**	33·0	3·0	80	23	**57**	59·8	19	67	17	12	1,420
3(a) Dry Campine	—	**17·0**	8·4	—	—	—	4·6	·2	—	1·4	6·0	·8	—	·6	31·6	30·8	73	30	42	58·1	5	35	8	8	622
(b) Improved Campine	1·9	**20·5**	5·6	·2	—	·1	**10·3**	1·2	—	3·8	3·9	2·0	—	1·3	28·1	20·2	71	19	52	71·5	4	36	16	10	648
4(a) Ypres–Courtrai Region	**15·5**	4·5	**15·4**	·7	**10·4**	**7·2**	7·6	1·2	·4	3·2	6·3	1·4	—	·9	23·6	1·6	73	38	35	48·3	9	37	14	14	1,134
(b) Tournai Region	**16·4**	3·5	**12·5**	1·0	2·0	**6·0**	3·6	·5	·6	5·6	6·9	1·5	—	2·8	29·9	8·6	78	41	37	47·6	13	11	3	14	828
(c) Southern Hainaut	**12·8**	1·9	**11·5**	4·1	1·6	**4·3**	2·7	·2	·4	5·7	5·3	1·8	—	3·4	38·0	8·7	66	31	35	53·7	18	11	2	17	566
(d) Central Limon Region	**15·3**	3·3	**18·5**	2·1	1·7	**11·8**	4·4	·4	·2	6·9	5·2	1·5	—	2·5	21·1	3·5	65	32	33	50·8	8	50	4	14	576
(e) Eastern Hesbaye	**13·7**	3·0	**19·3**	1·3	·2	**12·5**	5·1	·4	·1	4·1	4·0	1·1	—	**11·3**	26·3	1·9	68	33	35	50·9	9	88	2	16	761
(f) Central Mixed Farming Region	**14·0**	5·0	**11·1**	1·4	1·8	1·4	3·1	·5	·2	6·3	5·6	1·1	—	3·7	**41·4**	6·5	79	39	41	51·7	14	27	3	16	514
5(a) Brussels–Hoeylaert Region	7·2	8·5	8·2	·4	·2	·1	6·7	1·5	—	4·5	4·6	**20·2**	2·2	2·5	15·2	11·8	39	13	26	67·9	18	37	12	25	1,045
(b) Malines–Lierre Region	2·8	**19·8**	3·3	·2	·4	—	**13·7**	**10·4**	—	5·2	4·8	**6·1**	·2	1·1	23·5	5·1	82	9	**73**	89·0	4	19	49	21	790
(c) Antwerp Region	2·3	**12·5**	4·3	·4	·2	·1	**10·5**	2·1	·2	2·1	6·1	**5·7**	1·0	2·5	**43·6**	2·9	95	33	**61**	64·6	16	38	11	16	1,435
(d) Liége Region	3·1	·6	7·5	·8	—	3·1	4·5	·6	—	2·4	2·8	**19·3**	1·0	**12·8**	34·6	11·4	59	17	42	71·5	23	43	11	14	2,654
(e) Charleroi–Borinage Region	6·3	2·1	7·7	1·1	·5	·6	4·5	·5	·1	3·3	4·5	**9·0**	·2	3·2	**43·1**	12·2	62	19	43	69·5	25	26	21	12	1,831
6 The Condroz Region	4·6	1·9	**13·8**	2·3	·6	·7	1·8	·2	·4	5·3	3·4	·7	—	2·0	36·0	26·1	49	25	24	48·3	7	20	2	10	300
7 Herve Fruit and Dairying Region	1·5	·3	1·9	·3	—	—	·4	·1	—	·9	·8	1·0	—	**24·1**	**84·9**	6·5	179	55	**124**	69·1	3	**143**	1	8	1,002
8 Famenne–Chimay Pastoral Region	1·4	1·6	6·1	·7	—	—	1·4	—	·1	5·5	1·2	·3	—	·8	**41·7**	38·4	45	24	21	47·9	3	9	17	7	186
9(a) The High Ardenne	·1	1·0	3·5	·3	—	—	1·4	—	—	2·6	·1	·2	—	·4	**40·6**	**50·0**	58	26	32	54·9	3	13	—	4	225
(b) The Ardenne Forest and Grass Region	·5	2·1	7·2	1·0	—	—	2·8	—	—	8·4	·6	·2	—	·5	22·8	**53·1**	38	21	17	44·1	2	15	1	6	166
(c) The Bastogne Region	1·5	3·6	**13·3**	1·0	—	—	4·5	·1	—	14·5	·6	·3	—	·2	30·2	29·0	52	29	23	43·9	3	25	1	9	225
(d) The Semois Valley	·2	1·9	4·6	·4	·9	—	2·6	—	—	6·7	·5	—	—	·6	16·8	**64·3**	24	13	11	46·2	2	11	2	4	136
10(a) The Arlon Mixed Farming Region	3·0	1·9	**11·5**	1·2	—	—	5·7	·4	·3	5·6	2·0	·7	—	1·1	33·9	26·8	40	18	22	55·3	7	36	5	9	288
(b) The Virton Forest and Mixed Farming Region	1·7	—	5·4	·7	—	—	3·3	·1	·3	1·8	1·3	·3	—	·5	33·1	**48·5**	28	14	14	49·2	2	20	5	5	194

* Including also "Flowers," mostly under glass.

Regions	Wheat	Rye	Oats	Barley	Industrial Crops	Sugar Beet	Maincrop Potatoes	Early Potatoes	Horse Beans	Trefoils	Other Fodder Crops	Open-air Market Gardening	Glasshouses *	Fruit	Grasslands and Orchards	Woods	Total Cattle	Beef Cattle	Dairy Cows	% of Dairy Cows	Sheep	Pigs	Goats	Horses	Poultry
	% of "Exploited" Land																Livestock per 100 Hectares of "Exploited" Land								
1(a) The Coastal Dune Belt	—	—	—	—	—	—	—	—	—	—	—	—	—	—	—	—	—	—	—	—	—	—	—	—	—
(b) The Polders	**13·4**	3·2	7·5	7·7	2·3	1·7	4·4	2·2	**1·5**	1·6	8·6	1·2	—	·3	**40·4**	·5	80	45	35	43·3	32	28	6	15	114
2(a) Sandy East Flanders	5·7	**21·1**	5·8	·9	2·9	—	**12·9**	—	—	2·9	5·2	2·3	—	**6·7**	22·2	5·6	91	33	**58**	63·0	27	32	6	13	151
(b) The Roulers Region	**11·0**	**9·8**	8·3	1·1	**9·1**	1·1	**13·9**	1·2	—	·4	5·7	2·8	—	1·9	27·6	1·3	72	30	42	56·0	23	27	16	11	126
(c) Southern Flanders	**12·7**	**9·6**	3·8	2·5	1·9	·4	**8·1**	—	—	4·3	8·1	2·1	—	**7·1**	30·6	3·2	90	35	**55**	61·0	37	26	14	12	106
3(a) Dry Campine	1·6	**18·7**	6·7	·3	—	·1	3·7	—	·7	·3	5·2	·7	—	·7	29·2	29·9	69	30	39	57·0	9	19	9	8	121
(b) Improved Campine	3·0	**17·6**	4·1	·4	1·3	·1	**8·1**	·3	—	2·7	4·0	1·5	—	2·3	28·3	23·2	74	25	49	67·5	5	19	23	10	113
4(a) Ypres–Courtrai Region	**19·0**	3·1	9·5	4·3	**7·8**	**4·7**	**8·0**	·3	·7	1·3	7·3	1·4	—	·5	26·5	1·9	68	32	36	52·5	12	23	9	14	100
(b) Tournai Region	**18·2**	2·4	6·7	5·5	2·4	**6·7**	3·4	—	·2	2·1	7·0	1·1	—	2·0	31·5	8·9	73	37	36	50·0	8	8	2	12	83
(c) Southern Hainaut	**15·8**	2·8	4·8	4·8	2·2	**5·2**	2·0	—	·2	2·0	2·7	·6	—	2·8	38·6	9·0	72	34	38	52·5	12	5	2	14	66
(d) Central Limon Region	**19·8**	3·4	8·8	4·4	3·0	**13·0**	2·8	—	·1	4·7	2·9	1·4	—	2·2	22·2	7·2	61	29	32	52·5	6	22	3	11	109
(e) Eastern Hesbaye	**17·3**	2·1	8·5	4·3	2·1	**13·4**	3·1	—	—	2·8	2·6	1·2	—	**14·0**	19·0	2·9	66	26	40	60·0	16	52	2	16	124
(f) Central Mixed Farming Region	**15·2**	3·2	5·4	4·5	1·8	3·1	2·5	—	·2	4·1	3·5	·7	—	2·8	**41·0**	6·9	81	51	30	37·3	12	17	3	14	81
5(a) Brussels–Hoeylaert Region	**10·3**	6·5	3·7	·7	1·1	·3	**10·5**	—	·3	3·5	7·3	**10·4**	2·5	4·9	12·2	16·4	28	13	15	54·0	16	18	16	11	142
(b) Malines–Lierre Region	4·9	**16·3**	3·6	·3	1·2	—	**10·6**	**7·5**	—	2·1	5·2	**9·0**	·2	1·9	25·2	6·4	91	33	**58**	63·5	6	15	32	17	170
(c) Antwerp Region	5·9	**12·0**	4·9	1·8	3·5	·8	**12·4**	·9	—	·7	6·4	**8·2**	·2	3·1	34·0	2·1	72	33	39	54·0	17	17	19	13	186
(d) Liége Region	7·9	·5	3·7	2·8	5	3·4	**14·6**	—	—	·5	1·8	**8·2**	·1	**14·6**	28·1	8·0	61	19	42	69·0	21	17	11	7	92
(e) Charleroi–Borinage Region	8·4	3·8	2·8	3·3	1·4	2·1	4·8	—	—	1·4	3·3	**4·7**	·1	2·2	**43·5**	11·3	70	30	40	57·5	18	8	9	11	73
6 The Condroz Region	**10·8**	2·1	5·9	3·3	·9	1·0	1·2	—	·4	2·9	1·9	·4	—	1·1	36·5	27·6	56	28	28	50·0	5	9	2	7	42
7 Herve Fruit and Dairying Region	4·3	1·5	1·2	·9	—	·5	1·5	—	—	·4	·8	—	—	**30·0**	**41·0**	10·0	164	60	**104**	63·5	6	23	1	6	96
8 Famenne–Chimay Pastoral Region	3·8	1·7	3·1	2·0	·1	·1	·7	—	—	1·0	·7	—	—	·4	**39·6**	42·5	48	24	24	50·0	2	5	1	4	28
9(a) The High Ardenne	1·3	·9	·6	·6	—	—	·7	—	—	—	·1	—	—	·1	38·1	**56·5**	56	24	32	56·5	3	1	—	3	21
(b) The Ardenne Forest and Grass Region	1·4	2·0	4·5	1·9	—	—	1·9	—	—	1·0	·8	—	—	—	29·0	**53·8**	35	16	19	54·5	1	7	1	4	43
(c) The Bastogne Region	3·1	3·6	8·3	4·1	·1	—	3·9	—	—	·8	·9	—	—	—	**41·5**	31·5	49	22	27	55·5	2	15	—	8	75
(d) The Semois Valley	·8	1·8	2·8	·3	·9	—	1·7	—	—	1·0	·7	—	—	—	19·7	**66·7**	21	9	12	57·8	1	6	1	3	28
10(a) The Arlon Mixed Farming Region	3·3	2·1	8·6	1·3	—	—	4·3	—	·4	3·0	2·7	·5	—	·5	**39·5**	31·5	50	24	26	51·5	6	21	7	8	88
(b) The Virton Forest and Mixed Farming Region	2·3	2·4	3·7	·7	—	—	2·1	—	·4	·6	1·4	·2	—	·3	29·2	**55·0**	31	16	15	47·6	2	8	4	4	42

* Including also "Flowers," mostly under glass.

1945

Regions	% of "Exploited" Land																Livestock per 100 Hectares of "Exploited" Land								
	Wheat	Rye	Oats	Barley	Industrial Crops	Sugar Beet	Maincrop Potatoes	Early Potatoes	Horse Beans	Trefoils	Other Fodder Crops	Open-air Market Gardening	Glasshouses *	Fruit	Grasslands and Orchards	Woods	Total Cattle	Beef Cattle	Dairy Cows	% of Dairy Cows	Sheep	Pigs	Goats	Horses	Poultry
1(a) The Coastal Dune Belt	—	—	—	—	—	—	—	—	—	—	—	—	—	—	—	—	—	—	—	—	—	—	—	—	—
(b) The Polders	**10·3**	3·1	**10·7**	8·5	2·7	·5	4·2	·9	**2·4**	1·5	8·5	1·5	—	·5	**39·8**	·5	92	53	39	43·5	25	55	5	15	126
2(a) Sandy East Flanders	5·6	**17·6**	7·7	1·8	·7	—	**10·2**	·3	—	4·3	4·8	2·1	—	**7·1**	23·9	6·0	103	47	**56**	54·0	21	66	4	10	176
(b) The Roulers Region	9·3	**7·0**	**10·9**	2·5	**8·0**	·3	**12·7**	·1	1·3	3·7	6·1	3·7	—	1·9	28·2	1·3	73	35	38	51·5	16	44	14	14	151
(c) Southern Flanders	**13·7**	**7·6**	5·2	3·7	1·4	·3	6·8	—	·1	6·0	7·8	2·3	—	**7·3**	31·0	3·6	95	43	**51**	54·0	32	45	12	11	104
3(a) Dry Campine	1·6	**13·9**	8·7	·3	—	·2	3·3	—	—	1·3	6·1	·6	—	·7	32·6	32·0	76	39	37	48·6	8	33	5	9	162
(b) Improved Campine	3·8	**17·1**	5·6	·4	·2	·1	5·6	·6	—	4·0	4·3	1·6	—	2·7	35·0	28·4	94	40	54	58·0	4	35	23	14	165
4(a) Ypres–Courtrai Region	**15·4**	2·0	**12·0**	5·7	**8·3**	2·7	7·1	·1	·3	2·1	6·2	1·9	—	·4	26·6	1·9	73	39	34	46·0	11	38	7	16	118
(b) Tournai Region	**13·6**	1·7	**11·4**	4·8	4·3	**3·9**	2·8	—	·5	4·2	6·7	1·1	—	1·8	32·0	8·9	75	40	35	46·2	7	15	1	15	106
(c) Southern Hainaut	**12·9**	1·2	9·7	7·0	2·7	2·8	1·3	—	·3	2·7	4·8	1·1	—	2·2	**39·4**	9·1	73	38	35	49·1	11	10	1	14	91
(d) Central Limon Region	**16·8**	2·1	**13·5**	5·6	3·0	1·6	2·1	—	·2	6·0	4·5	1·4	—	2·1	22·3	7·2	54	23	31	58·0	6	31	3	17	131
(e) Eastern Hesbaye	**14·2**	2·6	**13·4**	4·9	·9	**10·4**	2·3	—	—	4·0	5·9	1·6	—	**16·4**	18·1	2·6	69	36	33	48·3	14	60	2	12	149
(f) Central Mixed Farming Region	**12·7**	3·4	9·3	3·6	1·2	**11·9**	1·9	—	·3	5·4	5·4	·9	—	2·4	**42·0**	6·8	90	48	42	46·2	13	26	2	16	104
5(a) Brussels–Hoeylaert Region	**11·6**	5·7	6·4	1·2	·6	·1	6·7	—	—	5·5	6·2	**16·0**	2·9	4·3	12·1	22·0	33	17	16	48·8	15	33	21	15	220
(b) Malines–Lierre Region	6·2	**12·4**	3·7	·4	·1	—	·3	**6·5**	—	—	7·5	**8·6**	·1	1·8	26·8	6·8	96	44	**52**	54·3	6	33	29	19	231
(c) Antwerp Region	4·7	**11·5**	6·7	1·9	—	—	**11·3**	·8	—	1·4	6·8	**5·9**	·2	3·8	38·0	3·6	83	40	43	51·8	12	34	16	15	166
(d) Liège Region	6·3	·4	5·7	3·4	·8	2·7	3·1	—	—	·8	2·3	**12·2**	·1	**16·9**	21·0	20·0	60	26	34	57·2	19	28	9	9	185
(e) Charleroi–Borinage Region	8·8	2·2	6·6	7·1	·5	1·0	2·3	—	—	3·7	5·3	2·8	·1	1·3	32·0	11·2	48	19	29	55·0	10	8	5	9	118
6 The Condroz Region	8·7	1·3	8·1	3·7	·5	·4	·8	—	·3	4·2	6·2	·4	—	1·2	36·7	27·8	61	34	27	44·0	6	13	1	9	53
7 Herve Fruit and Dairying Region	2·5	·4	2·2	1·2	—	—	·9	—	—	·3	1·0	·3	—	**22·6**	**56·0**	8·2	189	88	**101**	53·5	11	38	1	8	124
8 Famenne–Chimay Pastoral Region	3·1	·9	4·0	1·4	—	—	·5	—	—	1·3	·9	·1	—	·4	**40·5**	43·0	53	29	24	44·9	2	5	—	6	31
9(a) The High Ardenne	·7	·4	1·8	·8	—	—	·7	—	—	—	—	—	—	—	**53·6**	41·0	72	35	37	47·6	2	4	—	4	25
(b) The Ardenne Forest and Grass Region	1·0	1·2	4·1	1·2	—	—	1·3	—	—	1·2	·7	—	—	—	30·1	**56·5**	38	20	18	48·0	1	6	—	5	39
(c) The Bastogne Region	2·5	1·9	6·9	2·6	—	—	2·2	—	—	1·4	·9	—	—	—	**42·9**	37·0	47	25	22	47·0	2	8	—	7	50
(d) The Semois Valley	·4	1·2	2·6	·2	·8	—	1·3	—	—	1·3	·6	—	—	·1	20·0	**68·1**	23	11	12	51·8	1	7	1	3	29
10(a) The Arlon Mixed Farming Region	3·3	1·4	9·3	·6	—	—	3·0	—	·1	3·5	2·1	·3	—	·3	38·8	31·4	51	26	25	48·2	6	22	4	9	83
(b) The Virton Forest and Mixed Farming Region	1·9	1·7	4·0	·4	—	—	1·6	—	·2	·9	·8	·1	—	·3	29·0	**57·0**	33	18	15	46·0	3	9	3	4	44

* Including also "Flowers," mostly under glass.

XVI

SOME PRINCIPLES OF LAND-USE PLANNING [1]

By E. C. WILLATTS, B.Sc. (Econ.), Ph.D.

The concept of comprehensive land-use planning in this country is strangely new. This densely-populated country, with an average of more than one person per acre, has been slow to recognize the need for planning the use of its resources. Planning suggests an ordered scheme for the use of scarce resources, and a regard for the wider interests of the country as a whole and not merely short-term sectional interests. It requires a proper appreciation of the resources whose use must be planned, and it is essential that the public as a whole shall understand the need for planning and approve its broad principles.

There is little doubt that before Britain became a predominantly urban and industrial country its occupants regarded the land, that is to say, the soil from which they derived their wealth, " as a precious heritage, to be used but not destroyed, and to be handed on intact, improved if possible, for the use of those who came after." [2] But in the last century or more it was natural that an increasingly urban population, whose livelihood was heavily dependent on industry and overseas trade and whose larder was largely stocked with imported foods, should pay relatively little heed to rural problems. Land was more profitable and gave rise to more employment, if used for industry, commerce or other urban purposes than for food production, and the relative productivity of the agricultural worker was below that of the industrial worker. Professor S. R. Dennison has given a brilliant exposition of the logic of this point of view in his minority report as member of the Scott Committee.[3] The core of the argument is: ". . . we can produce more food by employing labour and capital in manufacturing goods which are exported, and importing food in return, than we can by using the same amount of labour and real capital in agriculture to produce the food directly."

[1] Written in 1947. The views expressed by Dr. E. C. Willatts are his own, and not necessarily those of the Ministry of Town and Country Planning in which he holds an appointment as Senior Research Officer.

[2] Sir E. John Russell, " Agricultural Restoration of Mining and Quarrying Sites," *Agriculture*, Vol. LIV, No. 2, 1947, p. 49.

[3] *Report of the Committee on Land Utilization in Rural Areas*, H.M.S.O., 1942, Cmd. 6378.

Such reasoning led, of course, only indirectly to a general disregard for land as such.

However, public opinion has, in the last decade or so, undergone a radical change of outlook on the subject of intrinsic land values. There have been two distinct stages. There was a first stage (exemplified by the Ribbon Development Act of 1935) of anxiety to preserve the amenities of the countryside from the threat of rapidly expanding suburban development. This was partly due to the natural and proper though very incomplete view of the countryside as providing recreation and amenity for the urban dweller, who was concerned to avoid spoiling a source of æsthetic satisfaction. The second stage was of concern for the loss of agricultural land, especially if of high intrinsic quality.

It is significant that the Town and Country Planning Act, which was passed in 1932, showed very little appreciation of agricultural land except as providing scope for urban development. The aim of the Act was to direct building into appropriate zones of varying density but its effect, though not its intention, was often to accelerate the rate of loss of agricultural land by zoning rural areas for housing at very low densities to the acre. Typical schemes prepared under it had no regard for the qualities of soil or the problems of agriculture.[4] The period was one which will be remembered for the rash of hoardings advertising land as "ripe for development." Houses were often the most profitable crop on farm land. Reasonable development there must always be, but how unreasonable was much of the inter-war development is clear even to the eye of quite inexpert observers, while the unreasonably expansionist nature of many of the official planning schemes of that time is exemplified by the single fact that by 1937, when not half the country was covered by such schemes, those already prepared had zoned enough land to accommodate an increase of nearly 300,000,000 people,[5] at a time when the country's population was becoming almost stationary. There were many reasons why such a fantastic position was achieved, and they have been adequately discussed elsewhere.[6]

With the rapid increase in population in the nineteenth and early twentieth centuries the majority of our towns expanded, and in the latter part of this period, with more generous standards of housing and of provision of land for such purposes as open spaces, the propor-

[4] E. C. Willatts, "Present Land Use as a Basis for Planning," *Geography*, June, 1938.

[5] *Report of the Committee on Land Utilization in Rural Areas*, H.M.S.O., 1942, Cmd. 6378, p. 41.

[6] Ibid., Chap. VI.

tion of urban land to population increased well beyond what had been customary during the rapid building development of the Industrial Revolution. This had tended to give rise to an assumption, widespread in pre-war physical planning, that all towns would grow, some fast and some slowly, but all requiring more and more room for expansion. At the same time there was a marked tendency for towns to have a high density of dwellings at the centre and to become less and less dense towards the periphery, where their thinly scattered houses merged into the countryside.

The former concept is now as untenable as the latter tendency is unacceptable to good planning practice. It is increasingly realized that towns do not all continue to grow, and some decline. Indeed, however wounding it may be to local pride, and however reluctant some local authorities may be to accept the facts, the truth is that large numbers of towns in recent years have lost rather than gained population. There are, of course, obvious reservations to be made in comparing towns, as commonly understood and recognizable physical entities, with administrative areas, but an examination of official figures published by the Registrar-General [7] shows that of the County Boroughs, Municipal Boroughs and Urban Districts of England and Wales, 35 per cent were declining in the period 1921–31 and 44 per cent in 1931–9. For Wales alone the figures are 59 per cent and 72 per cent.

In general, in this period of physical expansion, some of the difficulties of full land-use planning stemmed from an inadequate appreciation of the real nature of the problem by the general public and, as legislation is rarely in advance of public opinion, progress in the matter was not possible until public opinion had been modified. This was not achieved until, firstly, the results of relevant research work had been disseminated [8] and, secondly, the problem had become one of imperative national interest when the sudden shock of war threatened serious shortages of the products of the soil. The war, with its virtual cessation of normal building and its successful introduction of large-scale economic and physical planning, provided a specially favourable opportunity of studying the fundamental problems, of stock-taking and of discovering how quickly there had arisen a wide measure of agreement on the principles which should govern the use of our land. Within the space of three years the Barlow, Scott and Uthwatt reports had

[7] *Census of England and Wales, 1931,* Table 9B ; Registrar-General's Mid-1939 Estimates of Resident Population in *Statistical Review.*

[8] Sir George Stapledon, *The Land, Now and Tomorrow,* Faber and Faber, 1935, was a notable landmark in this respect, and it was less a coincidence than a sign of the times that the findings of the Land Utilisation Survey of Britain began to be published a few months after the first edition of Stapledon's book.

reviewed the broad range of problems and it was significantly the Coalition Government which described the agreed objectives of Planning in the 1944 White Paper on Control of Land Use.[9]

> New houses . . . the new layout of areas devastated by enemy action or blighted by reason of age or bad living conditions ; the new schools which will be required . . . the balanced distribution of industry . . . the requirements of sound nutrition and of a healthy and well-balanced agriculture ; the preservation of land for national parks and forests, and the assurance to the people of enjoyment of the sea and countryside in times of leisure ; a new and safer highway system . . . the proper provision of airfields—all these . . . involve the use of land, and it is essential that their various claims on land should be so harmonized as to ensure for the people of this country the greatest measure of individual well-being and national prosperity.

In 1943 the Ministry of Town and Country Planning was " charged with the duty of securing consistency and continuity in the framing and execution of a national policy with respect to the use and development of land throughout England and Wales." [10] The Department of Health was given comparable responsibility in Scotland.

Under the Town and Country Planning Act, 1947, reasonable development plans are required to be made for the whole country and it is claimed that means have been devised by transferring the development value in land to the community, for overcoming the major financial obstacle of compensation which has hitherto bedevilled almost all attempts to plan through fear of sectional interests holding the larger public interest to ransom. If the means are sound it will be possible for national interests, and not the " power of the purse," to decide major issues.

The new prospect has been expressed as follows :

> When the urban estates in and around our towns and cities were being developed the rent or price at which land was leased or sold was determined by the demand for the land in the area concerned. Furthermore, the type of development, for example whether residential or commercial, was also determined by the demand. But under the modern conception of positive planning, as demonstrated in recent legislation, the value of the land can be largely determined by the actions of the authorities as the plans and the developments evolve. The demand for the land can be regulated and steered to fit the plan. New Town Corporations or local authorities have the power to make or kill land values.[11]

(This, of course, is on the assumption that the Minister approves the plans.)

[9] Cmd. 6537, p. 3.

[10] *Minister of Town and Country Planning Act*, 1943, 6 and 7 Geo. 6, Chap. V, p. 1.

[11] Henry W. Wells, F.R.I.C.S., F.A.I., " The Management of Publicly Owned Land," *The Estates Gazette*, Vol. CL, No. 4291, Oct. 4, 1947.

The task of land-use planning is too all-embracing to be the concern of one department and machinery has been devised for co-ordination and harmonization with other departments, among which the Ministry of Agriculture has established a special organization concerned with Rural Land Utilization. Similarly, it is not the task of the professional planner, single handed, to survey his area and prepare his plan. He must inevitably seek help and advice from professional and technical colleagues and not least from practising geographers who can interpret for him, and advise him on, the contributions to be made by specialist experts of the land sciences and related professions including the geologist, geomorphologist, pedologist, agriculturalist, horticulturalist and botanist.

It is clear, of course, that in most of its aspects planning is not comparable with those exact sciences in which it is possible to observe the unfailing operation of rigid general laws but it is possible to discern certain major principles which, although by no means codified, are either accepted or whose acceptance is being strongly urged by disinterested students of the subject.[12] As, however, Britain is neither a virgin land nor a totalitarian state in which it might be possible to plan development without the hindrance imposed by the legacies of past planning—or the lack of it—the application of general principles must have regard to particular conditions which prevail in different areas. The British genius for compromise must inevitably extend to physical planning.

A first major principle, however, is the avoidance of the destruction of good agricultural land. It is a government decision that good agricultural land is not to be taken when less valuable land is available. The principle is a good guide and has been repeatedly stated, e.g. the Minister of Town and Country Planning, in steering his recent Bill through the House of Commons, declared that " one of the main purposes of planning is to ensure that agricultural land is preserved as far as possible."

The loss of some of the best land will not stop for, rare as that land is, its physical qualities, if it is suitably located, may render it subject to irresistible demands for other purposes than food production. In very recent years a large part of the only remaining extensive tract of very intensively cultivated market-garden land in Middlesex, an area of level, well-drained land at Heathrow, has been used for London's new airport and few would dispute the need for this modern facility, within easy reach of the Commonwealth's capital, even though it has

[12] An enunciation of " Principles in Planning " in relation to New Towns is made in the *Final Report of the New Towns Committee*, H.M.S.O., 1946, Cmd. 6876.

involved the loss of land particularly well adapted and situated for supplying fresh vegetables to the metropolitan market. The principle is nevertheless being applied, and there have been recent cases where planning consent to development in the countryside has been refused on purely agricultural grounds.

The question of what constitutes good agricultural land is a very large and complex one and in view of the importance of the subject it is astonishing how little attention had been paid to it until the last decade. In that period, however, it has received increasingly close attention.

The enquiries of the Barlow Commission [13] stimulated the Land Utilisation Survey to conduct a special investigation into the question of the quality of land.[14] The resulting map, destroyed in one of the worst enemy raids, was redrawn and published in 1944 by the Ordnance Survey on the scale of 1/625,000 as one of the series of national planning maps.[15] This affords a valuable small-scale picture of the broad position and suggests the need for detailed examination in every case of importance. Since its preparation much valuable work has been done. One of the earliest of a growing series of replanning reports, *A Plan For Plymouth*,[16] included a land-classification map showing the national scheme in greater detail, with accompanying reports. Under the auspices of the West Midland Group on Post-war Reconstruction, soil scientists, geographers and planners acting as a team devised a scheme of land classification based on factors of soil and site in an endeavour to provide a simple index to quality and have issued maps based on resultant surveys.[17] This work, however, suffers from the restriction to three major categories and an unwillingness to map in detail, while the scheme only aims at revealing the quality of the land and not its character. The ideal solution might be the completion of the detailed work of the Soil Survey of England and Wales, mapping series, textural types and phases on the scale of six inches to one mile. This should furnish a valuable, if complex, classification of soil char-

[13] *Report of the Royal Commission on the Geographical Distribution of the Industrial Population*, H.M.S.O., 1939.

[14] *Vide.*, *Nature*, Vol. CXLIII (No. 3620), Mar. 18, 1939, pp. 456–9 and L. D. Stamp, *Fertility, Productivity and Classification of Land in England and Wales*, The Land Utilisation Survey, 1941.

[15] Great Britain : 1/625,000 map of Land Classification (2 sheets), Ordnance Survey, 1944.

[16] J. Paton Watson, Patrick Abercrombie and others, *A Plan for Plymouth : The Report Prepared for the City Council*, 1943.

[17] The West Midland Group on Post-war Reconstruction and Planning ; *Land Classification in the West Midlands*, Faber and Faber, 1947. Reconstruction Research Group of the University of Bristol : *Gloucester and Wiltshire Land Classification*, Arrowsmith, Bristol, 1947.

acteristics, but would still require skilled interpretation. Soils do not easily lend themselves to reasonably simple classification because they are products of environment as well as of parent material and do not possess easily recognizable individuality. Unfortunately most soil-mapping in this country lags behind physical planning and current indications suggest that the work of a team of about twenty soil surveyors might not provide a complete map of the whole country in much less than half a century. It would, however, be possible to survey in much less time the much smaller part of the country in which the possibility of a change of land-use brings the question of soil qualities to the fore. There would still remain the very large problem of adjusting the type of rural land-use to the quality of the land. Any study of land to determine its relative quality requires also a study of the optimum use of each type of land.

A second principle is that the physical development of towns should be reasonably controlled. It has been mentioned above that the former concept of continuous growth is now outmoded. Henceforth planners will be required to prepare outline plans to meet the needs of their areas for no more than the foreseeable future. Given certain standards of density, it is clear that the clue to the amount of land required by a town depends on its future population and it may well be found that public opinion will support the concept of measuring the needs of each area by reference to an assumed future population, the approximate size of which will involve very elaborate estimation. Such a course would avoid the difficulty in which local planning authorities previously found themselves, of being tempted or forced to throw a shadow of uncertainty over more extensive areas of land than were likely to be required for " development." Now they will be required

> to set out on the plan (to " designate " as it was termed) such land as they would be requiring in the course of the following ten years. In the case of agricultural land they were required to decide as to the land they would require in the following seven years . . . if land is not designated at least there is some guarantee that it will not be required in the course of the next ten years.[18]

In choosing their areas for development they will be helped by the concept of the " Urban Fence " as devised by the Ministry of Agriculture's Planning Branch. The " Fence " is a line around a town, drawn to include areas of undeveloped land which are so situated as to be of little use for farming, and inside which the Ministry of Agriculture declares it has no further interest except in allotment gardens.

[18] *The Estates Gazette*, Vol. CL, No. 4295, Nov. 1, 1947. Report of an address by the Minister of Town and Country Planning to the Valuers' Association.

A related basic principle of planning is that there shall not merely be as little loss of agricultural land as possible but that there shall be due regard paid to the need to avoid severing farm units. It is only very recently that there has been any appreciation of the fact that "many farms are carefully balanced economic units, the disturbance of which would mean a sacrifice in both efficiency and productivity, resulting from the difficulty to the farmer of applying even elementary principles of both crop and stock rotations." The problem had been mentioned in the Scott Report [19] in 1942 and in the following year was expounded in detail by L. Dudley Stamp in an appendix to Watson and Abercrombie's *A Plan for Plymouth*,[20] from which the above quotation is taken. The authors of this report were able to make the claim that "Plymouth has the honour of being the first town and region in this country which has given agriculture its proper place in its scheme of planning. It is here declared to be one of the prime determinants of all forms of development. . . ." [21]

Thirdly, the sizes of settlements must take account of various factors, including the need to avoid the coalescence of urban nuclei and to preserve peripheral belts of open agricultural country. This should contribute to the maintenance of valuable contacts between town and country as well as help to avoid excessive daily travel to and from work. In deciding the size of settlements it is desirable on the one hand to avoid such excessive size as will tend to lessen a sense of civic consciousness and unity and on the other to ensure a well-balanced provision of occupations and services.

Fourthly, there are certain needs of the countryside which should be met by ensuring that country towns constitute adequate centres of commercial, social, cultural and educational life. Few would deny that too much of our physical planning has hitherto been concerned chiefly with towns and very little with the countryside and its villages. There have been all too few planning surveys of the complex regional inter-relationships of town and country and British geographers are partially to blame for this by their neglect of major regional studies of their homeland. There is, however, a growing awareness of the intimate relationships between most towns, especially the older ones, and their neighbouring rural settlements. Towns normally serve not merely those who live within their own boundaries but provide for many of the needs of those who live in the surrounding small settle-

[19] *Report of the Committee on Land Utilization in Rural Areas*, H.M.S.O., 1942, Cmd. 6378.

[20] J. Paton Watson and Patrick Abercrombie, *A Plan for Plymouth: The Report prepared for the City Council*, Underhill Ltd., 1943.

[21] Ibid., pp. 44–5.

ments, in what some have called the " rural neighbourhood " of the town, but which Smailes [22] has suggested should be called the Urban Field.

The function of a country town as a focusing point for the life of its surrounding area is a very important field of study for the planner who, if he fails to realize this function, will not plan his town aright and will misunderstand the life of the country. The two are integrated. The town serves as a collecting and marketing point and a distributing centre, it makes provision for the fuller social and cultural life of the surrounding villages, which look to the town as a centre for those health, recreational and educational services which cannot be sustained in the smaller centres. Adequate provision, especially transport, must be made to ensure that the life of the villages and smaller settlements is adequately linked to that of the towns in order that the countryman may obtain the fullest benefit of the functions which the town can and should fulfil as the hub of rural life. Sir Patrick Abercrombie has indeed crystallized the matter by insisting that towns should never be regarded as " more than the nodal points in a general pattern of national existence." [23] Planning for the full realization of this integration of town and country life is one way in which a positive contribution can be made to the welfare of the countryside. If the countryside, so long in relative decay, is to prosper, planners must understand its needs and cater for them. There is a great and urgent need to provide, in the countryside, a standard of housing and related services which, when other factors are taken into account, will not compare unfavourably with those of the towns. Cottages, with light, piped water and water-borne sanitation must be built where they are really required and within reasonable reach of the local centres of shopping and civic life. One of the major dilemmas of country planning is that there are daily distances to be covered by those who live there : either the farm worker has to travel to work or his children have to journey to school. The division of this journey is the problem which has to be solved and there is no single universal solution. But sympathetic planning of the provision of the various essential services which the countryside requires can make a great contribution to its life and thus to the full use of our land resources.

Another important principle is that care must be taken to avoid sterilizing valuable mineral resources. This again is a vast subject and much has been written on it, but it is obvious that plans for the

[22] A. E. Smailes, " The Analysis and Delimitation of Urban Fields," *Geography*, Vol. XXXII, Part 4, Dec. 1947.

[23] *Homes, Towns and Countryside*, edited by G. & E. McAllister, Batsford, 1945.

reasonable exploitation of mineral wealth must be carefully integrated with those for other forms of development, especially where minerals are of such a nature that their full exploitation on a particular site is essential. Vast deposits of two of our chief minerals, coal and gravel, have been "sterilized" by development and this tends, especially in the case of the latter, to heighten the competition for undeveloped mineral-bearing land. It is generally expected that the nationalization of the country's coal resources and the centralized planning of their exploitation by the National Coal Board will facilitate the co-ordination of plans for physical development on the surface with those for effective winning of coal below ground and the disposal above ground of the vast quantities of waste material brought to the surface as an essential part of the mining process.

It is only natural that a small country in which mineral working is so important should pay serious attention to the problems of restoring land from which mineral wealth has been won. Before the present century it was common practice to restore to agriculture land from which clay, gravel and iron ore had been worked. Numerous examples of former full-scale restoration over shallow workings of brickearth, gravel and ironstone can be seen today, but the greatly increased rate of surface mineral working in the last half century, the replacement of hand methods by complex mechanical excavators, and the general disregard of old ideas of land conservation, heightened by the agricultural depression, wrought great changes and great havoc. During the last decade, however, there has been a change of outlook. There has been increasing public concern over the accelerated loss of agricultural land and the unsightly devastation caused in areas well beyond the old industrial districts so long accustomed to land spoliation.

The Kennet Committee [24] enquiry and report first focused public attention on the extremely serious problem of the destruction of agricultural land in the Midlands as a result of excavation of iron ore, although it is interesting to record that in 1918, when there was a serious threat of food shortages, the Food Production Department of that day instituted an enquiry and initiated experiments on restoration in this area.[25]

The Kennet Committee, while revealing that over 100,000 acres were liable to be affected in future by the appalling devastation consequent upon modern methods of excavating a shallow stratum of ore under a considerable overburden, concluded that, because of excessive cost owing largely to no suitable machinery being available, restoration

[24] *Report of the Committee on the Restoration of Land affected by Iron Ore Working*, H.M.S.O., 1939.

[25] Sir J. Russell, op. cit.

to agriculture was not feasible and that afforestation was in general the only practicable remedy. This conclusion was soon challenged. The war brought amazing technical advances in earth-moving machinery and, as has been noted, a change in public attitude to problems of land conservation, summarized by the Scott Committee as " In principle it is wrong that any body or person should be allowed to work land for the extraction of minerals and leave it in a derelict condition."[26] It was not only the public conscience which was disturbed. Some members of the industry were becoming much alarmed at the devastation they were causing and the restriction and increased cost of town development resulting from peripheral gravel-workings where " the Prospectors' large cheque-book " always outbid the efforts of planners to acquire land.[27]

The ironstone problem was the subject of a further report by Mr. A. H. S. Waters,[28] after a special technical investigation. He reviewed the problem and found the cost of fully restoring land already left derelict would be prohibitive, but that " developments of excavating machinery since 1945, particularly of the walking drag line, are such that a great deal more restoration is possible, concurrently with the excavation of the ironstone, than is now effected."[29] He suggested that if legal obligations to restore were imposed, progress in restoration would be considerable. The cost of separating and replacing " top soil " tends to be even greater than of levelling the land, but costs, except of draining and fencing, vary, so that " the total cost of restoration to agriculture may vary from as little as perhaps £40 an acre to over £500 an acre."[30]

It is pertinent to recall that the Kennet Committee found that in 1938 royalties paid in respect of ore working averaged £264 per acre in Northamptonshire, £377 in Oxfordshire and up to £717 elsewhere, and a further cash payment was often made, of £20 to £80 per acre, in lieu of restoration.

Restoration problems at least as serious, though different in some respects, confront the gravel-working areas. By the very nature of their geographical situation, many of these, as in the Thames and Trent valleys, are on land very highly esteemed for agriculture and in great demand for other purposes. The deposit has normally only a very shallow overburden of soil which is stripped off and the whole of the

[26] Op cit., pp. 62 and 94.

[27] C. Maw, " Quarry Dereliction and Restoration," *Cement, Lime and Gravel*, Vol. XVII, Nos. 5 and 6, Nov. and Dec. 1942.

[28] *Report on the Restoration Problem in the Ironstone Industry in the Midlands*, Summary of Findings and Recommendations, H.M.S.O., 1946, Cmd. 6909.

[29] Ibid., p. 8.

[30] Ibid., p. 9.

gravel removed, so that when, as is frequently the case, the water table is high, large and unsightly lakes result and to restore the sites it is necessary to convey to them an amount of material equal to that which has been extracted. The economics of transport, as well as the relative shortage of filling material, tend to confine restoration to the pits in relatively close proximity to large towns. In this age of concrete, the rapid rise in the demand for gravel, whose output increased from 2½ million tons in 1921 to 35 million in 1939,[31] has produced an acute problem which should be capable of mitigation by careful integration of plans for the future exploitation and restoration of sites with those for other forms of physical development. The findings of a Committee, set up in 1946 by the Minister of Town and Country Planning under the chairmanship of Mr. A. H. S. Waters to report on various aspects of sand- and gravel-working, including problems of control and restoration are therefore awaited with interest.

An older type of universal spoliation is the unsightly heaps of waste material, spoilbanks and other derelict land in older mining areas. A recent survey [32] of the Black Country has shown that more than 12 per cent of the area consists of derelict land, rising in the case of several local authorities to more than a quarter of their area. Here and in similar areas much can be done by modern machinery to render such sites suitable for various urban uses which otherwise tend to encroach on adjacent agricultural land, as well as to improve the appearance of the landscape and to obtain some return from the land.

There is a vast field of investigation into the problems of restoration and re-creation of soil on the sites of mineral workings and tip heaps. Within reason, the community has a right to insist on the principle of " having its cake and eating it " : working the minerals and restoring the land for food production or other uses. It is significant that in all cases of open-cast coal-working, which is conducted by the Ministry of Works, the land has to be restored to agriculture after the coal has been extracted, the upper soil layers being separated, replaced and levelled. Some will see in this State action a hopeful sign of that national concern for soil conservation which has hitherto not found practical expression.

Restoration of open-cast coal-workings to full agricultural practice represents a more hopeful prospect than in many other cases because restoration, with the surface soil replaced in position, proceeds con-

[31] *First Annual Report of the Secretary of Mines*, 1922. Ministry of Fuel and Power, " Statistical Digest," 1945.

[32] Ministry of Town and Country Planning, *Report on Derelict Land in the Black Country*, by S. H. Beaver, 1946.

currently with working and is supervised by capable officers of the County Agricultural Executive Committees. The main difficulty is in restoring drainage, for full land drainage is not possible until after consolidation, which may take five years. The land is normally sown to grasses and clover, which are specially beneficial in building up good soil texture,[33] but it is too early to be certain that the land will finally be as good as before its disturbance. So far only moderate success has attended the growth of cereals on such land. In the case of other mineral workings the surface soil is normally lost in working and restoration involves the problem of creating a new soil by natural regeneration, especially by putting humus into it by the liberal use of leguminous plants. The problem is not hopeless, but, as with so many matters in which nature is concerned, man must be patient and not think too much in terms of immediate results. For housing, factories, playing-fields and uses other than the exacting one of food production, restoration of much derelict land is relatively easy.

Questions of land and soil should never be far from the mind of the planner, who must constantly have regard for such matters of relief, structure, soil and drainage. He must adapt his plans to local circumstances, he must avoid, if he can, developing water-logged land, or undrained land, and land liable to flood. He does not yet know as much as he would like to about the complex relations of soil and health, even though he is beginning to appreciate the significance of the related aspects of micro-climate to healthy living. He realizes that there are many advantages in developing sloping land for residential purposes (especially if it slopes the right way !) and of concentrating on the flat land the factories and playing-fields which require level areas. He realizes, however, that in so far as there is a call for good land for private gardening, especially from allotment holders, it is desirable to reserve, if possible, suitable soil for this purpose. Those who are prepared to undertake extensive spade work will be discouraged by, say, intractable clay or sterile sand. On the other hand, planners are beginning to doubt the fantastic assertion that on former agricultural land developed for residential purposes "the food production . . . is greater than it was when the sites were purely agricultural."[34] It is possible that this idea was spread by those who wished to believe that no harm was done by taking agricultural land, but the fallacy in it is easily exposed.[35]

[33] Sir John Russell, op. cit.
[34] F. J. Osborn, *New Towns after the War*, Dent, 1942, p. 49.
[35] L. Dudley Stamp, "The Production of Gardens," *Town and Country Planning*, Vol. XV, No. 59, Autumn, 1947.

Planning, indeed, must aim at maximizing the use made of our land, even though it is essential that non-agricultural uses shall extend. Forestry, National Parks, Military Training areas and school playing-fields all tend to make enormous demands on land. Fortunately there is a principle of multiple use which recognizes that much of the country should and must serve more than one use at the same time. The countryside is a national asset whose recreational qualities are only maintained if it continues to be restrained, by farming, from its natural impulse to achieve a climax vegetation. The National Parks Committee fully realized that the parks they proposed "must not be sterilized as museum pieces. Farming and essential rural industries must flourish, unhampered by unnecessary controls or restrictions." [36]

With careful landscape planning, commercial forestry areas can have a recreational value and military training areas by no means always precludes farming or even recreation. The proposals for post-war military training areas [37] total 702,000 acres, to 24 per cent of which the public will have unfettered access and substantial rights over a further equal amount. Over the greater part of the training land agriculture or rough grazing will continue, and only 8 per cent of the whole will be actually lost to agriculture.

Planning is now entering upon a new phase. For the first time there is to be full land-use planning and not merely partial planning. Agricultural uses will take their proper place with residential, industrial and other purposes. Land-use planning will, in fact, be at last comprehensive. It will, of course, be flexible, a continuous process, subject to modifications and not a rigid adherence to a fixed plan. Great opportunities confront those whose task it will be to prepare the plans for the future. Equally, they are faced with great responsibilities and not least that of realizing that land is an asset which is of limited extent, a heritage which cannot be created by man and in whose use it is easier to make mistakes than to rectify them. Wise planners will always know that it is a good principle to realize that posterity will be grateful to them if they can pass on that heritage improved for the use of future generations.

[36] *Report of the National Parks Committee* (*England and Wales*), H.M.S.O., 1947, p. 8, Cmd. 7121.

[37] *Needs of the Armed Forces for Land for Training and other Purposes*, H.M.S.O., 1947, Cmd. 7278.

XVII

AFRICA IN ANCIENT TIMES

By H. J. WOOD, B.Sc. (Econ.), Ph.D.

This essay treats of knowledge of Africa in ancient times, the methods, motives and achievements of explorers in that long period of human history, and their legacy in the field of geographical literature and cartography.[1]

The African continent is bounded by the Mediterranean Sea on the north, by the Atlantic Ocean on the west, by the Suez isthmus, Red Sea and Indian Ocean on the east. Let us look at implied relations with non-African areas.

In the north the Mediterranean coastal lands of the eastern basin and of the western basin, and of the entire sea in fact, have a large measure of geographical and historical unity, whether they lie in Europe, Asia, or Africa ; the Straits of Gibraltar are narrower than the Straits of Dover and nowhere is the Mediterranean too wide for easy transit between lands on its borders. The eastern basin was the scene of the very beginnings of seafaring in the Occident, with its clear weather and steady winds in summer, abundant natural harbours, and numerous landmarks in islands and promontories. The entire Mediterranean Sea was only in slightly less degree favourable to early maritime exploration. The story of the exploration of the coasts of north Africa is then part of a wider story, going so far back as to be based on fragmentary and controversial evidence, drawing on place names and archæological evidence as well as on the written record. There was contact between Crete and Egypt in the fourth millennium B.C. A phase of maritime activity in which the Minoans, based on Crete, took the lead had passed its zenith by 1000 B.C. Later came the much disputed and enigmatic achievements of the Phœnician traders, based on Tyre and Sidon and later on Carthage, which was founded about 800 B.C. ; they discovered the Straits of Gibraltar in the early twelfth century and probably followed the western Mediterranean coasts of Africa on their way to Spain. Greek colonial expansion in the eighth, seventh and sixth centuries B.C. preceded, and was followed by, exploration of Mediterranean coasts. Of particular

[1] The author is particularly indebted to the work of M. Cary and E. H. Warmington, *The Ancient Explorers*, London, 1929.

interest, because the Greeks gained some knowledge of its hinterland from subsequent contacts, is the foundation of Cyrene *c.* 640 B.C. At about this time also the Greek Colæus discovered the Straits of Gibraltar, the Pillars of Heracles; Greek stations were founded on the north African coast on the route to Tartessus in southern Spain. Exploration had been completed before the rise of Roman power but with its advent the political and geographical unity of the entire Mediterranean reached full expression after about 200 B.C. and the north African coastal fringe became more and more part of a wider sphere.

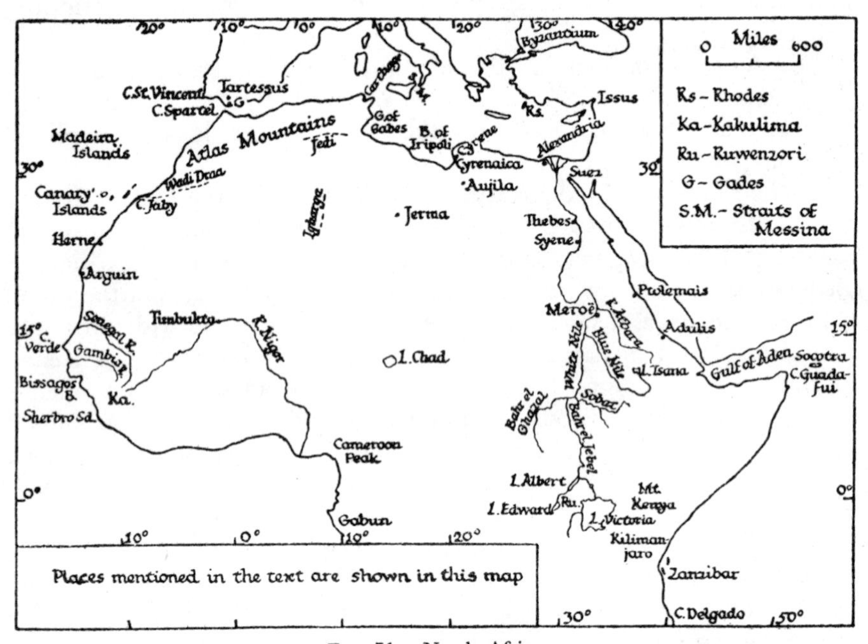

FIG. 71.—North Africa.

The isthmus of Suez presents no great barrier and has at times provided a causeway between Asia and Africa. The Red Sea lands of Arabia and Africa have geographically much in common and flank a trade route of regional but also of wider significance, linking the Indian Ocean and the Mediterranean. The modern traveller is aware of the oppressive heat of the Red Sea and it does not provide easy navigation conditions, yet in very early times it played a part as an outlet for Egypt and a meeting-place of men of African and Asiatic cultures. Early exploration is but dimly seen. Ancient Egypt, based on the Nile oasis, was linked by desert tracks to Red Sea harbours and at times by canal from the Nile to the Gulf of Suez. Craft plied on the Nile in antiquity and it is possible that Egyptian prototypes commenced the evolution of sea-going ships. In the Mediterranean the Egyptians were content to let the Minoans, Phœnicians and Greeks play the part of pioneers. They did, however, frequent the Red Sea as early as *c.* 3000 B.C.; their knowledge of its shores, largely based on trade, expanded and contracted with the passage of some thirty centuries and changing fortunes. In early times it sometimes reached to the frankincense lands of Punt, i.e. the Gulf of Aden lands including Somali coasts. There is record, for example, of a great expedition to these lands sent by Queen Hatshepsut in the period 1501–1479 B.C. Knowledge extended to Socotra. Under the Greek Ptolemies in Egypt the Red Sea coast of Africa was re-explored, in quest of ivory, of the cinnamon trade of the Somali coast, and for elephant hunting, and by 221 B.C. the coasts were known probably eastward to Guardafui.

A long stretch of the eastern coast of Africa must be considered too as part of a larger entity, that of the lands of the north-west Indian Ocean, bordering the Erythræan Sea of the Greeks. Here, regularity of favourable seasonal or monsoon winds provided encouragement to early seafaring. Just how early contact was made with the African coast, perhaps by Phœnicians, certainly by men of Arabia, is obscure, but no doubt it began by coasting. "Age-old coasting commerce" certainly existed between north-west India and the eastern coasts of Africa.[2] Ophir was the destination of Phœnicians of Tarshish sailing for King Solomon in the tenth century B.C. The view that Ophir is represented today by the great ruins of Zimbabwe in Southern Rhodesia conflicts with the generally held opinion that these may be ascribed to Bantu builders perhaps of the eighth or ninth centuries. Ophir has been conjectured to be in India, Ceylon, southern Arabia, as well as Africa, but the suggestion that Sofala may be its modern equivalent carries some weight, the port serving as an outlet for gold mined in inland districts, perhaps Havilah of the biblical texts, if this view is accepted. Warmington comments:

> The renewal of the voyage every three years and the varied nature of the wares which these Phœnicians brought back—silver, precious stones, Indian sandalwood and peacocks, apes, ivory and, above all, gold—suggest a trading voyage carried out by Sabæan Arabs of South Arabia like those which even today the Arab dhows make from the Red Sea by way of Muscat to Malabar and thence back to Africa and down the east coast to Madagascar and sometimes Sofala.

[2] See Cary and Warmington, op. cit., p. 60.

He concludes that "the question of early visits by civilized men to Southern Africa must remain unsolved." [3] As to the regions farther north, by the latter part of the first century A.D. it is known, from evidence provided by an anonymous Greek trader's manual (*Periplus Maris Erythræi*), describing coasts and exports and imports, that the African coast had been followed south from Guardafui (Cape of Spices) to the Zanzibar channel, to Menuthias Island, probably Zanzibar. In the second century A.D. a Greek, Dioscorus, reached Cape Delgado, the southern limit of modern Tanganyika.

Within the African continent there are great well-defined geographical sub-divisions that lend themselves to use as bases for discussion of the main features of the history of exploration. The north coastal margins, as we have seen, come within the Mediterranean sphere. West of the Gulf of Gabes the northern margins of the Atlas mountain region, by reason of relief, rainfall and position have always permitted a fair density of population based on crops and herds and trade and on the west these favourable factors extend south behind the Atlantic coast of Morocco to about 30° N. East of the Gulf of Gabes desert reaches the sea, with some amelioration in the region of the steppe of Cyrenaica. To the south of the Mediterranean margins extend a great zone of hot desert from the Atlantic coast, between about latitudes 30° and 15° N., eastward to the Red Sea and beyond, while it extends south on the African coast of the Red Sea, east along the southern shore of the Gulf of Aden, south again on the coast to the equator. This great zone has an average annual rainfall almost everywhere less than 10 in. and is a zone of sandy wastes, rock-strewn plateaux, providing little basis for human settlements save in occasional oases and scant pasture on its fringe. It is of course impossible to define the desert limits in rigid terms ; transition to areas of some winter rain in the north, with some summer rain in the south, in the Sudan, is gradual. In the latter zone steppe becomes savannah or parkland, the bases of human life more ample. A complication enters the task of reconstructing geographical environment of ancient times, indeed to a lesser extent that of any later period of history, i.e. the possibility of climatic change and consequent effect on vegetation and the bases of human settlement. In the main our sketch is no doubt valid, but there is a considerable consensus of opinion, despite the controversial nature of the problem, that conditions were rather less severe in later ancient times, at least on the desert margins. That the desert provided a formidable barrier there is no doubt, the writings of the period,

[3] See E. H. Warmington, *Africa in Ancient and Medieval Times* in *Cambridge History of the British Empire*, Vol. VIII, 1936, Chap. 3, pp. 54–5.

lacking in precise detail as they are, pointing to this. For example Herodotus (*c.* 440 B.C.) is categorical—" to the south and interior of Libya, the country is desert, without water, without animals, without rain and without wood ; and there is no kind of moisture in it." [4]

How far did travellers penetrate into the Sahara, and beyond, to the south ? The Carthaginians were believed to cross the Sahara in their search for trade, but on the whole the lure in ancient times was not strong enough to counter the obvious difficulties and such trade as may have existed in peaceful times was no doubt mainly carried on by intermediaries, such as the Garamantes, tribesmen based probably on Jerma, to the south of the Bay of Tripoli. It was through an indirect story that the Greeks of Cyrene heard of the first recorded crossing of the Sahara, an exploit of men from the Berber tribe of the Nasamones, a crossing probably made via the oasis of Aujila.

> When they had traversed much sandy ground, during a journey of many days, they at length saw some trees growing in a plain . . . they approached and began to gather the fruit that grew on the trees ; and while they were gathering, some diminutive men, less than men of middle stature, came up, and having seized them carried them away . . . However, they conducted them through vast morasses, and when they had passed these, they came to a city, in which all the inhabitants were the same size as their conductors, and black in colour ; and by the city flowed a great river running from the west to the east, and . . . crocodiles were seen in it.[5]

Thus the Nasamones would seem to have reached savannah and the east-flowing Niger, and possibly, as some conjecture, Timbuktu. The power of Rome on the Mediterranean fringe was sometimes challenged by men of the steppe and the desert but no great explorations were achieved by Roman armies in Africa, as they were in Europe. There is no record of Roman raids on the Gætulians to the south of Numidia (modern Algeria) but it is known that there was Roman reaction to forays by the Garamantes, although historical references are sparse. At an unknown date at least one Roman, Julius Maternus, reached the Sudan south of the Sahara—four months' journey from Jerma, to the district of Agisymba, but where this might be is quite conjectural. Possibly the route was via Lake Chad.

The most important of the desert oases is of course the valley of the Nile between the first cataract at Syene and the sea. Here civilization goes far back into antiquity, to rival that of the riverine civilizations of the Tigris–Euphrates and Indus valleys, perhaps to 5000 B.C. Navigation reached to Syene, and the annual inundations made agriculture possible in a desert region extending east and west

[4] Herodotus, iv, 185, Bohn edition, London, 1854.

[5] Ibid., ii, 32.

and south. Many were curious as to why the river flooded, as to the regions whence it came before entering Egypt. Not only does the Sudan region of savannah and steppe lie far to the south but the cataracts break the continuity of navigation and the great bend of the river makes the route devious. Long stretches of the Nile are unnavigable, but to the south the White Nile and a long stretch of the Blue Nile can be used by boats. Beyond the confluence of the White Nile and the Bahr-el-Ghazal the *sudd,* floating vegetation, tangled masses of aquatic plants in a sluggish stream, becomes a serious obstacle on the Bahr-el-Jebel and here was a limit never passed by ancient explorers. The Atbara and the Blue Nile show the way to the mountains of Abyssinia, a region of summer rains that provides the largest contribution to Nile flow downstream, but the approaches are formidable.

In ancient times the ramifications of Egyptian contacts extended for long periods into Nubia to the south, and into more remote and vague Ethiopia, extending to the Red Sea and the Gulf of Aden ; war and trade provided motives ; there were prizes of gold, incense and ivory. At greatest extent there was certainly penetration as far as the sources of the Blue Nile, from about 700 B.C. Egyptians may have reached the sources of the Atbara, and the Bahr-el-Ghazal confluence with the White Nile. Later there was contraction of knowledge, until the Greek dynasty of the Ptolemies ruled in Egypt and Greeks entered the field. From the Red Sea coast (modern Eritrea) a route was opened from Ptolemais to Meroë, and from Adulis exploration reached into the northern Abyssinian highlands, the region of the upper Atbara and of Lake Tsana. In the Roman period Nero, planning an expedition against the Ethiopians, sent pioneers to find the sources of the Nile ; they reached the *sudd* and could sail no farther. Pliny refers to the episode, but Seneca is the chief authority. He learned from the explorers that they

> came to immense marshes, the outcome of which neither the inhabitants knew nor can anyone hope to know, in such a way are the plants entangled with the waters, not to be struggled through on foot or in a boat, because the marsh, muddy and blocked up, does not admit any unless it is small and holding one person.[6]

A little later exploration extended up the Blue Nile to the mountains and Lake Tsana. Finally we must notice a surprising episode, early in the Christian era, an approach from a different direction. A Greek, Diogenes, returning from India, missed his course near

[6] Quoted from Cary and Warmington, op. cit., p. 175.

Guardafui, and was blown south as far as Rhapta (Dar-es-Salaam ?). Either from direct investigation, or hearsay, he reported, according to a story derived from Marinus of Tyre through Ptolemy, two large lakes to the north of a long east–west trending mountain range, snow covered, the " Mountains of the Moon." Streams fed by melting snow reached these lakes, and from each of these two streams flowed northward, uniting to form the Nile, which received a stream flowing from Lake Tsana. The mountains may be reasonably assumed to be a distorted version of the Ruwenzori range, the lakes to relate perhaps to lakes Victoria and Albert. Ptolemy's Pylæ mountains and Mount Maste may correspond to Mounts Kenya and Kilimanjaro respectively.[7]

We will now consider the extent of ancient knowledge of the western coast of Africa, of the Atlantic shore trending south-west from the Straits of Gibraltar, and the alleged circumnavigation of the continent in ancient times. There was no early regular contact with the ocean shores of North Africa although the Phœnicians probably knew the Straits of Gibraltar from about 1200 B.C., the Greeks from about 650 B.C., and Atlantic trade from bases in Spain was an early development. The attraction of tin drew the adventurous to the north, and control of the Straits meant this trade more than any other. There were, however, interesting episodes. At a conjectural date, possibly about mid sixth century B.C., a Greek, Euthymenes, claimed to have seen on the west African coast a crocodile-infested river, the waters of which were being driven back by an onshore wind. He himself believed this to be the upper reach of the Nile ; Cary identifies it with the Senegal. More important and interesting is the colonizing venture of the Carthaginian, Hanno, generally thought to have taken place *c.* 500 B.C., some twenty years after earlier Carthaginian contacts. Hanno's account was placed in a temple in Carthage and a fragmentary Greek version of the text is available for the modern scholar —Hanno's *Periplus*. Several students of ancient geography have brought scholarship and ingenuity to bear on the task of interpretation, but inevitably there can be no certainty. Modern place names have no relation to those in this text. Warmington has made a translation and it is so unusual to find an original narrative in the field of ancient exploration that we will quote from it extensively, giving Cary's identifications in brackets. Hanno " set sail, taking fifty-oared ships, sixty in number, with a multitude of men and women to the number of thirty thousand, supplies of corn, and the usual equipment also."

[7] *C.H.B.E.*, op. cit., pp. 67, 68.

After founding cities at certain places the expedition reached the Lixus (Wadi Draa). The narrative goes on :

We took interpreters . . . and sailed along the desert [Sahara] towards the midday for nine days, and then again towards the rising sun for a one day's course. There we found in the inner recess of a certain gulf a small island which had a circumference of five stades ; this we peopled with settlers, and named Cerne [Herne].[8] We judged from the distance of our coasting voyage that it lay in a straight line with Carthage ; for the voyage from Carthage to the Pillars and the voyage thence to Cerne seemed to be of equal length. From here onwards we sailed through the mouth of a big river named Chretes [one branch of the Senegal], and came to a lake ; and the lake contained three islands bigger than Cerne. We completed a voyage of one day's sail from there, and came to the inner recess of the lake, above which huge mountains reached up crowded with wild men [Guanches] . . . Sailing thence we came to another great and broad river [Senegal], swarming with crocodiles and hippopotamuses. Thence we turned back again and came back to Cerne. From here we sailed towards the midday sun for twelve days, coasting along the land, all of which was inhabited by Ethiopians . . . On the last day we made fast under high and thickly-wooded mountains [Cape Verde] ; the wood of the trees was sweet-smelling and mottled. We were two days in rounding these mountains and found ourselves in an immeasurable indentation of the sea [estuary of the Gambia], in one part of which there was a plain towards the land. From this we could see by night fire flaring up on all sides . . . we sailed thence onwards for five days along the shore until we came to a great gulf which, our interpreters said, was called the Horn of the West [Bissagos Bay]. In this was a big island, and on the island was a marine lake, and in this another island [Orang] ; we disembarked on this, and by day could see nothing but forest, though by night we saw many fires burning and heard the sound of pipes and cymbals, and a rolling of drums and endless shouting. . . . We sailed away in a hurry, and then passed along a country which was all ablaze and full of fragrant smoke ; and huge rushing streams of fire plunged into the sea ; and the land was inaccessible because of the heat. Hurriedly therefore we sailed away from thence also in fear ; and during four days as we were borne along we saw by night the land in a mass of flame ; and in the midst was a fire greater than the others, and mounting to an enormous height ; it seemed to touch the stars. This was the highest mountain which we saw, and it is called the Chariot of the Gods [Kakulima ?]. Having sailed thence along the blazing streams, on the third day we came to a gulf called the Horn of the South [Sherbro Sound]. In the recess was an island [Macauley] like the former one, containing a lake, and in this was another island full of wild people. By far the greater number were women with hairy bodies, which our interpreters called Gorillas [chimpanzees] . . . we sailed no farther than this, because our food gave out.[9]

Thus the narrative. A vital point in interpretation concerns the position of Cerne, placed variously by commentators near Cape Juby,

[8] 10 stades approximately equal 1 mile.

[9] Quoted from E. H. Warmington, *Greek Geography*, London, 1934, pp. 72–5.

at Goree near Cape Verde, and between these extremes at Arguin or, most probably, at Herne. Cary in making this identification finds it necessary to amend two days' sailing along the Sahara coast as recorded in the narrative to nine days' sailing, and this modification it will be noticed has been adopted by Warmington in his translation; it illustrates the difficulties of the investigator.[10] The "huge mountains" beyond the Chretes prove baffling. The somewhat lurid description of fires is generally accepted as referring to grass fires, and has a basis in fact. It may be noted that a very optimistic view places Hanno's farthest point in the Gabun estuary, just north of the equator; Cameroon Peak in eruption has been held to be the "Chariot of the Gods." The ancient text presents a challenge and the puzzle is fascinating, if insoluble.

The voyage of Hanno stands out as an isolated episode. Some trade, with gold as the lure, followed, Carthaginian traders probably reaching Cerne. Their reports of the Atlantic were designed to discourage interlopers. The Greek historian and explorer Polybius, shortly after 146 B.C., is thought by Cary to have reached the Senegal, but that year, in which Carthage fell, seems to have marked the end of traffic with West Africa.[11] It may be noted that the Canary Islands, probably visited earlier by the Carthaginians, were definitely explored by an expedition sent by Juba, a king of Morocco (*c.* 25 B.C.–A.D. 25). The Madeiras and the Canaries came to be identified with the legendary Isles of the Blest and were called the Fortunate Islands.

How much is known of circumnavigation of Africa in ancient times? Nordenskiöld asserts that "it is obvious that Africa had already been circumnavigated several times during the pre-Christian era."[12] The evidence for this statement is of great interest. Herodotus tells all too briefly a story of a Phœnician voyage, made when Necho ruled Egypt, *c.* 600 B.C. He says:

> Libya shows itself to be surrounded by water, except so much of it as borders on Asia. Necho, king of Egypt, was the first whom we know of that proved this; he, when he had ceased digging the canal leading from the Nile to the Arabian Gulf, sent certain Phœnicians in ships, with orders to sail back through the Pillars of Heracles into the Northern Sea, and so to return to Egypt. The Phœnicians accordingly, setting out from the Red Sea, navigated the Southern Sea; when autumn came, they went ashore,

---

[10] Cary and Warmington, op. cit., p. 50. A. E. Nordenskiöld, in his *Periplus*, Stockholm, 1897, translated by F. A. Bather, suggests 12 days (p. 112).

[11] Nordenskiöld, however, argues that navigation on these coasts seems to have been continuous from the time of Hanno until the second century A.D. See *Periplus*, p. 113.

[12] *Periplus*, p. 116.

and sowed the land, by whatever part of Libya they happened to be sailing, and waited for harvest ; then, having reaped the corn, they put to sea again. When two years had thus passed, in the third, having doubled the Pillars of Heracles, they arrived in Egypt, and related what to me does not seem credible, but may to others, that as they sailed round Libya, they had the sun on their right hand. Thus was Libya first known.[13]

Warmington points out that Herodotus heard the story some hundred and fifty years after the event, and that some Greek thinkers from early times had the idea of an Ocean river flowing endlessly round the world, so that a sceptical attitude to such a story would not be likely : further, that the Phœnicians produced not only some of the greatest explorers but also some of the greatest liars in history. However, he gives a possible reconstruction of such a coasting voyage of some 16,000 miles, taking the view that whether it actually took place is " not proven." [14]

Herodotus goes on to relate a story about Sataspes, a Persian, who was to have been impaled by order of Xerxes (485–465 B.C.), but his mother secured a change of sentence, " a greater punishment," namely, to sail round Libya. " Dreading the length of the voyage and the desolation " Sataspes returned, after many months of sailing south along the African coast, and was duly impaled. His excuse " that his ship could not proceed any farther, but was stopped " was not acceptable.[15]

At a later period, from the second century B.C. to the beginning of the Christian era, it was rumoured that ships from the Phœnician city of Gades (Cadiz) were sailing round Africa to its north-eastern coasts and to Arabia, to avoid the exactions of Egyptian intermediaries on the orthodox Red Sea route. Strabo gives an account, derived from Poseidonius, of the exploits of the Greek Eudoxus. Some time between 117 and 108 B.C. this mariner, returning from India, was allegedly carried down the coast of East Africa, and came across a ship's figurehead, which later in Alexandria was identified as part of the wreck of a ship from Gades. Eudoxus conceived the idea of sailing round Africa to India, to avoid handing over cargo to the Ptolemaic government. He sailed with a cargo from Gades, but came to grief on the coast, presumably of West Africa, and after many adventures regained Spain. Eudoxus set forth once more but the fate of this venture is unknown. Modern scholars are of the opinion that the Eudoxus tradition is worthy of belief in its essentials.[16]

We have glanced at some of the main features of exploration of Africa in ancient times. The information available is clearly very

[13] Herodotus, iv, 42, 43.
[14] Cary and Warmington, Chap. V. See also *C.H.B.E.*, op. cit., pp. 55, 56.
[15] Herodotus, iv, 43.
[16] See *C.H.B.E.*, op. cit., pp. 60–2.

incomplete. Fragments only of the story are available, and these often are derived so indirectly and in such garbled form that the historian in many instances would be contemptuous of their use, were better alternatives available. Referring to Greek exploration Warmington writes, " sources for the most part are unsatisfactory ; actual reports of travellers are very rare, and those writers who record matters concerning geography were at times ignorant or careless or both." [17] None the less the known achievements are not unimpressive, particularly if we bear in mind the poverty of technique of exploration by land and sea, for coping with adverse geographical environment in the realm of the unknown. There was no great superiority in weapons over those of newly-encountered peoples, whose first reactions were no doubt often hostile ; certainly tradition often endowed them with fabulous and uninviting characteristics. Compared with our own day progress both by land and by sea was very slow. At sea sailing-ships, seldom of more than 250 tons burden, were closely dependent on weather and currents. Writes Nordenskiöld :

> When criticizing the sea-voyages of olden times, we should remember that, before the invention of the compass, the navigator had at his disposal no means of determining the cardinal points when on the open sea in cloudy weather. The vessels of those days were moreover unfit both for cruising and for braving rough sea. . . . Trading-vessels, which could seldom be manned with several hundred rowers, would have had then, as was still the case in the middle of the nineteenth century, no other motive force than that of the wind. . . . Navigation along unknown coasts was absolutely dangerous.[18]

Of the Greeks Warmington writes : " The defects of their ships, as compared with the medieval and early modern vessels, lay not so much in small size and low speed as in the absence of good steering gear and rigging, which prevented full use of winds. . . ." [19] The ancients lacked not only precise means of measuring position but also time and distance. Distance and direction were but approximately known even in more familiar regions, on land, while distances at sea were computed by dead reckoning, making allowance for wind and current, and were of course often grossly inaccurate. For measurement of latitude a form of sundial, the gnomon derived from Babylonia, could be used under favourable circumstances, but probably was seldom employed. Longitudes were based on dead reckoning. Sun, stars and winds were used to find bearings.

We have seen that the ancient geographers are often the source of such accounts of exploration as survive ; their ideas reflect to some

[17] *Greek Geography*, p. xx. [18] *Periplus*, p. 4. [19] *Greek Geography*, p. xix.

extent the results of exploration as known to them, and are therefore deserving of some attention, particularly as they have influence on geographical exploration and ideas in later times. There is of course no reason to assume that their expositions embody all the results of prior discovery. Herodotus for example shows no knowledge of the voyage of Hanno. Moreover the exposition of such knowledge as each author succeeded in accumulating is highly variable in clarity and worth.

Hecatæus of Miletus, called by Tozer [20] the father of geography, wrote the first treatise on the subject about 520 B.C., although Anaximander (*c.* 580 B.C.) is given the credit of being the first map maker. His *Periodos* or Description of the Earth is preserved only indirectly and in fragments. His conception of the inhabited world, the *Oikoumene*, was the traditional Homeric one of a circular plane, surrounded by an ocean stream, and from this the Nile was thought to flow, making its way to the Mediterranean.

The geographical passages in Herodotus are in parenthesis, his work being primarily a history, written about 440 B.C. Lacking system and often perspective, his work none the less has fascination because, the product of a curious mind, it is largely based on travel and diligent enquiry. Strabo in his contempt for Herodotus overlooks important merits. In a reference to the Nile we see the Greek love of symmetry in the parallel he finds with the Ister (Danube) in length and direction. He writes " the Nile flows from Libya, and intersects it in the middle : and as I conjecture (inferring things unknown from things known), it sets out from a point corresponding with the Ister. For the Ister . . . divides Europe in its course." He adds : " no one is able to speak about the sources of the Nile, because Libya, through which it flows, is uninhabited and desolate." [21] Actually he does mention a story about the sources of the Nile, which he received from an official whom he suspects of trifling with him. This was to the effect that the river issued from bottomless fountains between the mountains of Crophi and Mophi, between Syene and Elephantine—half the water flowing north to Egypt, half south to Ethiopia. Herodotus makes the Nile flow from its sources to Syene in an easterly direction, incorporating the great stream discovered by the Nasamones flowing from west to east. On its banks, four months' journey from the border of Egypt, is placed the farthest known point, the region of the Automoli, the Sennar district of today. Midway between in terms of travelling time

[20] H. F. Tozer, *A History of Ancient Geography*, 2nd edition with notes by M. Cary, Cambridge, 1935, p. 70.
[21] Herodotus, ii, 33, 34.

he places Meroë. These views contrast with those held later by some Greeks, when " an astonishing theory arose that eastern Africa, in the unknown regions where the Nile took its source, was joined to North India so that the Nile and the Indus were one river and the Arabian Sea was a great lake." [22] This view was dispelled when Alexander sailed down the Indus to its sea outlets in 325 B.C.

The Nile often intrigued the geographers of ancient times, its great delta, its arrival as a great river from the south, above all its periodic summer inundations. The Mediterranean zone as a whole has a winter maximum of rainfall and river régime reflects this so that summer floods aroused much curiosity. Herodotus writes:

> . . . some of the Greeks . . . have attempted to account for these inundations in three different ways . . . One of them says that the Etesian winds are the cause of the swelling of the river. . . . But frequently the Etesian winds have not blown, yet the Nile produces the same effects . . . The second opinion . . . says that the Nile, flowing from the ocean, produces this effect: and that the ocean flows all round the earth. The third way of resolving this difficulty is by far the most specious, but most untrue. For by saying that the Nile flows from melted snow, it says nothing . . . how . . . since it runs from a very hot to a colder region, can it flow from snow?

He goes on to argue that hot winds blow from those regions, " that the country, destitute of rain, is always free from ice," that " the inhabitants become black from the excessive heat." He proceeds to give his own views on the cause of the inundations: " During the winter season, the sun, being driven by storms from his former course, retires to the upper parts of Libya; this in a few words comprehends the whole matter . . . for it is natural that the native river streams should be dried up." [23] Thus does Herodotus argue on a geographical topic; other thinkers in ancient times, we shall see, were nearer the truth.

Herodotus accepts a customary division of the known world into three continents, Europe, Asia and Libya. The Nile was usually the limit of Asia to the Greeks, but he objects to thus dividing Egypt between the continents, and includes Egypt in a tract of Asia, with Libya as an appendage. Beyond the Pillars of Heracles he names the promontory of Soloeis (probably Spartel), and refers to the dumb commerce carried on by Carthaginians on the coast to the south, i.e. exchange of goods without the use of spoken language. Africa from the Mediterranean coast southwards, he divides into three zones. A coastal fringe, mountainous, wooded and inhabited by sedentary peoples in the west, is in the east low and sandy and inhabited by nomads. A zone to the south is infested by wild beasts and further

[22] *C.H.B.E.*, op. cit., p. 59.

[23] Herodotus, ii, 20–4.

south a ridge of sand extends from Thebes to the Straits of Gibraltar. On this ridge of sand he places at regular intervals of ten days' travel, inhabited places, where fresh water gushes from hills of salt (of which all houses are made and where date palms grow—the oases of the Ammonians, of Augila, of the Garamantes, the Atarantes (who curse the sun for his scorching heat) and the Atlantes (who dwell near cloud-capped Atlas Mountain). Here we see a distorted picture of the oases of north Africa, made symmetrical in distribution. Herodotus has a catalogue of very miscellaneous remarks about the peoples of Africa ; the semi-fabulous Macrobians, far to the south-east near the ocean coast ; the Automoli, deserters from Egypt to Ethiopia, of the upper Nile ; the troglodyte Ethiopians, who are fleet of foot and screech like bats and are hunted in four-horse chariots by the Garamantes ; the Nasamones, bordering on Cyrenaica ; the Lotophagi (lotus eaters) to the west of the latter ; the groups of oasis dwellers, and many others.

The works of Eratosthenes (*c.* 276–196 B.C.) of Alexandria have perished and are known only indirectly. His great achievements were in the direction of determining the dimensions of the globe, of the inhabited world (about which the ancients argued a great deal), and in determining a main parallel and a main meridian. The notions underlying the fixing of positions were an early development in Greek thought, the sphericity of the earth being accepted by some at least before Aristotle produced a reasoned case. Eratosthenes believed in an Africa surrounded by sea and unlike Herodotus gives a reasonably correct reason for the Nile floods. It is interesting that Polybius (*c.* 204–122 B.C.) throws doubt on the sea-girt character of the African continent. The view is also attributed to him that the neighbourhood of the equator was less hot than bordering zones on either side, and was habitable and inhabited ; this is the converse of a view held by some both earlier and later, e.g. Ephorus in the fourth century B.C. and Strabo, that great heat formed a real barrier to the south.

Strabo is the most outstanding comprehensive geographer of antiquity, despite his shortcomings, and fulfils the important function of providing much information about his forerunners. His work was produced during the Augustan age, being written before A.D. 23. Strabo in mathematical geography is inferior to Eratosthenes ; but follows him in constructing his map in relation to one main parallel and one main meridian, based on what were thought to be known positions : the former passing through the Pillars of Heracles, Straits of Messina, and Rhodes to the Gulf of Issus, the latter through Meroë, Syene, Alexandria, Rhodes and Byzantium. Like other ancient geographers, however, he depicts the Mediterranean coast of Africa as almost a straight line. His rather brief descriptive account of Africa treats mainly of Egypt. His information on the upper Nile is based mainly on Eratosthenes. The great bend of the Nile between Syene and Meroë is known, and the fact of tributaries coming in to the upper river from the east, enclosing the great " island " of Meroë. Strabo mentions also a lake " Psebo " above Meroë which, it has been suggested, relates to Lake Tsana. On the subject of the Nile inundations he writes :

> Now the ancients depended mostly on conjecture, but the men of later time, having become eye-witnesses, perceived that the Nile was filled by summer rains, when upper Aethiopia was flooded, and particularly in the region of its farthermost mountains. . . . This fact was particularly clear to those who navigated the Arabian Gulf as far as the Cinnamon-bearing country, and to those who were sent out to hunt elephants. . . .

He refers later to " the question which even to this day is still being investigated. I mean why in the world rain falls in summer but not in winter, and on the southernmost parts but not in Thebais and the country round Syene." [24]

Strabo describes Libya as having the shape of a right-angled triangle with its base along the Mediterranean coast, its perpendicular side along the Nile as far as Ethiopia, continued on to the ocean and the hypotenuse between Ethiopia and Maurusia (Morocco zone). He writes : " Now as for the part at the very vertex of the above mentioned figure, which begins approximately with the torrid zone, I speak only from conjecture because it is inaccessible, so that I cannot tell even its maximum breadth. . . ." [25] Elsewhere he writes :

> All those who have made coasting voyages on the ocean along the shores of Libya, whether they started from the Red Sea or from the Pillars of Heracles, always turned back after they had advanced a certain distance, because they were hindered by many perplexing circumstances, and consequently they left in the minds of most people the conviction that the intervening space was blocked by an isthmus ; and yet the whole Atlantic Ocean is one unbroken body of water. . . .[26]

For Strabo the inhabited world is bounded on the south by a zone uninhabitable on account of heat. The parallel which runs through the country 3000 stades south of Meroë and also through the Cinnamon producing country " must be put down as the limit and the beginning of our inhabited world on the south." [27]

Of very great importance, both on its merits and because of its

[24] *The Geography of Strabo*, Loeb edition, Vol. VIII, pp. 17, 19.
[25] Ibid., Vol. VIII, p. 157.
[26] Ibid., Vol. I, pp. 119, 120.
[27] Ibid., Vol. I, p. 439.

influence in later times, is the work of Claudius Ptolemy, who probably worked in Alexandria, and published works on astronomy and geography, the latter *c.* A.D. 150.[28] He aspires to correct the work of a forerunner, Marinus of Tyre, lost to us except in so far as embodied in his own. Ptolemy produced a map of the known world with a network of parallels and meridians, which can be reconstructed from his data. However, the number of scientifically ascertained positions on which these were based was lamentably small, and most of

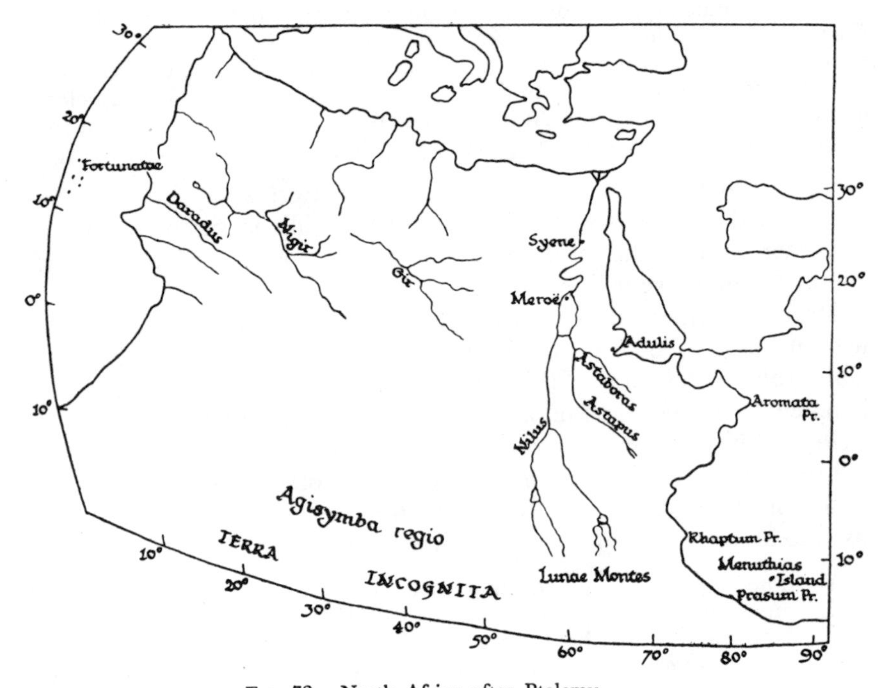

Fig. 72.—North Africa after Ptolemy.

the data used was based on distances as reported by travellers on land and sea, inaccurate not only as we have seen because of methods of computing distance but because units of measurement were not standardized. Despite his arbitrary corrections errors are numerous and in the direction of over-statement. Moreover, he divided the equator and other great circles into 360 degrees (herein following Hipparchus) with a degree equivalent of 50, instead of the correct 60, geographical miles. His longitudes were reckoned from the Fortunate Islands, the Canaries, placed only conjecturally two

[28] See E. L. Stevenson's edition of the *Geographia*, New York, 1932.

and a half degrees westward of the Sacred Promontory, Cape St. Vincent, while actually they are some nine degrees westward. Thus the known world was extended in terms of latitude and longitude, with important effects in later times, when the limitations of Ptolemaic data were forgotten. The idea of map projection is seen in his curving parallels and meridians. Thus the Ptolemaic map has a spurious appearance of accuracy. Let us look at its representation of Africa.[29]

Ptolemy departs from the ideas of his forerunners in many directions and the interpretation of his map in terms of known modern geography has been the subject of much controversy, partly because some have been misled by being too credulous of his latitudes. His limiting parallel on the south is that of 20° S. His " Agisymba regio " and " Lunæ Montes " are south of 10° S. : here is great distortion that gives a false impression of his range of knowledge. These features may correspond with the Chad region in about 15° N. and the Ruwenzori range on the equator. His " Gir " and " Nigir " rivers in the light of this fact have no necessary relation to the Sudan, neither is given an outlet to the sea, and a possible interpretation is that they refer to water courses to the south of the Atlas region—possibly, as Cary suggests, the Jedi and the Ighargar ; the words " *ghar* " and " *n'ghar*," he points out, are still used in modern Tuareg, and denote water. Ptolemy states that the Libyan desert bounds Africa to the south and moreover it should be noted that he places the Gir and Nigir in the same latitudes as the Fortunate Islands. His delineation of the upper Nile basin, by good fortune, achieves a schematic approximation to the truth, by indicating lake sources of the Nile. A striking delineation of the map is that of the east coast of Africa trending south-east from Rhaptum (on the coast near Zanzibar) to Prasum Promontory (probably the southern Cape Delgado), whence *terra incognita* extends south of the Indian Ocean to join a west facing coast of China. Thus the Indian Ocean becomes a lake, circumnavigation of Africa impossible, and we are reminded that the ancients were not unanimous in the belief that ships could sail round the continent. However, the Ptolemaic view was less popular than " the old Eratosthenic idea that the inhabited part of the world lay wholly within the north temperate zone, with a circumnavigable Africa wholly north of the Equator." [30]

Ptolemy was the last great geographer of ancient times and we have completed our outline of the outstanding contributions of ancient authors to African geography. Their works, significantly, were all written in Greek, the Romans showing little interest in geographical

[29] See *C.H.B.E.*, op. cit., pp. 65–70. [30] Ibid., p. 70.

topics. Very few geographical works were written in Latin and they were of inferior calibre ; only those of Pomponius Mela and Pliny are worthy of our attention, and this mainly because of their great influence in later times, for example through Solinus, a writer of the third century A.D.

Pomponius Mela wrote a popular compendium *c.* A.D. 43 and is much cited by Pliny ; much of his material is derived from Herodotus and even earlier writers, and he has a predilection for traditional fables. He writes of " antichthones " living in a southern temperate zone, unknown and beyond reach because of great heat in intervening regions. Of Nile theories he seems to favour a view that the great river originated in this zone and made its way to Ethiopia by means of a concealed channel. The sources would thus be satisfactorily explained as fed by orthodox winter rains, but of the southern hemisphere, giving floods in the Nile valley in the northern summer. However, he also cites the view that the Nile sources are in the region of the western Ethiopians and shows no hostility to this theory.

Pliny the Elder (A.D. 23–79) amassed in his *Natural History* a great number of facts, and part is devoted to geography ; he shows great interest in the bizarre. He has the common view of his time that the inhabited part of the earth is bounded on all sides by ocean, but goes further and asserts that it is established fact. In relation to Africa he writes :

> Hanno made the passage round from Gades to the borders of Arabia, and left a written account of his voyage ; as did also Hamilco, who was sent out at the same time to explore the outer coasts of Europe. Moreover, Cornelius Nepos states that within his own time a certain Eudoxus . . . set out from the Arabian Gulf and accomplished the passage to Gades, and long before him Caelius Antipater asserts that he had seen a merchant who had sailed from Spain to Ethiopia for the sake of trade.[31]

Bunbury, from whose work this quotation from Pliny is taken, writes of his assertions on navigation of the ocean surrounding the inhabited earth : " It would be difficult to find a stronger instance of the promiscuous manner in which Pliny raked together his materials, or of the total want of critical judgement, or even common accuracy with which he made use of them." It is a great misfortune that the only version of the West African voyage of Polybius that has survived is that given by Pliny because coherent reconstruction becomes impossible from reference to his garbled names and distances. Strabo fails to mention the voyage. Bunbury points out, however, that Pliny is not entirely without merit as, for example, in parts of his description

[31] E. H. Bunbury, *A History of Ancient Geography*, London, Vol. II, 1879, p. 383.

which are based on Juba's work on Africa. Bunbury sees in Pliny's reference, doubtless influenced by Juba, to the river Nigris "the first distinct notice of the great river that has attracted so much attention in later times under the name of the Niger," and he quotes Pliny as saying that it has "the same nature as the Nile; it produces reeds and papyrus and the same kinds of animals, and it becomes swollen at the same period." [32] The Nigris is a part of Juba's Nile which had its source in the west of Mauretania, flowing into a lake, thence underground and reappearing to form another lake; thence underground again and reappearing once more as the Nigris; it flowed on ultimately as the Astapus past Meroë and so to Egypt. It may be noticed that Strabo describes some as believing that the Nile sources are in far Maurusia, the same Moroccan zone.

In conclusion, let us emphasize that the modern estimation of the worth of these ancient geographers has no necessary correlation with that of contemporary or later times. Strabo, for example, is unmentioned by Pliny and by Ptolemy, and has to wait until the fifth century for a large degree of appreciation. Tozer comments on the slight extent to which Ptolemy influenced the ordinary Roman mind at any time, and Kimble, with reference to the Middle Ages and Christendom, writes: "After the fifth century and until the beginning of the fifteenth, we find only occasional references to Ptolemy and his school, while, as for his '*Geography*,' it is scarcely ever noticed." [33]

[32] Ibid., Vol. II, p. 435. For a recent consideration of various views as to identification, see J. O. Thomson, *History of Ancient Geography*, Cambridge, 1948, pp. 267–9.

[33] G. H. T. Kimble, *Geography in the Middle Ages*, London, 1938, p. 10.

UNIVERSITY OF LONDON

KING'S COLLEGE AND THE LONDON SCHOOL OF ECONOMICS JOINT SCHOOL OF GEOGRAPHY

LIST OF STUDENTS
1921–1947

The year 1921 was the first year in which examinations for Honours in Geography in the Faculties of Arts and Science were held in the University of London and in which one student, registered at King's College, had been trained partly at the London School of Economics.

KING'S COLLEGE

1921

B.A.

REA, ELSA C.
STAMP, LAURENCE DUDLEY

1923

B.A.

WOODS, MADALINE E.

LONDON SCHOOL OF ECONOMICS

1921

B.A.

Nil

B.Sc. (Econ.)

MARVEN, F. H.
POTTER, F.
RICHARDS, E. T.

1922

B.A.

PEILE, ALICE M.

B.Sc. (Econ.)

APPLETON, J. B.
CAIR, S. H.
GILES, F. C.
OLIVER, J. L.
PESTER, FRANCES E.
TREGEAR, T. R.

1923

B.A.

Nil

1923

1923

B.Sc.(Econ.)

Horniblow, E. C. T.
Hunt, Winefride
Pracy, Elsie M.
Pullen, H. G.
Saul, F.
Smith, D. H.
Snellgrove, A. A.
Tailby, M. A.
Thorogood, J. W.
Wilson, A. S.

1924

B.A.

Robinson, Margaret R.

1924

B.A.

Nil

B.Sc.(Econ.)

Marfell, Marie T. F.
Munday, F. G.
Overington, S.
Saunders, C. L.
Shepherd, O. D.
Thomas, F. L.
Walsh, C. E.

Ph.D.(Econ.)

Bryan, P. W.

1925

B.A.

Bell, G.
Cawley, Marjorie A.
Crinson, C. R.
Merewether, Jeanne A. M.
Warren, Lucy M.

1925

B.A.

Borgeaud, A. H.
Price, D. J.

B.Sc.(Econ.)

Burdett, Phyllis E.
Colls, Amy M. E.
Crowe, P. R.
Hookey, F. R.
Hum, S. G.
Wood, H. J.

1926

B.A.

Evans, R. J.
Letkey, Ethel W.

B.Sc.

Morgan, R. S.

1926

B.A.

Newson, Marion

B.Sc. (Econ.)

Ashlin, Ivy M.
McNamara, P. M.
Pooley, J. E.

M.Sc. (Econ.)

Uyehara, S.
Wilson, Martha E.

1927

B.A.

Allen, Emily F.
Dennis Kathleen F.
Saner, Beatrice R.
Stainer, D.
Stamp, L. M.
Tebbs, Vera C.

B.Sc.

Linton, D. L.

M.A.

Warren, Lucy M.

1927

B.A.

Nil

B.Sc. (Econ.)

Buchanan, R. O.
Jeans, Annie E.
Jones, R. E.
Kay, A.
Oakey, A. J. H.
Sweet, S. F.

M.Sc. (Econ.)

Mootham, O. H.

1928

B.A.

Brooks, N. C.
de Gruchy, Catherine R.
Higson, Lavinia E.
Lee, Christina E.
Lowe, Grace E.
Phillips, H. C.

1928

B.A.

Nil

1928

B.Sc.

Grove, Beryl E.
Rowsell, Eveline H.

1928

B.Sc. (Econ.)

Calver, Evelyn M.
Colclough, J. R.
Newell, F. M.
Thorpe, W. N.
Waite, P. C.
Williamson, Agnes F.
Willis, Gwenyth M.
Wolf, E. L. S.
Wood, R. J.

M.A.

Hassan, M. S.

1929

B.A.

Carter, Elizabeth P.
Dyer, H. F.
Hadlow, L. H.
Kimble, G. H. T.
Morrow, Mary
Murphy, A. J.
Wilson, E. E.
Young, M.

B.Sc.

Curtis, Dorothy M.
Pool, M.
Porter, W. McD.
Reynolds, Kathleen H.
Sears, Lily V.
Smetham, D. J.

M.A.

Cons, G. J. (Goldsmiths College)

1929

B.A.

Stephenson, J.
Walker, Evelyn L.

B.Sc. (Econ.)

Croome, J. L.
Hunt, C. J.
Morley, H. T.
Payne, W.
Tanner, L. S.
Thornley, J. A.
Wellstead, G. E.

Ph.D. (Econ.)

Robertson, C. J.

1930

B.A.

Clifton, Cecelia M.
Cooke, Jenny
Davis, Vivien M.
Edwards, Gwendolen A.

1930

B.A.

Clarkson, Kathleen N.

1930

B.A.

Flanagan, Una E.
Garrett, A. J.
Hadlow, F. L.
Jacobs, Rachel E.
Kay, Dorothy H.
MacDonald, K. R.
Rizvi, S. M. T.
Satterly, Dorothy L.

B.Sc.

Boon, E. P.
Burrows, R. P.
Heywood, C. L.
Horrocks, N. K.
O'Dell, A. C.
Sheppard, T. V.
Wright, Gertrude

1930

B.A.

B.Sc. (Econ.)

Green, H. A.
Hancock, T. W.
Lebon, J. H. G.
Maguire, W.
Tappenden, W. L.
Willatts, E. C.
Woolford, F. H.

M.A.

Ginige, A.

1931

B.A.

Best, A. E. J.
Burslem, Dora P.
Ciompka, Ella M.
Coles, R.
Garton, Winifred M.
Gosling, L. O.
Hewitt, W. G.
Sutton, Muriel F.

B.Sc.

Goddard, H. W.
Kelly, F. L.
Pace, W. A. G.
Price, Constance
Woodhouse, B.

1931

B.A.

Bailey, Ivy
Gahagan, Monica
Green, Edna C.

B.Sc. (Econ.)

Crouch, R.
Glass, D. V.
Hodge, W. H.
Michaels, M. I.
Ranger, A.
Rosenstein, S.
Smith, L. F. H.
Ward, E.
Winterton, C. A.

1931

M.A.

Kimble, G. H. T.

M.Sc.

Sargent, C. P.

1931

M.A.

Beaver, S. H.
Morris, F. G.
Price, D. J.

D.Sc.(Econ.)

Ormsby, H. R.

1932

B.A.

Bates, Florence E.
Forbes Leith, Ailsa G.
Gensbury, Frances M. D.
Innes, M. R. H.
Walkley, A. E. T.
Wilkinson, J.
Young, J. H.

B.Sc.

Fieldgate, Margaret H.
Kingsland, F. H.
Ram, R.
Wheelton, C.

M.A.

Gaught, A. S.

Ph.D.(Arts)

Saner, Beatrice R. M.

1932

B.A.

Nil

B.Sc.(Econ.)

Elcome, Barbara M.
Green, F. H. W.
Nāg, T. C.
Pacey, H. E.
Pierson, A. C.
Roberson, B. S.
Thomas, J. L.
Turrill, E. F.
Wallis, A. G.

Ph.D.(Econ.)

Buchanan, R. O.
McNamara, P. M.

1933

B.A.

Balch, Kathleen M.
Bull, R. G. F.
Dale, E.
Donithorn, W. C.
Driscoll, M. J. M.

1933

B.A.

Bloom, Daisy G.
Cotton, A. F. E.
Hopkins, F. P.

1933

B.A.

GRAY, G. D. B.
GRAY, MARGARET
HUMPHREY, SYLVIA
HUNT, MARY I.
KAY, H. F.
KING, S. H.
PARISH, DORIS M.
SATTERLY, MARJORIE S.
SHAW, G. W.
WANSBROUGH, KATHLEEN M.
YOUNG, MOIRA B. A.

B.Sc.

GALE, LUCY M. D.
HOLLIDAY, L. A.
JONES, MARJORIE
LAKE, H. C. G.
SMALL, EDITH
VAIZEY, J. D.

1933

B.A.

B.Sc. (Econ.)

BLAKE, R. A.
BROWN, C. H.
DIACK, P. G. A.
EDWARDS, T. J. I.
GRACE, S. G. LE C.
GROSSMAN, M.
HARRISON, L. J.
HONER, D. E.
JOY, B.
MARLOW, C. V.
MCCORMICK, R.
MILSON, H.
RICHARDSON, J. R.
SMITH, A. E.
YOUNG, H. E.

M.Sc.

O'DELL, A. C.

1934

B.A.

BARRELL, K. W.
FITCHETT, C. E.
HARRIES, P. A.
KEEBLE, R. W. J.
KING, F. H.
LEES, MARY W.
MORRIS, H. H.
PUCKETT, C. H.
SMITH, DOROTHY J.
WOODWARD, MARGARET G.

1934

B.A.

TEMKIN, NETTIE

1934

B.Sc.

Hines, G. M.
Mabbitt, Doris A.
Mills, E. F.
Purcell, T. S.
Rouse, W. J. C.
Webb, J. O. W.
Willson, J. B.
Wyatt, F. J.

Ph.D. (Arts)

Coles, R.

1934

B.Sc. (Econ.)

Ashby, R. A.
Bottoms, S. H.
Browne, S. E. D.
Dann, Violet
Hooper, E. N.
Lakin, F. H.
McIver, J. A.
Orsborn, K. C.
Ramsbottom, Margery E.
Phillips, L.
Pyke, Evelyn M.
Smith, G. E.
Tucker, R. L.
Woodlands, A. F.

1935

B.A.

Beck, Frances S.
Broughton, Margaret E.
Burden, G. H.
Clarke, May K.
Dewick, Nancy D.
Gabriel, Joyce R.
James, J. R.
Johnson, Ella
Morris, J. W.
Obendorf, Winifred
Smith, Stella M.

B.Sc.

Learner, C. H.
Smith, F. T.

1935

B.A.

Hartop, P. W.
Oram, A. E.

B.Sc. (Econ.)

Alexander, K. A. de W.
Cadman, D. J.
Cattle, A. J.
Cornish, E. A. E.
Dakin, C. J.
Hardy, A. V.
Jones, W. E.
Lewis, D. E. S.
Marfell, G. J. C.
Renshaw, H. G.
Tubbs, A. E.
Warren, Margaret N. N.

1935

B.Sc.

M.A.

Garrett, A. J.

M.Sc.

Sheppard, T. V.

1935

B.Sc. (Econ.)

Wilkins, H. F.
Wilkins, Phyllis M.

M.Sc. (Econ.)

Green, F. H. W.

1936

B.A.

Earl, R. S.
Jeffery, F.
Lyons, H. R.
Malcolm, Stella M.
Minch, D. J.
Muthukrishnan, S.

B.Sc.

Ahmed, A. M.
Batchelor, Elizabeth C.
Jennings, Winifred
Manning, G. R.
Millington, A. G. E.
Morse, Jennie E. D.
Pearson, Sybil W.
Thomas, A. M.
Titmarsh, A. B.
Tryhorn, Margaret S.
Turner, Evelyn M.

1936

B.A.

Hopkins, P. G. H.
Page, Mary P.

B.Sc. (Econ.)

Backhouse, A. F. D.
Cale, R. P.
Capsey, G. H.
Church, R. J.
Clarke, L. W.
Dewhurst, G. A.
Finden, H. S.
Fry, J. C.
Hobbs, G. R.
Hodd, R. E.
Hodge, C. J.
Lewis, G. E. D.
Muxworthy, D. C.
Neary, J.
Read, C. W. F.
Richards, G.
Rogers, R. H. E.
Round, P. H.
Shanahan, Dorothy R.
Singleton, W. R.
Stokes, E. C.
Weston, A. C. E.
Wilmoth, M. St. J.

M.A.

Stephenson, J.

1937

B.A.

Attrill, N. E. C.
Barham, Marjorie I. L.
Birchmore, B.
Gray, Martha H.
Hamer, S. M.
Hardy, Flora
Husted, G. J.
Jenkins, J. E. C.
Jones, Kathleen H.
McLaren, R. E. H.
Martindale, Eileen E. M.
Orpin, Margaret J.
Pike, D. F. B.
Powell, D. J.
Salinger, Madeline G.
Salmon, Alice M.
Smith, Winifred M.
Walker, Francesca M. T.
Walton, P. C.

B.Sc.

Bird, W. J.
Blackeby, D. H.
Dennett, Edith E.
Paterson, Jean W.
Stringer, K. V.
Tipping, Gladys E.
Wiltshire, Mary

1937

B.A.

May, J.
Wells, S. F.

B.Sc.(Econ.)

Adam, J. H.
Armitstead, A. R.
Brech, R. J.
Britnell, P. R. F.
Cary, Agnes M.
Cauter, T.
Clayton, W. B. G.
Coe, W. R. J.
Duke, C. H. A.
Gilbert, W. L.
Henderson, Daphne J.
Hipkins, G. J.
Hope, J. A.
Hostler, T.
Hutchinson, E. A.
Jackson, J. G. C.
Lawrence, J. C.
Lee, A. J.
McBain, F. C. A.
Melamid, A.
Milford, H. J.
Owen, A.
Radleigh, J. R. Y.
Savage, K. R.
Saville, J.
Sheppard, W. W.
Spooner, W. M.

1937

M.A.

King, S. H.

1937

M.A.

Arrowsmith, F. A.
Dudlyke, E. R.

Ph.D. (Arts)

Rizvi, S. M. T.

Ph.D. (Econ.)

Willatts, E. C.

1938

B.A.

Barker, Sylvia Y.
Burnett, F. T.
Clements, Elizabeth
Dobson, E. B.
Franklin, Doris M.
Hallworth, H. J.
Harris, A. H.
Rayns, A. W.
Sarkar, S. A.
Stribley, W. S.
Sutcliffe, Dorothy
Watkins, M. A. J.

B.Sc.

Ellis, Margaret J.
Evans, Lilian M.
Jameson, Margaret E.
Jones, Ivy M.
Lewis, A. D.
Lewis, Nesta J.
McKenzie, Mary
Owen, Marjorie B.
Rivers, F. A.
Senior, Winifred M.

1938

B.A.

Nil

B.Sc. (Econ.)

Askew, R. A. A.
Bourne, R. T.
Bridgeland, F. P.
Butler, L. G.
Chadwick, A.
Cooper, J.
Ensor, P. T.
Hargraves, Beatrice M.
Lucas, G. W.
Murison, I. G.
Pearson, S.
Rubin, L. J.
Shellabear, J. E.
Spink, P. J.
Tancock, D. R.
Try, C. G. A.
Vince, S. W. E.

1938

M.Sc.

Smith, F. T.

1938

M.Sc. (Econ.)

Coles, R.

Ph.D. (Arts)

Mosby, J. E. G.

1939

B.A.

Berrey, A. J.
Black, Winifred M.
Burke, J. J. N.
Hallam, J. H.
Mather, D. B.
Shepperd, Edith M. C.
Uplekar, G. S.
Ward, Phyllis M.

B.Sc.

Brailey, V. G.
Cornish, R. T.
Cowling, C.
Davies, Olwen E. G.
Erasmus, D. W. E.
Feakes, B. A.
Gilliland, J.
Hare, F. K.
Hawkin, C.
Heppel, B. C.
Miller, H.
Remmington, Winifred M.
Searles, H. J.
Sheshgiri, B. S.
Smith, R. H.
Storey, K. L.
Thompson, B. W.

M.A.

Jenkins, J. E. C.

1939

B.A.

Alsop, Gwyneth M. D.
Scott, Beryl J.

B.Sc. (Econ.)

Barker, E.
Bartholomew, Joan V. M.
Benson, C. O.
Bovingdon, Joy P.
Dancer, W. S.
Donald, A. S.
Dunley, C. A. L.
Holleyoak, W. F.
Hughes, J. M. C.
Ker, D. C. A.
Lovatt, G. W.
Pugh, G. P.
Rees, H.
Richmond, Eileen J.

M.A.

Durrani, M. A. K.
Luke, Winifred R.

M.Sc. (Econ.)

Bryant, R. W. G.
Mead, W. R.

1940

B.A.

Allen, D. A.
Bard, Joyce E.
Collinson, Gwendolen E.
Crawford, Margaret C.
Ford, V. C. R.
Hallum, Lesley M.
Henton, R. G.
Hillier, Ruth C.
Lee, G. W.
Molloy, Mary T.
Puddephatt, R. C.
Rahman, M. H.

B.Sc.

Sandry, Kathleen F. M. (née Ainsworth)
Allchin, L. J.
Bailey, Kathleen M. F. (née Williams)
Clements, Marie F.
Loretto, Jean C.
Rodgers, Evelyn F.
Wisdom, Joan M.

1940

B.A.

Pardoe, C. H.
Peiris, W. M. D.

B.Sc.(Econ.)

Bland, J. C.
Bornstein, Joyce
Bosdet, Joan M.
Dalling, Joyce M.
Fryer, D. W.
Giddy, H. P.
Gledhill, E.
Hegan, Marjorie J.
Laidlaw, Patricia
Peacock, D. L.

Ph.D.(Econ.)

Bao, C.

1941

B.A.

Barraclough, Marian C.
Basu, B. B.
Kirk, W.
Stephens, C. H.
Stevens, Margery K.
Stevenson, G. A.
Taylor, Joan M.
Tristram, Edith B.

B.Sc.

Dean, Muriel E. (née Bradforth)
Douglas, I. H.
Love, W. A. L.
Price, Mary S.
Russell, Ruth M. H.
Saunders, Margaret J.

1941

B.A.

Nil

B.Sc.(Econ.)

Allen, D. M.
Brookes, A.
Fry, F. J.
Robilliard, A. H. S.

1941

M.Sc.

Boon, E. P.

1941

Ph.D. (Science)

Howell, E. J.

1942

B.A.

Barfoot, R. W.
Lascelles, H. B.
Martin, Diana M.
Partridge, Joy E.
Pearce, Margaret J.
Seymour, Eileen N.
White, Frances
Whitehead, Mavis J.
Williams, Margaret J.
Youldon, Irene G.

B.Sc.

Camp, Ruth E.
Carr, Rosemary F.
Connolly, Mary P.
Knowles, Barbara P.
Owen, Irene I.
Spill, Mary

1942

B.A.

Newell, Dulcie M.

B.Sc. (Econ.)

Cassar, Josy
James, Dorothy C.
Scola, P.
Scott, P.
Seidmann, M.

M.Sc. (Econ.)

Fryer, D. W.

1943

B.A.

Clark, Kathleen M.
Garrett, Joan M.
Skene, J. S.
Sutcliffe, Eileen C.
Swales, Evelyn M.
Thorpe, Beryl M.

B.Sc.

Gifford, Cynthia J. (née Alexander),
Liddle, Ellen E.
Roberts, Joan
Salmond, K. D.

1943

B.A.

Burns, Edith M.
East, Jean B.
Kent, Mollie

B.Sc. (Econ.)

Nil

1943

1943

M.A.

Hindia, M. E. A. A.

M.Sc.(Econ.)

Wilkins, Phyllis M.

Ph.D.(Econ.)

Harrison-Church, R. J.

1944

B.A.

Cook, Betty D.
Musto, Suzanne M.
Royston, Audrey P.
Smith, Margaret J.
Watkins, Irene A.
Weber, Irene I.

B.Sc.

Clough-Ormiston, Kathleen M.
Colleypriest, Jean S.
Dawson, Clarice N.
Markes, Margaret I. F.

1944

B.A.

Shute, Gwendolyn E.

B.Sc.(Econ.)

Jenkin, Jean R.
O'Brien, Gonzaga M.
Osborn, Olive M.
Thatcher, D. C.

M.Sc.(Econ.)

Scola, P.

1945

B.A.

Mercer, Agnes

B.Sc.

Collins, Sylvia M.
Coppinger, Mary M.
Fleming, Frances J.
Harrison, W. T.
Prince, Margaret

1945

B.A.

Brookfield, H. C.
Cannon, Catherine E.

B.Sc.(Econ.)

Auty, R. M.
Averill, C. M.
Bransgrove, Beatrice M.
Creamer, D. J.
Lord, Marian
Moore, Mabel T.
Taupin, J. C.
Tomlinson, Katherine M.
Turner, Barbara L.

1945

M.A.

Youldon, Irene G.

Ph.D. (Arts)

King, S. H.

1946

B.A.

Burbidge, Cecilia M.
Burgess, Patricia M.
Grant, R. F.
McCallum, Marjory J.
McDanell, Maureen S.
Mills, Nancy S.
Saunderson, Stella M.
Scott, Mary S.
Smyth, Geraldine M.

B.Sc.

Crossley-Meates, Stella
Francis, Margaret L.
King, Margaret B. M.
Laundon, E. W.
Tabberer, Pamela A.

M.A.

Gray, G. D. B.

M.Sc.

Heywood, C. L.
Morris, W. D.

1947

B.A.

Alner, Marion G. C.
Blight, Muriel
Fletcher, D. M.
Halliday, A. J.
Harris, Phyllis M.

1945

Ph.D. (Arts)

Pounds, N. J. G.

1946

B.A.

O'Connor, Joan
Ross, Pamela E.

B.Sc. (Econ.)

Carter, P. M. G.
Davies, H. G.
McCarley, Rowena O.
Turner, Kathleen

M.Sc. (Econ.)

Napolitan, L.
Rees, H.

Ph.D. (Science)

Davis, S. G.

Ph.D. (Arts)

East, Jean B.

1947

B.A.

Brandon, J. J.
Evans, Ruth E.
Moss, Helen

1947

M.A.

Hill, Joyce I.
Johnson, B. L. C.
Kidson, C.
Lingwood, Barbara J.
Lowdon, Joan
Miller, Maureen E.
Parratt, Frances M.
Poulson, Kathleen I.
Sara, Margaret M.
Singer, Patricia M.
Stevens, Joan
Sullivan, Brenda E.
Thomas, A.
Watts, Jean L.
Willett, Pamela

B.Sc.

Brawer, M.
Britten, Menna
Brown, E. H.
Dass, Betty M. E.
Fleming, H. L.
Restall, Barbara M.
Stokes, Margaret P.

1947

B.Sc. (Econ.)

Arkinstall, Mary
Bath, A.
Buckland, M. P.
Davies, H.
Dennell, E.
Dennison, V. D.
Gledhill, A.
Hadley, L.
Hankinson, L. C.
Hooley, Megan
Lewis, Carol L.
Parsons, Patricia M.
Roberts, V. H. P.
Rodd, Ursula M.
Ruoff, F. H. G.
Sandall, R. St. C.
Sherwin, Donne
Wilkins, R. E.
Woods, Janet M.

M.A.

Dutt, P. K.

M.Sc. (Econ.)

Fielding, J. T.

Ph.D. (Arts)

Dayal, P.

INDEX